FLORA ZAMBESIACA

Flora terrarum Zambesii aquis conjunctarum

VOLUME EIGHT: PART SEVEN

FLORA ZAMBESIACA

MOZAMBIQUE

MALAWI, ZAMBIA, ZIMBABWE

BOTSWANA

VOLUME EIGHT: PART SEVEN

Edited by
G.V. POPE & E.S. MARTINS

on behalf of the Editorial Board:

S.J. OWENS
Royal Botanic Gardens, Kew

M.A. DINIZ
Centro de Botânica, Instituto de Investigação
Científica Tropical, Lisboa

G.V. POPE
Royal Botanic Gardens, Kew

Published by the Royal Botanic Gardens, Kew,
for the Flora Zambesiaca Managing Committee
2005

Typesetting and page make-up by Media Resources, Information Services Department,
Royal Botanic Gardens, Kew

Printed in the European Union by
The Cromwell Press

ISBN 1 84246 026 9

CONTENTS

FAMILIES INCLUDED IN VOLUME VIII, PART 7

Avicenniaceae, Nesogenaceae, Verbenaceae and
Lamiaceae (subfams. Viticoideae and Ajugoideae)

NEW COMBINATION PUBLISHED IN THIS PART

vi

127. AVICENNIACEAE

By R. Fernandes

Glabrous (or canescent) evergreen small trees of mangrove swamps, with extensive widely-spreading root systems bearing pneumatophores. Leaves simple, decussate, petiolate, without stipules; lamina entire, pulverulent (powdery)-puberulent or felted-tomentellous on lower surface, ± coriaceous. Inflorescences terminal or axillary, of contracted cymes at the end of short peduncles. Flowers actinomorphic, hermaphrodite, sessile, subtended by an involucre formed by a bract and two bracteoles. Calyx gamosepalous, deeply 5-lobed with lobes imbricate. Corolla gamopetalous, campanulate or campanulate-rotate; tube short; limb 4(5)-lobed, spreading, with subequal lobes or the posterior lobe slightly broader. Stamens 4, inserted in the corolla throat and alternating with the lobes, equal or sub-didynamous; filaments ± short, filiform; anthers ovoid or suborbicular, subexserted. Ovary superior, 2-carpellate, but appearing 4-locular due to ovary wall intrusions (false septa); placenta free, axile; style 1 short, bifid or bilobed, with apical stigmas; ovules 4, with 1 in each apparent locule, pendent, orthotropous. Fruit a capsule, l-seeded (by abortion of 3 ovules), usually asymmetric, compressed, opening by 2(4) thickened valves. Seed without a testa; embryo with large longitudinally folded cotyledons and a villous radicle, viviparous; endosperm fleshy.

A monogeneric family of the Old and New World tropics and subtropics. Usually treated as a subfamily of the *Verbenaceae*, it is here recognized as a separate family readily distinguished by the involucre of 1 bract and 2 bracteoles which subtend each flower, the free-central placentation with a c. 4-winged placenta, the orthotropous ovules and by the viviparous embryo and fleshy endosperm. Molecular evidence now places it near the more basal *Acanthaceae*, with which it shares few essential morphological characters. [A.E. Schwarzbach & L.A. Mc Dade. Phylogenetic relationships of the mangrove family *Avicenniaceae* based on chloroplast and nuclear chromosomal DNA sequences. Systematic Botany **27**: 84–98 (2002)].

AVICENNIA L.

Avicennia L., Sp. Pl.: 110 (1753); Gen. Pl., ed. 5: 49 (1754). —Moldenke in Ann. Missouri Bot. Gard. **60**: 149–154 (1973); in Rev. Fl. Ceylon **4**: 126 (1983). — Verdcourt in F.T.E.A., Verbenaceae: 144, fig. 21 (1992).

Characters as for the family.

A widespread genus with 3–6(12) species (fide *Verdcourt*) in tropical Africa, Asia, Oceania and America. It is represented by a single species along the east coast of Africa, including the Seychelles, the Comoros, the Mozambique Channel Islands, Madagascar and the Mascareignes, where it is a constituent of coastal mangroves.

Avicennia marina (Forssk.) Vierh. in Akad. Wiss. Wien. Math.-Naturwiss. Kl., Denkschr. **71**: 435 (1907). —Brenan, Check-list For. Trees Shrubs Tang. Terr.: 629 (1949). —J.G. Garcia in Estud. Ensaios Doc. Junta Invest. Ci. Ultramar [in Mendonça, Contrib. Conhec. Fl. Moçamb., II] **12**: 177 (1954). —F.W. Andrews, Fl. Pl. Anglo-Egypt. Sudan **3**: 193 (1956). — Moldenke in Fl. Madag., Avicenniaceae: 2, fig. 1 (1956); in Phytologia **7**, 4: 210 (1960). — Mogg in Macnae & Kalk, Nat. Hist. Inhaca Isl., Moçamb.: 13, fig. 5d, plate III(6) (1958). —Dale & Greenway, Kenya Trees & Shrubs: 581, fig. 106 (1961). —Gomes e Sousa, Dendrol. Moçamb.: 660, t. 214 (1967). —J.H. Ross, Fl. Natal: 300 (1972). —E.A. Weiss in Econ. Bot. **27**, 2: 189 (1973). —Palmer & Pitman, Trees Southern Africa **3**: 1971, photo facing p. 1956, photo p. 1972 & 1974, fig. p. 1975 (1973). —K. Coates Palgrave, Trees South. Africa, ed. 2: 816 (1988). —Verdcourt in F.T.E.A., Verbenaceae: 144, fig. 21 (1992). —M. Coates Palgrave, ed. 3 of K. Coates Palgrave, Trees South. Africa: 975 (2002). TAB. **1**. Type from "islands and shores of Red Sea (Al Luhaygah).*

---

*For further synonymy and bibliography see Moldenke, in Phytologia **7**, 4: 210–213 (1960).

Tab. 1. AVICENNIA MARINA. 1, flowering branch (× ²/₃); 2, flower (× 6); 3, outer bract (× 6); 4, middle bract (× 6); 5, inner bract (× 6); 6, corolla, opened out (× 6); 7, ovary (× 6), 1–7 from *Semsei* 1813; 8, fruit (× ²/₃); 9, seed (× ²/₃); 8 & 9 from *Renvoize* 762. Drawn by Pat Halliday. From F.T.E.A.

Sceura marina Forssk., Fl. Aegypt.-Arab.: 37 (1775).

Avicennia tomentosa var. *arabica* Walp., Repert. Bot. Syst. **4**: 133 (1845).

Avicennia officinalis auct. praecipue Fl. Afr. Orient.: —sensu Schauer in A. de Candolle, Prodr. **11**: 700 (1847) quoad synon. pro parte et distr. geogr. pro parte. —sensu A. Richard, Tent. Fl. Abyss. **2**: 173 (1850). —sensu Klotzsch in Peters, Naturw. Reise Mossambique **6**, part 1: 266 (1861). —sensu J.G. Baker, Fl. Mauritius & Seychelles: 257 (1877); in F.T.A. **5**: 332 (1900). —sensu Gürke in Engler, Pflanzenw. Ost-Afrikas **C**: 342 (1895). —sensu Sim, For. Fl. Port. E. Africa: 94, 125 et 139, t. 83 (1909), non L. (1753). —sensu H. Pearson in F.C. **5**, 1: 225–226 (1910).

Usually a small evergreen tree 3–4 m tall, sometimes to 10(12) m tall, with a much branched rounded crown and an extensive widely spreading horizontal root system, less often a shrub; roots, particularly in muddy places, giving rise to numerous erect pencil-like pneumatophores, 10–38 cm tall. Trunk ± stout, up to c. 25 cm in diameter, sometimes with aerial 'breathing' roots (not reaching the ground) on the lower portion; bark smooth and somewhat powdery, whitish or yellow-green; branchlets decussate, slender, subterete or quadrangular, with a densely pulverulent (powdery)-puberulent whitish or greyish thin outer bark, nodes swollen. Leaves 3.5–12 × 1.3–5 cm, ovate or lanceolate to oblong-lanceolate or elliptic, acute or subacute at the apex, entire with margins slightly revolute, cuneate at the base or ± tapering into the petiole, glabrous or ± obscurely powdery on upper surface, densely pulverulent (powdery)-puberulent or felted-tomentellous on lower surface, coriaceous, usually light green on both surfaces or discolorous and silvery-greyish or greenish-yellow beneath, becoming dark olive-green to blackish above and sordid grey to yellowish beneath on drying; midrib slender and ± prominent on both surfaces, reticulation slightly raised above and obscured beneath; secondary nerves 8–15 on each side of midrib; petiole 3–14 mm long, flattened above, densely powdery-puberulent or tomentellous, rarely glabrescent. Cymes usually solitary in leaf axils towards the end of the branchlets, and usually with 3 at the apex of the branchlets, few- to many-flowered, capitate or subcapitate; the heads 7–15 mm in diameter; peduncles 0.6–4 cm long, quadrangular, sulcate when dry, densely thinly powdery-puberulent. Bracts and bracteoles 2–3 mm long, ovate to broadly ovate, concave, obtuse or rounded at the apex, ciliate, glabrous on the inner face, densely silvery-tomentose or sericeous on the outside, finally glabrescent, closely appressed to the calyx. Flowers fragrant. Calyx green; lobes 2–4 mm long, ovate or broadly ovate to elliptic, rounded at the apex, densely tomentellous outside, fimbriate-ciliate on the margins, persistent. Corolla white at first, turning yellow or orange to dark orange or reddish-orange, becoming blackish when dry, rigid, caducous; tube 1–2 mm long, glabrous; lobes 3–4 mm long, ovate, densely tomentellous or sericeous outside, glabrous inside. Filaments erect; anthers subequalling the filaments, suborbicular, sulphur-yellow, turning black. Capsule 1.2–3 × 0.7–2 cm, subglobose, broadly ellipsoid or ovoid, somewhat asymmetric and abruptly tapering into a short narrow apical beak at least when young, yellowish or pale to greyish-green densely powdery-puberulent, the calyx and involucre persisting unchanged. Seed large, compressed, germinating within the fruit while still on the tree. Cotyledons reniform, light green.

Mozambique. N: Mogincual, mouth of Mogincual R., 30.iii.1964, *Torre & Paiva* 11484 (COI; LISC; LMU; SRGH). Z: Pebane, veget. 5.x.1946, *Pedro* 2115 (LMA). MS: Marromeu, fl. ix.1973, *W. Bond* 10881 (SRGH). GI: Daimane, adjoining Inhambane, iii.1938, *Gomes e Sousa* 2098 (COI; K; LISC; SRGH); Inhassoro Distr., Bazaruto Island, Ponta Gengarene, 21.x.1958, *Mogg* 28729 (K; LISC; LMA; LMU; SRGH). M: Matutuíne Distr., margem esquerda do rio Maputo, a juzante de Matutuíne (Bela Vista), fl. 17.viii.1948, *Gomes e Sousa* 3793 (COI; K; LISC; LMA; SRGH); Maputo Distr., Inhaca Island, south of the Marine Biological Station, 29.xii.1956, *Mogg* 27030 (FHO; K; LMA; PRE; SRGH).

Widely distributed along the shores of the western Indian Ocean (east Africa from Egypt to the Cape including the Seychelles, the Comoros, the Mozambique Channel Islands, Madagascar and the Mascareignes); also along the coasts of the Red Sea, Persian Gulf, Arabian Sea, Bay of Bengal, the northern and eastern Indian Ocean, the South China Sea north to Hong Kong and Taiwan, and the islands of the Philippine Sea, the Coral Sea and South Pacific to New Zealand. Intertidal zone mudflats and sandy shores, and the estuaries and tidal river banks (brackish water). It is a common and often dominant constituent of mangrove swamps (usually the inland fringes of mangrove associations), and is also a pioneer of new mud banks; sea level. The bark is rich in tannin and yields a brown dye. The wood is durable and is used in boat building in Maputo, and for the ribs of dhows in East Africa.

127a. NESOGENACEAE

By M.A. Diniz

Annual or perennial herbs with intermixed simple and multicellular unbranched hairs, and usually also with gland-tipped hairs. Leaves opposite, simple, toothed or entire, exstipulate. Flowers axillary, solitary or clustered in cymes (bracteate in one species) or in bracteate spikes. Pedicels short (sometimes absent). Calyx (4)5-lobed, accrescent. Corolla gamopetalous slightly zygomorphic, funnel-shaped, unequally (4)5-lobed; lobes imbricate, either the lower or the two lateral lobes outermost. Stamens didynamous, epipetalous, inserted low down in the corolla tube, included; anthers 2-locular, the thecae pendulous, from a very short connective, tailed, ± divergent. Ovary superior, 2-locular; ovules 1 per locule, axial at the base of the locule, erect. Style filiform; stigma small, capitate. Fruit with pericarp crustaceous, indehiscent or at least tardily dehiscent, oblong-ovoid or obovoid, compressed, grooved at the base between the locules, pubescent at the apex around the thickened base of the style. Seeds oblong, foveolate; endosperm present.

A monogeneric family, previously included in the *Verbenaceae* until separated as a family in 1981.

NESOGENES A. DC.

Nesogenes A. DC., Prodr. **11**: 703 (1847). —Marais in Kew Bull. **35**: 799 (1981).
   Acharitea Benth. in Bentham & J.D. Hooker, Gen. Pl. **2**: 1142 (1876).

Features as for the family.

A genus of 7 species, 6 known from islands in the S Pacific and SW Indian Oceans and 1 from east Africa (Kenya, Tanzania and Mozambique).

Nesogenes madagascariensis (Bonati) Marais in Kew Bull. **38**: 37 (1983). TAB. **2**. Type from Madagascar.
   Stemodiopsis madagascariensis Bonati in Bull. Soc. Bot. Genève, ser. 2, **15**: 97 (1924).
   Nesogenes africanus G. Taylor in J. Bot. **68**: 84 (1930). —Marais in Kew Bull. **35**: 802, fig. 1 (1981); in F.T.E.A., Nesogenaceae: 1, fig. 1 (1984). Type from Tanzania.
   Nesogenes mansfeldianus Mildbr. in Notizbl. Bot. Gart. Berlin **11**: 820 (1933). Type from Tanzania.

Erect, slender or bushy, annual herb up to 45 cm tall. Stem terete, with short internodes and usually 2 branches at each node, or stem sparsely branched, pilose with spreading ± crisped hairs intermixed with multicellular gland-tipped hairs; branches 10–33 cm long, slender, ascending or sometimes procumbent. Leaves opposite, becoming ± scaberulous; petiole short, slender; lamina up to 35(45) × 10 mm, ovate-lanceolate to narrowly lanceolate, tapering to an acute apex, cuneate-attenuate at the base, and narrowly decurrent onto the petiole, entire or crenate-serrate with (2)4–5(6) teeth on each side, the upper leaves smaller and entire; indumentum of small acicular antrorse hairs on both surfaces intermixed with gland-tipped hairs, more numerous on the veins, with several rows of acicular antrorse hairs on the margins; veins ± obscure, in (2)4–5(6) pairs. Flowers in groups of 2–3 in each leaf axil, or solitary in upper axils; pedicels 0.3 mm long. Calyx narrowly campanulate with (4)5 deltoid lobes, hispid and ciliate with antrorse hairs and sometimes intermixed with pilose gland-tipped hairs; calyx tube 2–3 mm long; lobes 1–2 mm long. Corolla white, cream, pink or mauve, tubular-infundibuliform; tube c. 4 mm long, lobes c. 1 mm long, rounded. Filaments 1–2 mm long, glabrous; anther thecae c. 0.3 mm long, unequal, divaricate, inverted, tailed. Ovary 2-locular with 1 ovule per locule; style 2 mm long, filiform. Fruit 2 mm long, obovoid. Seeds 2, oblong, foveolate, with the foveoles in longitudinal rows.

Mozambique. N: Nampula Distr., serra de Mesa, c. 6 km from Nampula, 500 m, fl. & fr. 3.iv.1964, *Torre & Paiva* 11603 (COI; EA; K; LISC; LMU; PRE; SRGI I; WAG).
   Also in Kenya and southern Tanzania. On granitic rocky outcrops; 500 m.

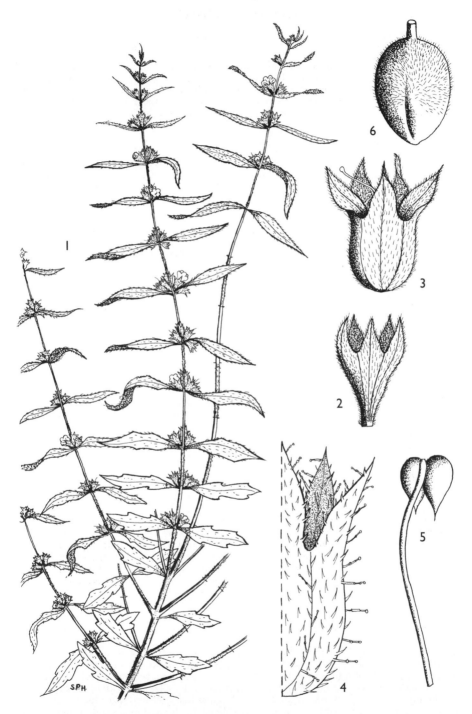

Tab. 2. NESOGENES MADAGASCARIENSIS. 1, habit (× ²/₃); 2, flowering calyx (× 8); 3, fruiting calyx (× 6); 4, detail of fruiting calyx (× 12); 5, stamen (× 18); 6, fruit (× 10), 1–6 from *Vollesen* 3705. Drawn by Susan Hillier. From Kew Bull.

VERBENACEAE and LAMIACEAE

Notes on the family delimitation of the Verbenaceae and Lamiaceae*

The traditional division of Verbenaceae and Lamiaceae, e.g., Bentham & Hooker in Gen. Pl. **2**: 1131–1223 (1876); Baker & Stapf in F.T.A. **5**: 273–502 (1900), is far from satisfactory. The delimitation of these families was based on whether the taxa were mostly woody with a terminal or subterminal style (Verbenaceae) or mostly herbaceous with a gynobasic style (Lamiaceae). The problems with this classification were pointed out by Junell in Symb. Bot. Upsal. **4**: 1–219 (1934), and elaborated upon by Cantino in Ann. Missouri Bot. Gard. **79**: 361–379 (1992); and Cantino et al. in Harley, R.M. & Reynolds, T. eds, Advances in Labiatae Science: 27–37 (1992). The traditional classification is difficult to implement, there being many intermediates, and also does not represent phylogenetically natural taxa.

A new circumscription of the families has been proposed which delimits groups which are both easily communicated and which are probably monophyletic. The classification is outlined in Cantino [in Ann. Missouri Bot. Gard. **79**: 361–379 (1992); in Harley, R.M. & Reynolds, T. eds, Advances in Labiatae Science: 27–37 (1992)]; Cantino, Harley & Wagstaff [in Harley, R.M. & Reynolds, T. eds, Advances in Labiatae Science: 511–522 (1992)]; and Thorne [in Bot. Rev. (Lancaster) **58**: 225–348 (1992)]. This delimitation of the Verbenaceae and Lamiaceae is now being accepted and used in the most recent textbooks (e.g., Zomlefer, Flowering Plant Families (1994) and is followed here. Wagstaff & Olmstead in Syst. Bot. **22**: 165–179 (1997) & Olmstead et al. in Am. J. Bot. **88**: 348–361 (2001) suggest that the Lamiaceae and Verbenaceae together do not form a monophyletic group, but are each related to different groups within the Lamiales.

The Verbenaceae is now restricted to subfamily Verbenoideae of traditional classifications (e.g., Briquet in Emgler & Prantl, Nat. Pflanzenfam. **4**, 3a: 132–182 (1895)) which has an indeterminate racemose inflorescence and a hypocrateriform corolla with stamens included; whereas in the Lamiaceae the inflorescence is cymose with determinate, usually opposite cymes and the corollas tubular and usually bilabiate with stamens usually exserted from the tube, but can be held within the lobes. The cymes in Lamiaceae are arranged usually in opposite pairs along an indeterminate axis, forming a thyrse. In some Lamiaceae the cymes are reduced to single flowers though bracteoles are often present below the flower in these cases indicating the cymose, rather than racemose, nature of the inflorescence. These gross morphological differences are supported by anatomical and pollen characters: the Verbenaceae have their ovules attached marginally on the carpel margin (the false septa separating the two carpels into 4 locules) and thickened pollen exine near the apertures; the Lamiaceae have the ovules attached on the face of the false septa (submarginally) and an unthickened exine. Good general accounts of the Verbenaceae can be found in Sanders in Harvard Papers in Botany **5**: 303–358 (2001); and S. Atkins in Kadereit (ed.), Fam. Gen. Vasc. Pl. (Kubitzki, ed. in chief) **VII**: 449–468 (2004). The Verbenaceae as recognized here includes 34 genera and around 1200 species, and is most abundant in temperate South America. Within the Flora Zambesiaca area, the family is represented by 9 genera and 35 species.

The classification of the Lamiaceae as used here follows the account of Harley et al. in Kadereit (ed.), Fam. Gen. Vasc. Pl. (Kubitzki, ed. in chief) **VII**: 167–275 (2004) which is generally consistent with the phylogeny presented by Wagstaff et al. [An ordinal Classification for the Families of Flowering Plants] in Ann. Missouri Bot. Gard. **85**: 531–553 (1998). Subfamilies Ajugoideae (including *Clerodendrum, Teucrium, Rotheca, Kalaharia, Karomia,* and *Caryopteris*) and Viticoideae (including *Premna, Vitex, Gmelina, Tectona* and *Callicarpa*) are dealt with in this part of Flora Zambesiaca. Both of these subfamilies have fruits which either do not split into 4 distinct mericarps or rarely do split late in development to produce 4 nutlets. These nutlets, when they occur, for example in *Teucrium*, have a conspicuous areole or scar which covers at least two-thirds of the nutlet's length. The other subfamilies present in the Flora Zambesiaca region, the *Lamioideae, Scutellarioideae* and *Nepetoideae*, have 4 distinct nutlets with small scars, rarely as long as half the nutlet's length. These subfamilies will be treated in volume 8 part 8 of Flora Zambesiaca.

128. VERBENACEAE

By R. Fernandes

Annual or perennial herbs or shrubs, rarely trees or vines, often aromatic, sometimes armed with prickles and/or spines. Stems frequently square in cross section. Leaves simple, opposite or sometimes whorled; lamina entire or serrate to lobed; stipules absent. Inflorescence indeterminate, terminal or axillary, lax or condensed, racemose, spicate or capitate. Flowers zygomorphic, hermaphrodite, hypogynous, usually with an inconspicuous nectariferous disk, each solitary in the axil of a single bract, sessile

---

* by A Paton.

or pedicellate. Calyx gamosepalous, persistent, campanulate or tubular, 5-lobed or 5-toothed, sometimes expanded or inflated in fruit. Corolla gamopetalous, hypocrateriform (with very narrow tube and abruptly spreading limb), funnel-shaped in *Verbena*, 5-lobed, often somewhat bilabiate (with a 2-lobed upper lip and 3-lobed lower lip), variously coloured. Stamens 4, didynamous, epipetalous; filaments distinct; anthers dorsifixed, dehiscing longitudinally, introrse. Gynoecium of 1 pistil, 2-carpellate, though sometimes reduced to one; ovary superior, 2-locular but typically appearing 4-locular due to ovary wall intrusions (false septa), usually slightly to moderately 4-lobed; ovules 1 in each apparent locule, usually anatropous, erect, placentation axile with ovules attached to the margin of the false septum (directly to carpel margins, not sub terminal); style 1, terminal; stigma 1, capitate or lobed. Fruit a drupe with 2 or 4 pyrenes, or a schizocarp splitting into 2 or 4 nutlets (mericarps), enclosed or subtended by persistent calyx; endosperm absent; embryo straight, oily.

A family of c. 36 genera, mainly tropical and subtropical, with only a limited number of representatives (usually herbs) in temperate regions.

Ornamental shrubs and trees which have been introduced into the Flora Zambesiaca region, and which are not otherwise dealt with in the text, are treated briefly below and included in the generic key that follows.

Citharexylum fruticosum L., Syst. Nat., ed. 10, **2**: 1115 (1759), as "*fruticus*". —O.E. Schulz in Urban, Symb. Antill. **6**: 61 (1909). —Moldenke in Lilloa **4**: 310 (1939). —H. Moldenke & A. Moldenke in Rev. Fl. Ceylon **4**: 284 (1983). Type from Barbados.
    Citharexylum cinereum L., Sp. Pl., ed. 2: 872 (1763). —Schauer in A. de Candolle, Prodr. **11**: 611 (1847).

A shrub or tree up to 16 m tall, without spines; branches and branchlets 4-angled, glabrous. Leaves opposite, petiolate; lamina 4.5–21 × 1.5–8 cm oblong, elliptic to obovate, acute or very shortly acuminate to obtuse or rounded and sometimes emarginate at the apex, more or less cuneate towards the base and there with 1–3 pairs of glands, entire, chartaceous to coriaceous, shiny and glabrous on both faces or shortly pubescent beneath. Racemes mostly terminal, 3.5–26.5 cm long, usually lax, nutant or pendulous, many-flowered. Flowers odoriferous, usually functionally dioecious. Calyx 2.5–4 mm long, cup-shaped, thin, with the apex very shortly 5-dentate or truncate. Corolla white or greenish-white; corolla tube 2–6 mm long. Drupe up to 13 × 12 mm, oblong or subglobose, pink or red to orange or purple when immature, then purplish or black at maturity, surrounded at the base by the persistent, indurated calyx.
    Mozambique. M: Maputo (Lourenço Marques), Jardim Tunduru (Vasco da Gama), 20.iv.1971, *Balsinhas* 1836 (K; LISC; LMA, as *C. spinosum*).
    Native of Florida, southern United States, West Indies and northern South America. Cultivated in some countries.
    Balsinhas 1836 has leaves shortly and finely pubescent beneath and perhaps belongs to var. *subvillosum* Moldenke. However, considering the great variation of the species, this variety is not recognized here.
    Brummitt 9222 (K), from a plant in cultivation on Maone Estate, Blantyre Distr., Malawi, is perhaps *C. spinosum* L.

Petrea volubilis L., Sp. Pl.: 626 (1753). —Moldenke in Repert. Spec. Nov. Regni Veg. **43**: 32 (1938); in Phytologia **7**: 446 (1961); in Ann. Missouri Bot. Gard. **60**: 86 (1973). —Biegel, Check-list Ornam. Pl. Rhod. Parks & Gard. [Rhodesia Agric. J., Research Report No. 3]: 83 (1977). —Puff in J. S. African Bot. **44**, 2: 119 (1978). —H. Moldenke & A. Moldenke in Rev. Fl. Ceylon **4**: 274 (1983). —Verdcourt in F.T.E.A., Verbenaceae: 2 (1992). Type a specimen in Hortus Siccus Cliforttianus (BM).

Shrub, small tree or woody climber to 13 m tall. Leaves 3–21 × 1.4–10.6 cm, oblong or elliptic, obtuse or shortly acuminate, entire but ± undulate on the margin, coriaceous, rather scabrid on both surfaces, nerves prominent; petiole 4–13 mm long. Racemes axillary and solitary, very abundant, erect to pendulous, lax, 8–29 cm long, many-flowered; pedicels c. 8 mm long at anthesis. Flowers purple, blue or mauve. Fruiting calyx of accrescent, linear-oblong venose, stiffly divergent lobes up to 22 × 7 mm, persisting after the corolla has fallen.
    Zimbabwe. N: Zvimba Distr., Rainbow Nurseries, Banket, c. 950 m, 27.x.1976, *Biegel* 5401 (SRGH). **Malawi**. S: Zomba, Zomba Botanical Garden, below Government Hostel, 14.ix.1979, *Masiye & Tawakali* 193 (SRGH). **Mozambique**. M: Maputo, Municipal Garden, 17.vi.1941, *Hornby* 4107 (LISC; LMA).
    Native to Mexico, Central America and Antilles, it was introduced in many countries of tropical and subtropical America, Oceania and Asia. In Africa, it is cultivated in some gardens.
    J. Scott s.n. (SRGH), Rainbow Nurseries, Banket, Zimbabwe, was determined as *P. racemosa* Nees by Moldenke, but is here considered to be the same as *P. volubilis*.

Key to the genera in the Flora Zambesiaca area

1. Fertile stamens 2; each flower ± sunk into an excavation of the rhachis; fruit of 2 mericarps, shortly beaked, striate on the back and usually pitted near the apex · · · 6. **Stachytarpheta**
 – Fertile stamens 4; flowers not sunk in excavations of the rhachis; fruit not as above · · · · 2
2. Flowers sessile or shortly pedicellate, arranged in spicate inflorescences (the inflorescences sometimes contracted and head-like at anthesis becoming ± elongate in fruit) or in lax racemes; fruit usually dry, (drupaceous in *Lantana*); plants annual or perennial herbs, undershrubs or shrubs · 3
 – Flowers distinctly pedicellate, arranged in elongate racemes; fruit drupaceous; plants shrubs, trees or woody vines · 8
3. Fruit drupaceous, purple-black or lavender in colour, usually with a thick, fleshy exocarp and hard endocarp · 2. **Lantana**
 – Fruit dry, dehiscing into 2 or 4 portions (mericarps), occasionally in *Chascanum* the mericarps remaining united · 4
4. Fruit dehiscing into 4 mericarps, each 1-seeded; plants naturalized (with the possible exception of *Verbena officinalis* subsp. *africana*) · 1. **Verbena**
 – Fruit dehiscing into 2 mericarps, occasionally these remaining united (*Chascanum*) · · · · 5
5. Mericarps usually 2-locular and 2-seeded (rarely 1-locular and 1-seeded), each mericarp with 2 combs of spines on the dorsal face; calyx covered with hooked hairs, persistent and becoming inflated and enveloping the mature fruit · 7. **Priva**
 – Mericarps usually 1-locular and 1-seeded, not echinate-spinulose on the dorsal face; calyx without hooked hairs and not becoming inflated · 6
6. Flowers in long terminal spikes or short spike-like racemes; mericarps ± strongly reticulately ribbed, separating from one another or remaining united when mature; calyx tubular, conspicuous · 5. **Chascanum**
 – Flowers mostly in axillary cylindric or subcapitate pedunculate spikes, these often elongating in fruit; mericarps smooth, separating from one another; calyx campanulate or compressed, inconspicuous · 7
7. Erect perennial herbs, undershrubs or shrubs · 3. **Lippia**
 – Trailing herb, rooting at the nodes · 4. **Phyla**
8. Calyx 5-lobed, the lobes much longer than the calyx tube, blue or violet, net-veined, very accrescent and stiffly diverging; fruit with 2 pyrenes, each 1-locular and 1-seeded; woody climber with entire leaves (cultivated) · **Petrea**
 – Calyx shortly 5-dentate (the teeth shorter than the tube), or truncate; fruit with 2–4 pyrenes, each 2-seeded; trees or shrubs, often spinose · 9
9. Leaves with 1–3 pairs of prominent glands at the base of the lamina; calyx cupuliform or patelliform, shorter than the fruit and partially surrounding it; fruit of two 2-seeded pyrenes (cultivated) · **Citharexylum**
 – Leaves without glands at the base of the lamina; calyx cylindric-tubular or subcampanulate, very accrescent, completely enclosing the fruit; fruit of four 2-seeded pyrenes · · 8. **Duranta**

1. VERBENA L.

Verbena L., Sp. Pl.: 18 (1753); Gen. Pl., ed. 5: 12 (1754). —Verdcourt in F.T.E.A., Verbenaceae: 6–10 (1992). —Atkins in Kadereit (ed.), Fam. Gen. Vasc. Pl. (Kubitzki, ed. in chief) **VII**: 463 (2004).

Annual or perennial herbs, small shrubs, woody climbers or trees. Stems 4-angled, erect ascending or procumbent, glabrous or hairy. Leaves opposite and usually decussate, rarely 3-whorled or alternate, dentate, or often incised or variously lobed or partite, rarely entire. Inflorescences spicate with the spikes terminal, sometimes axillary, simple or corymbosely or paniculately arranged; spikes densely crowded, or elongate with distant flowers. Flowers small or medium-sized, sessile, each solitary in the axil of a narrow bract. Calyx tubular, 5-angled and ribbed on the angles, unequally 5-toothed, ± dilated at the base in the fruiting stage and then ventricose, eventually splitting. Corolla funnel-shaped or hypocrateriform; tube straight or curved, cylindric or widening slightly towards the apex, usually villous within particularly at the mouth and about the insertion of the filaments; limb spreading, oblique, weakly 2-lipped, with unequal obtuse lobes, rounded or emarginate at the apex. Stamens 4, included, didynamous; filaments inserted about or above the middle of the corolla tube; anthers ovate, with parallel or diverging thecae without appendages, or the connectives of upper 2 stamens each provided with a clavate or glanduliform appendage. Ovary 2-carpellate, entire or slightly lobed at the apex, 4-locular at anthesis, each locule l-ovulate; ovules anatropous, erect, attached laterally at or near the base; style usually equalling the stamens, divided at the apex into a short, stigmatic anterior lobe and an acute horny sterile posterior tooth. Fruit a schizocarp, enclosed within the persistent calyx, separating when ripe into four 1-seeded mericarps; mericarps linear or oblong, with dry pericarp, longitudinally ribbed with a raised reticulation toward the apex. Seeds oblong; endosperm lacking or scanty.

A genus of c. 350 taxa (including species and infraspecific taxa), mostly native to temperate and tropical America, with 2 species in Europe, Asia and N Africa. Several species, and their hybrids, are introduced in the Old World as garden ornamentals.

The concept of *Verbena* employed here includes *Glandularia* J.F. Gmel. which has often been recognized as a separate genus, e.g. by Sanders, R.W., Harvard Papers in Botany **5**, 2: 303–358 (2001). The treatment of *V. bonariensis* L., *V. brasiliensis* Vell. and *V. litoralis* Kunth by Dr. P. Yeo, in Kew Bull. **45**: 101–120 (1990), is followed here.

The species found in the Flora Zambesiaca area, with the exception of *Verbena officinalis* subsp. *africana*, are all introduced and most have become naturalised. Taxa which occur in the Flora Zambesiaca area only as cultivated garden ornamentals include: *Verbena wrightii* A. Gray, *Verbena* × *hybrida* Groenl. & Rümpler and *Verbena* × *teasii* Moldenke.

Cultivated taxa

Verbena wrightii A. Gray, Syn. Fl. N. Amer. **2**: 337 (1878).

Branches and peduncles hispid-hirsute with long whitish hairs. Leaves 2–4 cm long, bipinnatifid or trifid with segments deeply incised or divided, the ultimate segments short, 1.5–2.5 mm broad, oblong, acute; blades hirtellous or hirsute on both surfaces, mainly beneath on the nerves; petiole short or absent. Spikes solitary on short peduncles, rather compact and head-like in flower, elongating slightly in fruit. Calyx 7–9 mm long, angular with green angles, densely long hispid-ciliate and with gland-tipped hairs; calyx teeth unequal, 1–1.5 mm long, triangular at the base and linear-subulate at the apex. Corolla pink or rose to magenta; tube 10.5–12 mm long, exceeding the calyx by 4–7 mm, villous outside; limb 6–8 mm wide, villous at the base; lobes retuse.

Zimbabwe. C: Harare, c. 1440 m, v.1924, *D. King* in *Eyles* 4349 (K; SRGH).

Native of N America (west, central and east of U.S.A. and Mexico), cultivated as a garden ornamental in some countries.

Verbena × hybrida Groenl. & Rümpler. —Biegel, Check-list Ornam. Pl. Rhod. Parks & Gard. [Rhodesia Agric. J., Research Report No. 3]: 106 (1977).

This is the common garden verbena with white, pink, red, blue, purple or variegated flowers and is easily distinguished by the leaves lobed or toothed and corolla limb up to c. 2 cm wide.

Zambia. C: Chilanga, Mount Makulu Research Station.

Verbena × teasii Moldenke

Stems decumbent or ascending, much branched; branches ascending, slender. Leaves variable in shape and size, ± deeply incised-pinnatifid, often ± 3-partite the lowest divisions usually again pinnatifid; body of blade and lowest lobes relatively broad; lobes sharply acute. Spikes flattened and almost capitate, later elongating to c. 15 cm, densely many-flowered. Bractlets c. one third as long as the calyx. Calyx 7–13 mm long. Corolla showy, blue, purple, pink or white; tube 15–20 mm long; limb 5–11 mm wide.

Mozambique: M: Maputo, Umbelúzi Experimental Station.

Key to the species in the Flora Zambesiaca area

1. Leaves undivided, the margins at most only serrate or incised-serrate · · · · · · · · · · · · · · 2
- Leaves ± deeply divided · 5
2. Leaves petiolate, or if sessile then not amplexicaul, attenuate towards the base; spikes not dense (except at the top), elongate, up to 20 cm long, lax below in fruiting specimens, slender · 4. *litoralis*
- Leaves sessile, at least the median ones widening or subcordate at the base, amplexicaul; spikes dense and shorter, even in fruiting specimens · 3
3. Corolla showy, tube 9–10 mm long, limb 7–10 mm in diameter; floral bracts 2–3 times as long

as the calyx; plant a creeping rhizomatous herb with aerial stems 15–60 cm tall ·· 1. *rigida*
– Corolla not showy, tube up to 5.5 mm long; limb up to 3 mm in diameter; floral bracts ±
   equalling the calyx; plants not creeping rhizomatous herbs, stems erect usually robust and
   up to 2.5 m tall ·· 4
4. Uppermost corolla limbs overtopping the spike; floral bracts mostly slightly shorter than
   the calyces, the bracts at the spike apex exceeded by the uppermost corolla limbs; hairs on
   the leaf lower surface short and ± closely crowded on the venation, somewhat hyaline ···
   ··· 2. *bonariensis*
– Uppermost corolla limbs not overtopping the spike; floral bracts equalling or slightly
   longer than the calyces, the uppermost bracts forming a short tufted apex to the spike
   below which the uppermost corolla limbs form a ring; hairs on the leaf lower surface
   venation longer and less crowded, whitish ···························· 3. *brasiliensis*
5. Corolla small, tube c. 2.5 mm long, limb c. 2 mm in diameter, not showy; calyx 2–2.5 mm long;
   style c. 1 mm; mericarps 1.5–2 mm long; spikes slender, becoming very elongate in fruit;
   median leaves tripartite to pinnatifid, up to 10.5 × 8 cm ······ 5. *officinalis* subsp. *africana*
– Corolla larger, tube 10–13 mm long, limb more than 6 mm in diameter, showy; calyx 7–9
   mm long; style 9–13 mm long; mericarps 3–3.5 mm long; spikes at first ± condensed or
   head-like with flowers arranged on a contracted axis; leaves smaller, more deeply divided,
   with narrower segments ·· 6
6. Stems strigose with short appressed antrorse hairs; calyx strigose and with some patelliform
   glands; leaves 2-pinnatisect or 2-trisect, with the ultimate segments linear, c. 1 mm broad ·
   ·· 6. *aristigera*
– Stems densely hirsute; calyx hirsute with glandular hairs interspersed, without patelliform
   glands; leaves 2-pinnatifid or 3-fid, with the ultimate segments oblong, a little broader
   (cultivated) ··· *V. wrightii*

1. **Verbena rigida** Spreng., Syst. Veg., ed. 16, **4**, 2: 230 (1827). —Cabrera, Man. Fl. Alreded.
   Buenos Aires: 395 (1953). —Moldenke in Phytologia **11**: 62, 80 (1964). —Backer &
   Bakhuizen van den Brink Jr., Fl. Java **2**: 596 (1965). —Leon & Alain, Fl. Cuba **4**: 282, fig.
   121B (1965). —Adams, Fl. Pl. Jamaica: 628 (1972). —Franco in Fl. Europaea **3**: 123 (1972).
   —H. Moldenke & A. Moldenke in Rev. Fl. Ceylon **4**: 203 (1983). —Verdcourt in F.T.E.A.,
   Verbenaceae: 8 (1992). Type from Uruguay.
        Verbena venosa Gillies & Hook. in Hooker, Bot. Misc. **1**: 167 (1829). —W.J. Hooker in
   Bot. Mag. **59**: t. 3127 (1832). —Walpers, Repert. Bot. Syst. **4**: 27 (1845). —Schauer in A.
   de Candolle, Prodr. **11**: 541 (1847). —Briquet in Engler & Prantl, Nat. Pflanzenfam. **4**, 3a:
   147 (1897). —H. Pearson in F.C. **5**, 1: 208 (1901). —J.H. Ross, Fl. Natal: 299 (1972). Type
   from Argentina.
        Verbena bonariensis var. *rigida* (Spreng.) Kuntze, Revis. Gen. Pl., part 3: 255 (1898). —
   Briquet in Annuaire Conserv. Jard. Bot. Genève **7–8**: 291 (1904). Type as for *Verbena rigida*.

   Further synonymy and bibliography can be seen in Moldenke & A. Moldenke, loc. cit.
(1983).

   Perennial mat-forming herb from a creeping rhizome, with aerial stems 15–60 cm
tall. Stems simple or little-branched, ascending-erect, decumbent or creeping at the
base, c. 3 mm in diameter at the base, sharply 4-angled, scabrous-pubescent to hispid
with short bristle-hairs intermixed with glandular-headed hairs, sulcate; internodes
usually shorter than the leaves, sometimes elongate. Leaves sessile, 3–12 × 1–2.5(4)
cm, oblong, narrowly obovate, or oblong-lanceolate, acute at the apex, ± cuneate in
the lower leaves and subcordate to amplexicaul in upper leaves, unequally and
coarsely incised-serrate with rather distant teeth, ± stiff, scabrous, ± hispidulous on
both surfaces or callose-strigose above and shortly hispid on the nerves beneath, pale
green; midrib and lateral nerves impressed above, prominent beneath.
Inflorescences of solitary terminal spikes or 3–5 spikes in dichotomous cymes; spikes
0.5–2.5 cm long, up to 5 cm long in fruit, cylindrical, dense, elongating in fruit;
peduncles up to 6 cm long; bracts ± closely imbricate until fruiting, 2–3-times as long
as the calyx, lanceolate to long attenuate-subulate, acute at the apex, mostly glandular-
pubescent, ciliate on the margins, purple toward the apex; flowers showy. Calyx
hidden by a subtending bract (3.5)4–6 mm long, cylindric at flowering time, widening
below in fruit, glandular hairy, red or green. Corolla tube 8–10 mm long, 2–4 times
as long as the calyx, exceeding the calyx teeth by 5–8 mm, slender, cylindric, curved,
pubescent outside and inside in upper part, purple to violet, blue-violet or mauve in
the exposed portion, white in the lower part; limb 7–10 mm in diameter, slightly hairy

at the mouth, with the lobes broad, emarginate or bilobed, the median one longer. Stamens inserted near middle of the tube. Style about half as long as the corolla tube. Mericarps c. 2 mm long, raised-reticulate in the upper part, with 5 slender costae towards the base, minutely densely whitish tuberculate at the commissural face.

Zimbabwe. C: Harare Distr., Dombashawa road, c. 17 km from Harare (Salisbury), 20.ii.1976, *Pope & Biegel* 1541 (BR; SRGH).
Native to South America, from central Brazil to northern Argentina; introduced in the West Indies and southern United States of America, and in Kenya, Tanzania, South Africa (Transvaal, KwaZulu-Natal, Cape Provinces) Swaziland and Madagascar. A garden escape, now naturalized, grassland on rocky dry slope in doleritic soil; c. 1600 m.

2. **Verbena bonariensis** L., Sp. Pl.: 20 (1753). —Moldenke in Phytologia **8**: 246 (1962). — Franco in Fl. Europaea **3**: 123 (1972) pro parte. —Yeo in Kew Bull. **45**: 105, fig. 2 (1990). —Verdcourt in F.T.E.A., Verbenaceae: 9 (1992). Lectotype from Argentina (= *Linnaean Herb.* 35.11, chosen by Yeo).
   Verbena trichotoma Moench, Suppl. Meth. Pl.: 131 (1802) *nom. superfl., illegit.*
   Verbena bonariensis var. *conglomerata* Briq., Arkiv. Bot. **2** (4) no. 10: 11, 3B (1904). Type from Brazil.

Stiffly erect perennial herb up to 2.5 m tall. Stem simple or little-branched, ascending-erect, strongly 4-angled, with pale ribbed angles and flattish faces, somewhat constricted just below the nodes, scabrid to hispid or hirsute; branches mostly 5–9 cm long; indumentum sparse, consisting of spreading bristle-like hairs, sometimes intermixed with short glandular hairs. Leaves (3)5–12(17.5) cm long and up to 3 cm wide, ovate, ovate-lanceolate or oblong, acute at the apex, narrowed above the base and enlarged below into a subauriculate, semiamplexicaul base, serrate to double serrate at the margin, the base sometimes entire; lamina stiff, scabrid with patent acicular tubercle-based hairs, the indumentum densest beneath particularly on the nervation; midrib and nervation impressed above, raised beneath; the upper leaves 7–10.5 cm long and narrowly oblong to linear-lanceolate, the uppermost becoming bract like, sometimes the uppermost entire. Inflorescence of numerous spikes in a dense flat-topped or rounded cluster, the spikes cymosely arranged and mostly in dense groups of triads; spikes densely flowered, up to 4.5 cm long and 4–7 mm thick, ovoid becoming cylindric in fruit, the central one sessile, the lateral ones subsessile; bracts usually a little shorter than the calyx, 2.5–3(4) mm long, lanceolate, attenuate towards the acute apex, pilose with some minute glandular hairs intermixed, ciliate on the margins. Calyx (2.5)3–3.5 mm long, hispidulous outside with small glandular hairs intermixed. Corollas deep purple, the uppermost overtopping the spike; tube (3.25)3–4.75(5–5.5) mm, exserted above the calyx teeth for (0.5)0.75–1.5 mm; limb c. 3 mm in diameter. Stamens inserted near middle of the corolla tube; anthers 0.5 mm long. Style (1.25)1.5 mm long. Mericarps 1.5–2.1 mm long.

Zambia. S: Choma Distr., Muckle Neuk, c. 19 km north of Choma, 1344 m, l0.x.1954, *E.A. Robinson* 905 (K; SRGH). **Zimbabwe**. N: Shamva, 22.xii.1941, *Pardy* in *GHS* 8488 (SRGH). C: Harare, *Eyles* 6948 (K; SRGH). E: Mutare Distr., Vumba, 1756 m, 23.ii.1960, *Head* 47 (BM). **Mozambique**. M: Moamba Distr., between Moamba and Sábiè, banks of Incomáti R., 4.xii.1940, *Torre* 2239 (COI; LISC).
Native to Argentina, Uruguay and Chile. Now widely naturalized in southern North America, Europe, Asia and Australia, and in Africa in Kenya, Tanzania, Dem. Rep. Congo, South Africa (KwaZulu-Natal, Transvaal, Free State and Cape Province), Lesotho and Swaziland; also in Madagascar and Réunion. In damp soil in grassland, in headwaters of streams and on river banks; also in old cultivation and on roadsides; 65–1500 m.

3. **Verbena brasiliensis** Vell., Fl. Flum.: 17 (1829); t. 40 (1831). —Yeo in Kew Bull. **45**: 111, fig. 3 (1990). —Verdcourt in F.T.E.A., Verbenaceae: 9 (1992). Type from Brazil (lectotype Fl. Flum., t. 40, chosen by Verdcourt).
   Verbena quadrangularis Vell., Fl. Flum.: 16 (1829); t. 39 (1831). Type from Brazil (lectotype Fl. Flum., t. 39, chosen by Verdcourt).
   Verbena litoralis var. *brasiliensis* (Vell.) Briq. in Annuaire Conserv. Jard. Bot. Genève **7–8**: 292 (1904).
   Verbena bonariensis sens. auct. plur., non L. (1753).

Perennial herb up to c. 2.5 m tall. Stem stiff, branched above, 4-angled, with whitish angles and flat faces, up to 7 mm wide at the base, somewhat constricted just below each node, scabrid and sparsely to densely hispid with tubercle-based hairs up to 1.5 mm long, the bristles whitish spreading and denser along the angles; branches 4–9 cm long, suberect or forming a wide V with the stem. Leaves (1.4)5–8(12) cm long and 1.2–3.5(4.5) cm wide, oblong or ovate-oblong to lanceolate, acute at the apex, semiamplexicaul at the base, serrate to double serrate on the margin, the teeth ± acute, the base sometimes entire; lamina stiff, scabrid with patent acicular tubercle-based hairs, the indumentum densest beneath particularly on the nervation; midrib and nervation impressed above, raised beneath; the upper leaves smaller and relatively narrower, sometimes linear-lanceolate, sometimes entire. Inflorescences much branched, consisting of numerous spikes in dense flat-topped or rounded clusters, the spikes cymosely arranged and mostly in dense groups of triads, of which the median spike is usually longer and subsessile, and the lateral ones on peduncles up to 12 mm long; spikes densely flowered, 0.5–1 cm long and conical when in flower, 3.5–8(10–12) cm long and 0.3–0.6 cm thick and cylindric when in fruit, sometimes arched; bracts usually equalling or slightly overtopping the calyx, the upper ones distinctly longer than the buds and forming a short tuft at the top of the spike, 2.5–3(3.5–4) mm long and 0.75–1 mm wide at the base, lanceolate, attenuate towards an acute sometimes ± filiform apex, pilose without glandular hairs intermixed, ciliate on the margins. Calyx (1.75)2–2.75(3) mm long, with acute teeth, hispidulous outside. Corolla mauve, purple or bluish-purple, pubescent outside, the uppermost not overtopping the spike except sometimes at the end of flowering; tube (2.5)2.75–3(3.5) mm long, (0)0.5–0.75(1–1.25) mm exserted above the calyx teeth; limb 2.75–3 mm in diameter. Stamens inserted near middle of the corolla tube; anthers 0.25–4 mm long. Style (0.75)1–1.25 mm long. Mericarps 1.2–1.7 mm long, with 3–5 ribs on the back, and slightly raised-reticulate in the upper part.

Botswana. SE: South East Distr., Gaborone Dam, 24°42'S, 25°54'E, 1050 m, 20.xii.1978, *O.J. Hansen* 3556 (K). **Zimbabwe**. W: Umzingwane Distr., Esigodini (Essexvale), Ncema River, c. 1088 m, 28.xii.1965, *Simon* 601 (K; LISC; SRGH). C: Harare, University campus, 21.iv.1963, *Loveridge* 675 (K; SRGH). E: Mutasa Distr., Penhalonga, 26.viii.1937, *McGregor* 82 (BR; K; SRGH). S: Masvingo Distr., Kyle Ferry, Lake Mutirikwi, 12.ii.1970, *Chiparawasha* 142 (SRGH). **Mozambique**. MS: Sussundenga Distr., Serra Zuira, on Chimoio (Vila Pery) road, 2100 m, 2.iv.1966, *Torre & Correia* 15578 (LISC). GI: Massingir Distr., rio dos Elefantes right bank, on way to Albufeira de Massingir (Lagoa Nova), 19.i.1973, *Lousã & Rosa* 300 (LMA). M: Matutuíne Distr., Catuane, banks of Maputo R., 15.v.1964, *A. Moura* 69 (COI; LMU).

Native of tropical South America, now widely naturalized in the Malaysian Peninsula, Indonesia, tropical Australia, etc., and in Africa in Kenya, Tanzania, South Africa (KwaZulu-Natal and Transvaal), Madagascar and Réunion.

In damp soil in grassland, in headwaters of streams and on river banks; also in old cultivation and on roadsides; 37–2270 m.

This widespread species has often been confused with *V. bonariensis* L. However, it may be distinguished from this species by its spikes in which the uppermost floral bracts form a short tufted apex not overtopped by the uppermost corolla limbs, and by its floral bracts being equal to or slightly longer than the calyces (mostly shorter in *V. bonariensis*), and in its leaf indumentum in which the hairs on the leaf lower surface venation are longer and less crowded than in *V. bonariensis*, whitish not hyaline, and by its anthers smaller and inserted higher up in the corolla tube than in *V. bonariensis*.

4. **Verbena litoralis** Kunth in Humboldt, Bonpland and Kunth, Nov. Gen. Sp. **2**: 276 (1818); t. 137 (1817). —Yeo in Kew Bull. **45**: 115, fig. 4 (1990). Type from Peru.

    Verbena caracasana Kunth in Humboldt, Bonpland and Kunth, Nov. Gen. Sp. **2**: 275 (1818). Type from Venezuela (Caracas).

    Verbena brasiliensis sensu auct. plur. non Vell. (1829).

Perennial herb or sometimes a shrub 0.2–2(2.10 m) tall. Stem erect or suberect, branched above, 4-angled, with whitish angles and flat faces, somewhat constricted just below each node, glabrous or sparsely hispid with short tubercle-based hairs, sometimes ± hirsute in the upper part. Leaves 2–5(11) cm long and 1–2 cm wide, ovate, rhombic-ovate, spathulate or oblanceolate, acute at the apex, tapering to the base into a narrow petiole, or subsessile, not amplexicaul, serrate to double serrate on the margin, the teeth sometimes obtuse; lamina stiff, scabrid with sparse short appressed antrorse tubercle-based hairs on both surfaces, the bristles more numerous on the lower surface particularly on the venation; midrib and nervation slightly impressed above, raised beneath; the upper leaves up to 5 × 1 cm, oblanceolate or lanceolate to linear, tapering towards the base, sometimes petiolate. Inflorescences much branched, with several to numerous corymbose-paniculately arranged spikes, these mostly in lax groups of triads of which the median spike is usually longer and subsessile and the lateral ones on peduncles up to 5–8 mm long, these sometimes replaced by branches bearing also triads; spikes slender or filiform, up to 20 cm long in fruit, the youngest flowers closely spaced at the spike apex and the maturing flowers becoming more widely spaced toward the base of the spike as the rhachis elongates; bracts green, (1.25)2–3 mm long, slightly shorter than or equalling the calyx, lanceolate, ± attenuate towards the acute apex, hispidulous outside and shortly closely ciliate. Calyx (1.5)2–3 mm long in flower, hirsute, hispidulous or tomentose. Corolla lavender or lilac to violet or violet-blue, 3–3.5 mm long, not overtopping the apex of the spike, pubescent outside; tube exserted from the calyx for 0.5–1 mm; limb inconspicuous, 2.5–3 mm wide, with subequal, obtuse lobes. Style c. 1 mm long. Mericarps 1.2–1.7 mm long, longitudinally striate on the back and raised-reticulate in the upper part, minutely densely whitish tuberculate at the commissural face.

Zambia. W: Kitwe, 27.iv.1966, *Mutimushi* 1398 (K; SRGH). **Zimbabwe**. W: Bulawayo Distr., Matschemshlope R., near Bulawayo Golf Course, 1370 m, 26.xi.1972, *Simon* 2289 (K; LISC; SRGH).

Native of South America and now widely naturalized as a subtropical and tropical weed. In damp soil in grassland, in headwaters of streams and on river banks; also in old cultivation and on roadsides; c. 1300 m.

5. **Verbena officinalis** L., Sp. Pl.: 20 (1753). —Franco in Fl. Europaea **3**: 123 (1972). — Townsend in Fl. Iraq **4**: 651 (1980); in Fl. Turkey **7**: 33, t. 117/8–10 (1982). —Meikle, Fl. Cyprus: 1251 (1985). —Verdcourt in F.T.E.A., Verbenaceae: 6 (1992). Syntypes from the Mediterranean.

Subsp. **africana** R. Fern. & Verdc. in Bol. Soc. Brot., sér. 2, **62**: 305, fig. 1 (1990). —Verdcourt in F.T.E.A., Verbenaceae: 8, fig. 1 (1992). TAB. **3**. Type: Zimbabwe, Harare, between Avondale West and Mabelreign, 1480 m, *Drummond* 4858 (K, holotype; B; BR; LISC; S; SRGH).

   Verbena officinalis var. *natalensis* Krauss in Flora **28**: 68 (1845). —Moldenke in Phytologia **36**: 279 (1977) *nom. nud.*

   Verbena officinalis sensu auct. Fl. Afr. pro parte, non L. (1753).

An erect perennial, rarely annual, herb 60–160 cm tall. Stem leafy, unbranched below and with a few long inflorescence branches above, ± woody and subterete at the base, 4-angled above with angles hispid and faces flat or slightly sulcate, the faces sparsely hispid with short tubercle-based hairs intermixed with short glandular hairs. Leaves 4–10.5 cm long and 2.5–8 cm wide, ± ovate in outline, 3-partite to pinnatifid, acute or obtuse at the apex, attenuate or narrowly cuneate at the base into a slightly winged petiole, semiamplexicaul; terminal leaf-segment up to 4.5 cm long, deeply lobed with lobes dentate; lateral segments up to 3.2 cm long and bluntly serrate, narrower and more deeply divided and with more numerous teeth in mid-stem leaves; lamina scabridulous, shortly appressed hispidulous and drying dark green on the upper surface, sparsely shortly spreading setose on the lower surface, with the midrib and main lateral nerves prominent and densely hispid below, slightly revolute on the margin; leaves of upper stem and branches often lanceolate or oblanceolate to linear in outline, acute, subsessile, sparsely toothed to subentire. Inflorescences of numerous elongate terminal and axillary slender spikes 5.5–26 cm long, rhachis and floral bracts glandular-pubescent, the youngest flowers closely spaced at the spike apex and the maturing flowers becoming more widely spaced toward the base of the spike as the rhachis elongates; bracts ± equalling the calyx in length, lanceolate or ovate-lanceolate, attenuate-acuminate, covered with short whitish hairs interspersed with sparse glandular hairs on the outside, ciliate on the margins. Calyx 2–2.5 mm long in flower, subcylindric, glandular pubescent outside, teeth c. 0.4 mm long, fruiting calyx slightly overtopping the bracts. Corolla small, lilac or mauve to magenta, purple, violet or blue, glabrous; tube c. 2.5 mm long, narrowly cylindric, constricted below the limb; limb subequally 5-lobed, with lobes 0.75–2 mm long,

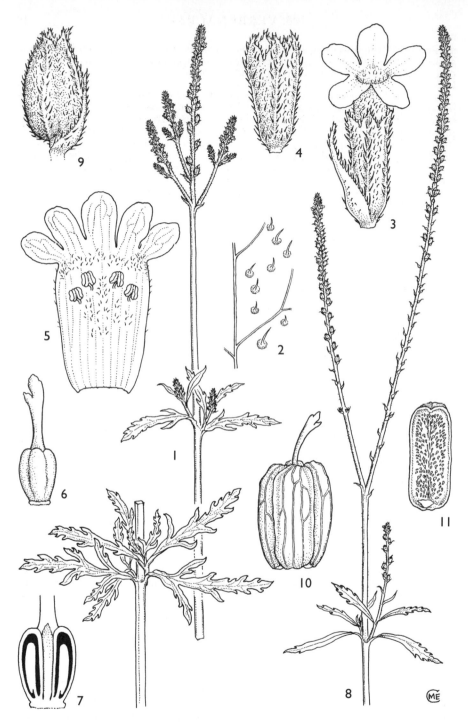

Tab. 3. VERBENA OFFICINALIS subsp. AFRICANA. 1, part of flowering stem (× 2/3); 2, detail of upper leaf surface (× 8); 3, flower (× 14); 4, calyx (× 14); 5, corolla, opened out (× 14); 6, ovary (× 20); 7, longitudinal section of ovary (× 30), 1–7 from *Gillett* 16559; 8, fruiting stem (× 2/3); 9, fruiting calyx, enclosing the fruit (× 12); 10, fruit (× 16); 11, nutlet (× 16), 8–11 from *Tweedie* 1529. Drawn by M.E. Church. From F.T.E.A.

lower lobe 1.75 mm wide, upper lobe 1.25 mm wide. Style c. 1 mm long. Mericarps 1.5–2 mm long, 3–4-striate, longitudinally raised-reticulate in upper part, minutely densely whitish tuberculate at the commissural face.

Zambia. C: Lusaka, 14.viii.1963, *E.A. Robinson* 5596 (K; LISC; M; S). S: c. 19 km north of Choma, c. 1350 m, 25.x.1958, *E.A. Robinson* 2903 (K; SRGH). **Zimbabwe**. N: Mazowe Distr., 5.xii.1961, *Whellan* 1878 (BR; SRGH). C: Harare, Marlborough suburb, xii.1957, *Lennon* 52 (BR; K; SRGH). E: c. 5 km from Mutare (Umtali), c. 1300 m, 29.xii.1930, *Fries, Norlindh & Weimarck* 4002 (K; LISU; M).

Also in Dem. Rep. Congo, Sudan, Ethiopia, Uganda, Kenya, Tanzania, South Africa (KwaZulu-Natal, Transvaal) and possibly elsewhere in tropical Africa; also in Nepal and Taiwan. Very local then frequent in damp soil, in dambos and stream margins and in wooded grassland; also in old cultivation and on roadsides; 1300–1500 m.

Although this taxon has been called *V. officinalis* by some African botanists and collectors it is nevertheless distinct from typical *V. officinalis*. It is distinguished by the leaves more deeply divided, and the lobes more acutely serrate; by the floral bracts which are relatively longer and narrower, being lanceolate or ovate-lanceolate, attenuate-acuminate, up to 2.5 mm long and equalling the calyx (in *V. officinalis* these bracts are ovate, up to 1.5 mm long and shorter than the calyx). It is also distinguished by the flowering calyx which is cylindric, (not ellipsoid as in *V. officinalis*); by the corolla tube which is cylindric becoming constricted just below the limb, (in *V. officinalis* it widens continuously towards the limb); by the style slightly longer (1 mm versus 0.75 mm), and more exserted than in *V. officinalis*; and by the mericarps slightly longer.

The typical taxon occurs in Europe, the Middle East, Asia to China and Indonesia, Canary Islands and North Africa, now introduced almost throughout the world. What is called *V. officinalis* shows a great deal of variation throughout the Old World, and Verdcourt, in F.T.E.A., Verbenaceae: 8 (1992), considers it possible that the East African and Flora Zambesiaca taxon represents an extreme variant introduced into Africa, where it has become naturalized. It is also possible that subsp. *africana* corresponds to an American taxon, because it also has affinities with *V. riparia* Rafin. ex Small & Heller, *V. ehrenbergiana* Schauer and *V. hallei* Small.

6. **Verbena aristigera** S. Moore in Trans. Linn. Soc. London, Bot., **4**: 439 (1895). — Moldenke in Phytologia **8**: 189 (1952). —Verdcourt in F.T.E.A., Verbenaceae: 10 (1992). Type from Brazil.

    Verbena tenuisecta Briq. in Annuaire Conserv. Jard. Bot. Genève **7–8**: 294 (1904); Dict. Gard. **4**: 2209 et 2213 (1951). —Moldenke in Phytologia **11**, 4: 280 (1965); op. cit. **16**: 208 (1968); op. cit. **41**: 403 (1979). —J.H. Ross, Fl. Natal: 299 (1972). —H. Moldenke & A. Moldenke in Rev. Fl. Ceylon **4**: 208 (1983). Type from Paraguay.

    Glandularia tenuisecta (Briq.) Small, Man. Southeast. Fl.: 1139 (1933). —Cabrera, Man. Fl. Alreded. Buenos Aires: 397 (1953). —Troncoso in Darwiniana **13**: 477, fig. 4 (1964). Type as above.

    Glandularia aristigera (S. Moore) Tronc. in Darwiniana **14**: 636 (1968).

    Verbena tenera sensu Binns, First Check List Herb. Fl. Malawi: 104 (1968) et auct. plur. non Spreng. (1825).

    Verbena tenuisecta var. *alba* Moldenke in Phytologia **16**: 210 (1968).

    Verbena tenuisecta forma *rubella* Moldenke, loc. cit. (1968); loc. cit. (1971).*

Perennial prostrate trailing herb 10–80 cm high. Stems often woody at the base, decumbent, sometimes rooting at the lower nodes, tetragonal, sparsely pilose to pubescent or glabrescent. Leaves petiolate, with short sterile densely leafy branchlets or fascicles of smaller leaves in the axils, 2–3.5 × 2–3 cm, ± triangular or ovate in outline, 2-pinnatisect or 3-sect with the segments also pinnatisect or 3-sect; the ultimate segments up to 1 cm long and c. 1 mm wide, linear, obtuse or acute; lamina somewhat thick, revolute on the margins, strigose-pubescent or ± glabrous on both surfaces, sometimes with brownish glands, entire or toothed; petiole up to 2 cm long. Inflorescences of solitary terminal spikes on stems and branches; spikes c. 1.5 × 2 cm, ovoid and dense in flower, elongating to 5–12 cm in fruit; peduncles 1.5–5 cm long, elongating to 10 cm or more in fruit; bracts green, 1/3 to 1/2 as long as the calyx, ovate-lanceolate, densely canescent-puberulous or strigose and sometimes glandular; flowers numerous showy, the upper ones dense and ascending, the lower ones spreading. Calyx 7.5–9 mm long, tubular, becoming wider below and constricted above in fruit, densely appressed whitish strigose-pubescent with some intermixed

---

*A very extensive bibliography and synonymy of *V. tenuisecta* can be seen in Phytology **11**, 4: 280–283 (1965).

darker glands; teeth ovate at the base, and abruptly contracted into a long seta 0.75–2 mm long, unequal. Corolla mauve, lilac, mauve-blue, purple, purple-red, deep pink or sometimes white; tube 10–11 mm long, exserted for 3–5 mm from the calyx mouth, glabrous outside, hairy inside at the throat, with two longitudinal dense bands of white setae below the insertions of the lower stamens; limb c. 10 mm wide, the lobes 3–4 × 3–4 mm, broadly obcordate, emarginate. Stamens included, or with glands just exserted. Style 9–13 mm long, included, persistent. Mericarps c. 3.5 mm long, black, striate, longitudinally raised-reticulate in upper part.

Botswana. SE: 13 km northeast of Gaborone, 24°31'S, 25°58'E, 1050 m, *O.J. Hansen* 3536 (K; SRGH). **Zambia**: W: Mufulira, 10.i.1948, *Cruse* 114 A (K). **Zimbabwe**. W: Bulawayo, Hillside, Weir Avenue, on roadside, iii.l954, *O.B. Miller* 2239 (K; LISC; SRGH). C: Harare, industrial site, 12.xii.1968, *Mavi* 790 (SRGH). E: Mutare Distr., 12.ii.1948, *Chase* 1339 (K). **Malawi**. S: Ntcheu Distr., Gochi Rest House, Kirk Range, c. 1600 m, 30.i.1959, *Robson* 1348 (BM; K); Blantyre Distr., Chichiri, Kamuzu Highway, 11.iii.1969, *Msinkhu* 1 (SRGH). **Mozambique**. GI: Homoíne, Experimental Camp of C.I.C.A., 10.x.1945, *Pedro* 310 (LMA). M: Maputo, Jardim Tunduru (Vasco da Gama Garden) (cult.), 22.xi.1950, *Carvalho* 68 (LMA).

Native of South America (Brazil, Uruguay, Argentina, Paraguay, etc.), now widely cultivated as a garden ornamental, often becoming naturalized. In Africa it also occurs in Egypt and Nigeria (fide *Moldenke*), in Dem. Rep. Congo, South Africa (KwaZulu-Natal and Transvaal) and Swaziland. In *Brachystegia* woodland and grassland; also in old cultivation and on roadsides; 100–1520 m.

Although Moldenke, in Phytologia **8**: 190 (1962) and Phytologia **11**: 284 (1965), treated *V. tenuisecta* Briq. and *V. aristigera* S. Moore as separate species, it is here considered best to follow Troncoso who, in Darwiniana **14**: 636 (1968), considered them to be synonymous.

2. LANTANA L.

Lantana L., Sp. Pl.: 626 (1753); Gen. Pl., ed. 5: 275 (1754). —R. Fernandes in Bol. Soc. Brot., sér. 2, **61**: 125–214 (1989). —Verdcourt in F.T.E.A., Verbenaceae: 37–47 (1992). —Atkins in Kadcrcit (ed.), Fam. Gen. Vasc. Pl. (Kubitzki, ed. in chief) **VII**: 461 (2004).

Shrubs, undershrubs or perennial herbs, usually erect, sometimes scandent, rarely procumbent. Stems and branches tomentose, scabrid hairy, or sparsely to densely armed with prickles. Indumentum usually of simple appressed or spreading hairs, gland-tipped hairs sometimes also present; young parts and leaf lower surface usually glandular-punctate. Leaves opposite or in whorls of 3(4), usually decussate, simple, usually petiolate, mostly aromatic; lamina usually serrate or crenate, often ± bullate or rugose. Inflorescences usually axillary, pedunculate, spicate; spikes densely cylindrical or subspherical; bracts small. Flowers small, each sessile in the axil of an ovate to lanceolate usually acuminate bract. Calyx small, ± tubular and shorter than the corolla tube, truncate to entire or sinuate-dentate, membranous, whitish. Corolla small, hypocrateriform, red, yellow, purple, blue, mauve or white, sometimes bicoloured with the throat yellow to orange, often changing colour after fertilization; tube slender, cylindrical or enlarging above middle; limb spreading, obscurely 2-lipped, 4–5-lobed with lobes obtuse to emarginate. Stamens 4, included; filaments short, inserted ± at middle of corolla tube; anthers ovate, with parallel thecae. Ovary 1-carpellate; carpel 2-locular; ovules 1 per locule, basal, erect or attached laterally near the base; style undivided with a somewhat thick, oblique or sublateral stigma. Fruit drupaceous; mesocarp ± fleshy, rarely rather dry; endocarp hard, 2-locular, not splitting into 2 pyrenes. Seeds without endosperm.

A genus of about 270 specific and infraspecific taxa (fide *Moldenke*), mostly native to subtropical and tropical America, with a few in the Old World tropics. Several species are widely cultivated as garden ornamentals and there are numerous cultivars. Some introduced species have become invasive weeds the most aggressive of which is *L. camara* L.

Species are notoriously difficult to determine because of the great variation exhibited, and hybridization is apparently very widespread. See R. Fernandes in Bol. Soc. Brot., sér. 2, **61**: 125–214 (1989), for discussion of problematic taxa and also of misidentifications by Moldenke (who subsequently misidentified taxa he had previously described from single specimens at intervals).

Biegel, Check-list Ornam. Pl. Rhod. Parks & Gard. [Rhodesia Agric. J., Research Report No. 3]: 69 (1977), has recorded the following as cultivated garden ornamentals in Zimbabwe:

Lantana montevidensis (Spreng.) Briq. (syn. *L. sellowiana* Link & Otto). A slender, much branched trailing or procumbent subshrub with rosy-lilac or violet flowers in flattish spikes on peduncles much exceeding the leaves.

Usually grown as a bedding plant or ground cover. No specimens have been seen from the Flora Zambesiaca area.

Drap d'Or, a sterile cultivar of *L. camara*, with golden yellow flowers, grown in Zimbabwe gardens. It has not become naturalized. No specimens have been seen from the Flora Zambesiaca area.

Each dichotomy of the key below should be followed, and the descriptions carefully read in doubtful cases.

1. Bracts linear or linear-lanceolate, up to 2 mm wide; spikes not elongating after anthesis; leaves all opposite · 2
 – Bracts broader, the lower ones more than 3 mm wide, not linear or linear-lanceolate; spikes ± elongating after anthesis; leaves opposite or 3-whorled, rarely 4-whorled · · · · · · · · · · 3
2. Indumentum of branches, petioles and peduncles without gland-tipped hairs; bracts shortly strigose, 4–13 × 1–1.5 mm; leaf lamina somewhat rigid on drying; stems often with recurved prickles · 1. *camara*
 – Indumentum of branches, petioles and peduncles with gland-tipped and bristle-like hairs intermixed; outer bracts ± densely hispid, up to 9 × 2 mm; leaf lamina more softly membranous on drying; stems without pickles · 2. *tiliifolia*
3. Leaves less than 2.8 × 2 cm, subrhombic to ovate, with 3–10 crenate teeth on each side; flowering peduncles slender, 1–2.8 cm long, subequalling or exceeding the subtending leaf · 4. *moldenkei*
 – Characters not combined as above · 4
4. Flowering peduncles shorter than the subtending leaf, usually not exceeding the petiole, up to 2(4) cm long; corolla rose-coloured to purple · 5
 – Flowering peduncles relatively longer, usually up to or exceeding the middle of the leaf, sometimes exceeding the subtending leaf; corolla rose-coloured to purple or white · · · · 6
5. Lowermost bracts more than (8.5)10 mm long and (5)7 mm wide, often purple or reddish-brown, conspicuously nerved; leaves usually 3-whorled; plant a many-stemmed perennial herb to 90(120) cm tall, rarely a shrub · 8. *rhodesiensis*
 – Lowermost bracts less than 10(13.5) mm long and 6.5 mm wide, usually green, inconspicuously nerved; leaves usually opposite; plant a ± branched shrub or sometimes a herb to c. 70 cm tall · 9. *angolensis*
6. Peduncles 1.5–2.5 times longer than the subtending leaf; leaves 3-whorled, laminas less than 5.5(7.5) cm long; lowermost bracts 5–8(10) × 3–5(6) mm, not or shortly acuminate, flat; spike very soon elongating, with the upper flowers much exceeding the bracts · 6. *swynnertonii*
 – Characters not combined as above · 7
7. Corolla rose-coloured to purple · 8
 – Corolla white · 9
8. Leaves usually opposite; leaf lamina less than 5.5(7.5) × 4 cm, not or slightly attenuate, rugose with nervation ± impressed above, with 9–15(18–21) marginal teeth; lowermost bracts up to 7.5 × 4.2 mm, contracted above into a subulate acumen, with margins revolute toward the apex · 3. *rugosa*
 – Leaves usually 3-whorled, sometimes opposite; leaf lamina usually larger, rather attenuate, smooth not rugose, with 24–34(39) small marginal teeth; lowermost bracts 8–12(14) × 3–5(6.5) mm, flat · 5. *trifolia*
9. Leaves usually 3-whorled, sometimes opposite; leaf lamina usually 2.5–3-times longer than wide, attenuate and acute, with 24–34(39) small marginal teeth; median and upper bracts ending in a ± filiform point, the lowermost 3–5(6.5) mm wide · · · · · · · · · · · · · · 5. *trifolia*
 – Leaves opposite; leaf lamina relatively shorter and broader, not or slightly attenuate, less acute with marginal teeth fewer and broader; median and upper bracts not ending in a filiform point, the lower bracts (3)4–8 mm wide · · · · · · · · · · 7. *viburnoides subsp. richardii*

1. **Lantana camara** L., Sp. Pl.: 627 (1753). —Schauer in A. de Candolle, Prodr. **11**: 598 (1847). —Briquet in Engler & Prantl, Nat. Pflanzenfam. **4**, 3a: 151 (1897). —J.G. Baker in F.T.A. **5**: 275 (1900). —Hiern, Cat. Afr. Pl. Welw. **1**, 4: 827 (1900). —H. Pearson in F.C. **5**, 1: 191 (1901). —De Wildeman, Études Fl. Bas-Moyen-Congo **1**: 309 (1906); in Bull. Jard. Bot. État **7**: 38 (1920). —T. & H. Durand, Syll. Fl. Congol.: 434 (1909). —R. Good in J. Bot. **68**,

Suppl. 2 (Gamopet.): 139 (1930). —Hutchinson & Dalziel, F.W.T.A. **2**: 269 (1931). —
Moldenke in Lilloa **4**: 288 (1939); Résumé Verbenaceae: 458 (Taxon. Index, 1959); in Ann.
Missouri Bot. Gard. **60**: 57 (1973). —A. Meeuse in Blumea **5**: 69 (1942). —Brenan, Check-
list For. Trees Shrubs Tang. Terr.: 639 (1949). —Martineau, Rhodesia Wild Fl.: 65 (1953).
—Garcia in Estud. Ensaios Doc. Junta Invest. Ci. Ultramar [in Mendonça, Contrib.
Conhec. Fl. Moçamb., II] **12**: 166 (1954). —White, For. Fl. N. Rhod.: 368 (1962) excl. spec.
White 1807 pro parte. —Meikle in F.W.T.A., ed. 2, **2**: 435 (1963). —Myre, Algumas Plantas
nocivas [Inst. Inv. Cient. Mocambique]: 9, fig. 1–3 (1964). —Backer & Bakhuizen van den
Brink Jr., Fl. Java **2**: 597 (1965). —Friedrich-Holzhammer et al. in Merxmüller, Prodr. Fl.
SW. Afrika, fam. 122: 7 (1967). —van der Schijff & Schoonraad in Bothalia **10**, 3: 496
(1971). —Adams, Fl. Pl. Jamaica: 629 (1972). —J.H. Ross, Fl. Natal: 299 (1972). —
Verdcourt in F.T.E.A., Verbenaceae: 39 (1992). —M. Coates Palgrave, ed. 3 of K. Coates
Palgrave, Trees South. Africa: 976 (2002). Type a cultivated specimen, *Linnaean Herbarium*
no. 783/4 (LINN, lectotype).

Lantana aculeata L., Sp. Pl.: 627 (1753). Type a cultivated specimen grown in Uppsala
Botanic Garden, *Linnaean Herbarium* No. 783/6 (LINN, lectotype).

Lantana mixta L., Syst. Nat., ed. 12, **2**: 417 (1767), as *"mista"*. —Willdenow, Sp. Pl. **3**: 315
(1800). —Persoon, Syn. Pl. **2**: 140 (1806). —Walpers, Repert. Bot. Syst. **4**: 60 (1845). —
Schauer in A. de Candolle, Prodr. **11**: 600 (1847).

Lantana scabrida Sol. in Aiton, Hort. Kew. **2**: 352 (1789); in W.T. Aiton, Hort. Kew., ed.
2, **4**: 44 (1812). —Walpers, Repert. Bot. Syst. **4**: 60 (1845). Type a cultivated plant (in
Hortus Kew.?).

Lantana antidotalis Schumach. & Thonn. [in Schumacher, Beskr. Guin. Pl.: 276 (1827)]
in Kongel. Danske Vidensk. Selsk. Naturvidensk. Math. Afh. **4**: 50 (1829). —Schauer in A.
de Candolle, Prodr. **11**: 598 (1847). —J.G. Baker in F.T.A. **5**: 276 (1900). —A. Chevalier,
Expl. Bot. Afr. Occ. Fr. **1**: 502 (1920). —Moldenke, Résumé Verbenaceae: 458 (Taxon.
Index, 1959). —R. Fernandes in Bol. Soc. Brot., sér. 2, **61**: 131 (1989), in adnot. Type from
Ghana (Guinea), *Thonning* 125 (C, syn., P. Yu, isosyn.), fide *Verdcourt*, loc. cit.

Camara vulgaris Benth., Bot. Voy. Sulphur: 154 (1846).

Camara aculeata (L.) Kuntze, Revis. Gen. Pl., part 2: 503 (1891).

Camara aculeata var. *subinermis* Kuntze, Revis. Gen. Pl., part 2: 503 (1891). Type as for
Lantana camara.

Camara aculeata var. *subinermis* forma *mixta* (L.) Kuntze, Revis. Gen. Pl., part 2: 503
(1891).

Camara aculeata var. *normalis* Kuntze, Revis. Gen. Pl., part 2: 503 (1891). Type as for
Lantana aculeata.

Lantana camara var. *mixta* (L.) L.H. Bailey, Cyclop. Amer. Hort.: 884 (1900). —Moldenke
in Ann. Missouri Bot. Gard. **60**: 58 (1973). —H. Moldenke & A. Moldenke in Rev. Fl.
Ceylon **4**: 229 (1983).

Lantana camara var. *aculeata* (L.) Moldenke in Torreya **34**: 9 (1934); in Lilloa **4**: 289
(1939); in Fl. Madag. **174**: 11 (1956); Résumé Verbenaceae: 458 (Taxon. Index, 1959); in
Ann. Missouri Bot. Gard. **60**: 58 (1973). —Leon & Alain, Fl. Cuba **4**: 287 (1957). Type as
for *L. aculeata*.

Lantana brittonii Moldenke in Phytologia **2**: 52 (1941); Résumé Verbenaceae: 55 (1958).
Type from Jamaica.

Lantana camara var. *camara* —Leon & Alain, Fl. Cuba **4**: 287, fig. 123B (1957)*.

Mostly a spreading shrub 0.35–5 m tall, forming dense thickets, or scandent to c.
10 m (or a tree to 8 m tall), strongly aromatic. Stem and branches 4-angled,
unarmed to slightly or copiously covered with recurved prickles, ± scabridulous and
sparsely to densely hispid or strigose, sometimes with short gland-tipped hairs
intermixed on peduncles and branchlets, ± densely sessile glandular towards the end
of peduncles and branches. Leaves opposite, petiolate; (3)4–11 × (2)2.8–6(7) cm,
broadly ovate to oblong-ovate, acute or ± acuminate at the apex, rounded to broadly
or narrowly cuneate at the base or sometimes subcordate, coarsely to closely crenate
or crenulate on the margins, scabrid with coarse tubercle-based hairs on the upper

---

* Many other species have been described which have since been treated as synonyms or
infraspecific taxa of *L. camara*. In addition numerous cultivars have been described, based
mainly on corolla colour and its variation during flower development and position within the
inflorescence spikes, (e.g., Kuntze, Revis. Gen. Pl., part 2 (1891); Moldenke in Phytologia, **45**:
296, 1980; L.H. Bailey, Stand. Cycl. Hort. N. ed. 2: 1819, 1927). See also Stirton's comments on
the naturalized *Lantana camara* complex in KwaZulu-Natal [in Andrews S, Leslie A, Alexander
C, (eds). Taxonomy of cultivated plants: Third International Symposium. Proceedings of the
meeting held in Edinburgh, Scotland 20–26 July 1998. Kew: Royal Botanic Gardens, Kew, 1999
publ. 2000].

surface ± pubescent or glabrescent beneath but scabrid on the raised venation, rugulose with impressed venation above, raised reticulate beneath; petiole 0.7–2 cm long, slender. Peduncles axillary, erect or ascending, 2–10 cm long at anthesis, becoming rigid, thickening toward the apex, hardly elongating in fruit, hispid to strigose and with small sessile glands. Flowers in ± flat corymbiform heads, 2–5 cm in diameter with axis 6–12 mm long, hardly elongating in fruit; bracts 4–13 × 1–1.5 mm, linear-lanceolate or oblanceolate, attenuate and acute or subobtuse at apex, shortly strigose. Calyx c. 3 mm long, very thin. Corolla variously coloured, orange, red, pink, mauve, pure yellow or white, often with a yellow or orange centre, young flowers pale changing to a deeper or a different colour with age, sometimes the inner flowers a different colour from the outer; tube 10–12 mm long, narrow in the lower half, widening slightly above, slightly curved, puberulous outside; limb 6–9 mm in diameter, lobes up to 5 mm wide and oblong to rounded. Drupes 3–6 mm in diameter, purple-black to black on ripening.

Botswana. N: Chobe Distr., Kasane, c. 960 m, 3.iv.1966, *Mutakela* 66/3 (SRGH). SE: Mahalapye, 16.iii.1977, *Camerik* 101 (K; PRE). Zambia. N: Chinsali Distr., Mbesuma Ranch, c. 1280 m, 10.x.1961, *Astle* 922 (K; SRGH). W: Kabompo Distr., 26 km WSW of Kabompo near Kabompo R., 24.iii.1961, *Drummond & Rutherford-Smith* 7280 (K; SRGH). C: Lusaka, waste ground near Woodgate House, 19.i.1958, *Angus* 1839 (B; BR; K; LISC; S; SRGH). S: Kafue, 26.ii.1963, *van Rensburg* 1450 (K; SRGH). Zimbabwe. N: Guruve Distr., Nyamunyeche (Nyamayanetsi) Estate, 7.ix.1978, *Nyariri* 332 (SRGH). W: Bulilima Mangwe Distr., by the road from Plumtree to Dombadema, c. 100 m west of Makwa R., 1300 m, 8.iv.1972, *Norrgrann* 128 (BR; E; K; SRGH). C: Sable Park, 8 km NE of Kwekwe (Que-Que), 15.xi.1976, *Chipunga* 87 (BR; K; SRGH). E: Mutare Distr., near Mutare (Umtali) town, c. 1200 m, 11.xi.1930, *Fries, Norlindh & Weimarck* 2845 (K; LD; LISU; M). Malawi. N: Mzimba Distr., east of Mzuzu, 13.iv.1977, *E. Phillips* 1985 (Z). C: Dedza Mountain Forest, 23.vi.1969, *Salubeni* 1360 (K; SRGH). S: Blantyre Distr., Ndirande Forestry Plantation Area near the Office, 25.iii.1968, *J. Phiri* 23 (SRGH). Mozambique. Z: Morrumbala, 23.v.1972, *Howard* c-120a (SRGH). MS: between Garuso and Chimoio (Vila Pery), 29.iii.1948, *Barbosa* 1273 (BM; BR; LISC; LMU). GI: Zavala Distr., between the lagoons Massava and Quissico, c. 7 km from Quissico, 22.xi.1968, *A. Fernandes, R. Fernandes & A. Pereira* 277 (COI; LMU). M: Maputo (Lourenço Marques), 18.xii.1940, *Torre* 2430 (BM; LISC; LMA).

Native of tropical and subtropical America, and introduced in the tropics where it is cultivated as a garden ornamental and hedge plant. It is now naturalized throughout the tropics and subtropics of both hemispheres. In dense thickets by roadsides, in forest clearings, pine plantations and old cultivations, occasional in deciduous woodlands and grassland; 0–2000 m.

In some parts it forms widespread uncontrollable thickets, and for this reason has been declared a noxious weed in Zambia and Zimbabwe. Certain cultivars are sterile and innocuous.

In the Flora Zambesiaca area a range of forms based on corolla colour may be recognized, and the following corolla colour variations occur: yellow and orange (in the same head), yellow-orange fading to red, yellow and red, orange and pink, yellow in the outer whorls and pink in the inner, white suffused with pale or mauve-pink towards edges and with an orange-yellow throat, all bright orange, all yellow, all pink with yellow throat, all mauve, all white.

Var. *mixta* (L.) Bailey is distinguished by some authors by the long whitish bristle-hairs toward the apices of branches and peduncles. While such plants are frequent in the Flora Zambesiaca area they do not differ in other characters and this variety is not formally recognized in this treatment. Furthermore, since most specimens are aculeate, with prickles varying in density from sparse to abundant, the separation of an aculeate variety [var. *aculeata* (L.) Moldenke] does not seem to be justified.

2. **Lantana tiliifolia** Cham. forma **glandulosa** (Schauer) R. Fern. in Bol. Soc. Brot., sér. 2, **61**: 179 (1989). Type from Brazil.

Lantana tiliifolia var. *glandulosa* Schauer in Martius, Fl. Bras. **9**: 257 (1851).

Lantana glutinosa Poepp. in Otto & Dietrich, Allg. Gartenzeitung **10**: 315 (1842). Type from Peru.

Lantana camara sensu F. White, For. Fl. N. Rhod.: 368 (1962) quoad *White* 1807 pro parte, non L. (1753).

A vigorous scandent shrub. Stem and branches subterete, not aculeate, glabrescent; young branches 4-angled, sparsely to densely hispid; indumentum comprised of tubercle-based gland-tipped hairs intermixed with scattered longer soft bristle-like hairs, densest at the nodes and toward the branch apices; tubercle-bases occasionally accrescent making older branchlets slightly scabridulous to the touch. Leaves opposite, petiolate, (3)3.5–12 × (1.5)2–6.5 cm, ovate or oblong-ovate, ±

acuminate at the apex, truncate or rounded at the base and cuneate to slightly decurrent into the petiole, closely crenate or crenulate on the margins, ± densely hispid on the upper surface in young leaves becoming scabrid with the tubercle-bases persisting in older leaves, pubescent-hispid to almost tomentose beneath in young leaves with hairs patent and not markedly tubercle-based, densest on the raised venation, with ± impressed nerves and reticulation above, raised reticulate beneath, drying dark above and paler beneath; petiole 0.7–2.5 cm long, slender, with indumentum similar to that on the branchlets. Peduncles solitary in leaf axils, ascending-erect to ± spreading, up to 10 cm long at anthesis, up to 13 cm long and becoming rigid and somewhat thickened below the spike in fruit; indumentum similar to that on the branchlets. Flowers in ± hemispherical heads 2–3 cm in diameter, spikes not elongating in fruit; bracts linear-lanceolate, the outer up to c. 9 × 2 mm, acute, hispid with long white bristle-like hairs densest on the margins. Calyx c. 2 mm long, thin membranous. Corollas ± exceeding the bracts, variously coloured, the outer deep pink, the inner orange, red magenta or pale yellow with the innermost deep yellow; tube c. 10 mm long, narrowly cylindric ± swollen about the middle, puberulous outside; limb 2.5–4 mm in diameter, the median lobe of lower lip larger than the lateral lobes. Drupes c. 5 mm in diameter, steel-blue on ripening.

The description above applies only to the African specimens seen by me.

Zambia. N: Mbala Distr., Mbulu R., on road to Sand Pits, 21.i.1970, *Sanane* 995 (K). W: Ndola, town park, 24.xii.1951, *White* 1807A (FHO; K). C: Lusaka, Woodlands (suburb of Lusaka), 19.i.1958, *Angus* 1825 (K; SRGH).
Native of south tropical America. Cultivated as a garden ornamental and as a hedge plant, it has become locally naturalized; 1250–1630 m.
Material of *White* 1807 in FHO is a mixture of *L. camara* and *L. tiliifolia* var. *glandulosa*. The latter element has been indicated by myself as 1807A, a duplicate of which is in K.

3. **Lantana rugosa** Thunb., Prodr. Pl. Cap., part 2: 98 (1800); Fl. Cap., ed. 2, **2**: 459 (1823). — Mogg in Macnae & Kalk, Nat. Hist. Inhaca Isl., Moçamb.: 152 (1958). —Friedrich-Holzhammer et al. in Merxmüller, Prodr. Fl. SW. Afrika, fam. 122: 7 (1967). —J.H. Ross, Fl. Natal: 299 (1972). —R. Fernandes in Bol. Soc. Brot., sér. 2, **61**: 162, t. 12 (1989). Type from South Africa (Cape Prov.).
  Lantana salviifolia Jacq., Pl. Hort. Schoenbr. **3**: 18, t. 285 (1798). —Thunberg, Prodr. Pl. Cap., part 2: 98 (1800). —Willdenow, Sp. Pl. **3**: 319 (1800). —Walpers, Repert. Bot. Syst. **4**: 64 (1845). —Schauer in A. de Candolle, Prodr. **11**: 605 (1847). —Gürke in Engler, Pflanzenw. Ost-Afrikas **C**: 337 (1895) pro min. parte. —J.G. Baker in F.T.A. **5**: 277 (1900), as "*salvifolia*" pro parte, quoad specimen *Curror* 20. —H. Pearson in F.C. **5**, 1: 190 (1901) pro max. parte. —Hutchinson, Botanist South. Africa: 301 (1946), as "*salvifolia*". — Martineau, Rhodesia Wild Fl.: 65 (1953), as "*salvifolia*". —R. Fernandes in Bol. Soc. Brot., sér. 2, **61**: 164 (1989) in adnot. Type: tab. 285, Hort. Schoenbr, non L. (1759).
  Lippia caffra Sond. in Linnaea, **23**: 88 (1850). —Moldenke in Phytologia **12**: 107 (1965). —R. Fernandes in Bol. Soc. Brot., sér. 2, **61**: 164 (1989) in adnot. Type from South Africa (Cape Prov.).
  Camara salviifolia (Jacq.) Kuntze, Revis. Gen. Pl., part 2: 504 (1891).
  Lippia lupuliformis Moldenke in Phytologia **2**: 470 (1948); in Phytologia **12**: 265 (1965), pro parte quoad specim. *Rudatis* 1145. —R. Fernandes in Bol. Soc. Brot., sér. 2, **59**: 256 (1986). Type from South Africa (KwaZulu-Natal).
  Lantana rugosa var. *tomentosa* Moldenke in Phytologia **3**: 38 (1948). —R. Fernandes in Bol. Soc. Brot., sér. 2, **61**: 164 (1989) in adnot. Type from South Africa (Cape Prov.).
  Lantana viburnoides sensu Moldenke Fifth Summ. Verbenaceae: saltem pags. 252 et 256 (1971) non (Forssk.) Vahl (1790).

A much branched shrub up to 2 m tall, or sometimes a woody perennial 45–90 cm tall. Stems and branches 4-angled, ± softly strigose and more densely so toward the apices, the indumentum with hairs ± tubercle-based and usually curving appressed antrorse, sometimes somewhat spreading, with sessile glands intermixed. Leaves opposite, rarely 3-whorled, petiolate, mostly shorter than the internodes; 0.7–5.5(7.5) × 0.5–4 cm, ovate, oblong-ovate or lanceolate, sometimes broadly ovate, acute, rounded at the base and cuneate into the petiole, closely crenate or crenulate on the margins with 9–15(18–21) shallow teeth on each side, strigose with whitish tubercle-based hairs on the upper surface, becoming scabrid with the tubercle-bases persisting in older leaves, densely shortly whitish hispid beneath and with minute sessile reddish glands on the lower surface, rugose with impressed venation above, ± strongly raised-reticulate beneath; petiole up to 8 mm long. Peduncles solitary in leaf axils, ascending-erect to ± spreading, 1.5–6 cm long at anthesis, 7(10) cm long and becoming rigid in fruit, mostly more than half as long as the subtending leaf or up to c. 3 cm longer than it, slender; indumentum similar to that on the branchlets. Flowering spikes up to 20 × 20 mm, subspherical or ovoid, with the lower bracts at first equalling the young spikes in length, spikes elongating to 3 cm in fruit becoming laxly conical or oblong-cylindric; lower bracts of flowering spikes up to 7.5 × 4.2 mm, broadly ovate to ovate-lanceolate, attenuate-cuspidate into a somewhat subulate apex, ± strongly contracted at the base, with margins revolute in the upper half, green, appressed pilose on both surfaces but more densely so and with sessile reddish glands on the outer surface, accrescent and up to 8–9 × c. 5 mm, usually not more than half as long as the spike, with nerves ± prominent in fruiting spikes; upper bracts smaller, ± tapering to the apex and less contracted at the base; spike-axes up to 4 cm long, with scattered scars. Calyx ± as long as the corolla tube, appressed-pilose, ciliate on the margins, thinly membranous. Corollas ± exceeding the bracts, pale rose, lilac to red-purple or violet with yellow throat, rarely white; tube (3)3.75–4 mm long, ± swollen about the middle, shortly puberulous; lower lip c. 3.5 mm broad, 3-lobed, the lobes rounded, the median one c. 2 × 1.5 mm, larger than the lateral ones. Stamens inserted ± at the middle of corolla tube. Drupes 2.5–3.5 mm long, purple or wine-coloured on ripening.

Botswana. N: Ngamiland Distr., Khwebe (Kwebe) Hills, c. 1088 m, 25.xii.1897, *Mrs. E.J. Lugard* 68 (K). SE: Central Distr., Letlhakane (Lotlakani), near Orapa, 11.v.1974, *A. Allen* 104 (PRE; S; SRGH). SW: Ghanzi Pan, 26.i.1969, *R.C. Brown* s.n. (K). **Zambia**. N: Mbala Distr., Lungu (Ulungu), 6.xi.1948, *Glover* in *Bredo* 6178 (BR). W: Solwezi Distr., c. 19 km from Solwezi along road to Mwinilunga, 20.xi.1972, *Strid* 2500 (K). **Zimbabwe**. W: Bulawayo, Hillside, c. 1440 m, xii.1956, *O.B. Miller* 4020 (K; LISC). E: Chipinge Distr., 2 km west of Musirizwi (Umzilizwe) R. near road from Mt. Selinda to Save (Sabi) R., c. 490 m, 22.xi.1974, *Müller* 2235 (SRGH). S: Chiredzi Distr., Gonarezhou National Park (Gona-Re-Zhou Game Reserve), 18.i.1971, *Kelly* 518 (K; SRGH). **Mozambique**. GI: Funhalouro Distr., 20 km from Funhalouro on Vilankulo (Vilanculos) road, 21.v.1941, *Torre* 2716 (COI; LISC; M). M: Namaacha Distr., Goba, proximidados do rio Maiuáua, 4.xi.1960, *Balsinhas* 189 (BM; COI; K; LISC; LMA; LMU; SRGH).

Also in South Africa (KwaZulu-Natal, Transvaal, Free State, Lesotho and Cape), Swaziland, Namibia and Angola. Often in moist soil in riverine vegetation and on dam and dambo margins; also in mopane and miombo woodlands, grassland, in alluvial soils, Kalahari Sands and on rocky outcrops; 100–1550 m.

Some authors have treated *L. kisi* A. Rich. as a synonym of *L. rugosa*. However, I consider *L. kisi* to be a separate species, mainly Ethiopian in distribution and not occurring in southern Africa (R. Fernandes in Bol. Soc. Brot., sér. 2, **61**: 139–141 (1989)). In my opinion *L. rugosa* occurs mostly south of the Zambezi River.

Well grown plants, from sheltered localities in moist soil, usually have larger more membranous leaves with a looser indumentum, and a nervation not markedly impressed above nor so prominent beneath. Such specimens, from the Flora Zambesiaca area, were sometimes incorrectly determined in herbaria as *L. viburnoides* (Forssk.) Vahl, which is a distinct species described from the Arabian Peninsula and only represented in Africa by the subsp. *richardii* R. Fern. (cf. R. Fernandes in Bol. Soc. Brot., sér. 2, **61**: 196–200, t. 17 & 18 (1989).

Some specimens with very small leaves and small spikes may be confused with *L. moldenkei*. However, this species is distinguished from *L. rugosa* by its relatively broader less rigid leaves, obtuse at the apex, with longer petioles, and by the smaller less acuminate bracts.

The branchlet and peduncle indumentum, usually strigose (appressed, antrorse), can sometimes consist of patent or spreading bristle-like hairs. However, intermediate forms occur, with subappressed or subspreading hairs, and it is considered therefore that indumentum type is not taxonomically significant, and plants with patent hairs on branches and peduncles are not recognized as separate from *L. rugosa*. Similar variations in indumentum are seen in *Lippia caffra* Sond., *Lantana angolensis*, *L. rhodesiensis* and *L. trifolia*.

4. **Lantana moldenkei** R. Fern. in Bol. Soc. Brot., sér. 2, **61**: 147, t. 5 & 6 (1989). TAB. **4**. Type: Zimbabwe, Goromonzi Distr., Chinhamhora Communal Land (Chindamora Reserve), Nyamasatsi R., 28.i.1964, *T. Müller & A. Smith* 22 (SRGH, holotype; K).

A much branched straggling shrub 0.6–1.6 m tall. Stems with a slightly fissured pale bark, glabrous; branchlets slender, brownish, with internodes 0.3–2.5(4) cm long and leaf scars somewhat raised, ± softly strigose and more densely so toward the apices, the indumentum of whitish ± tubercle-based curved-appressed antrorse hairs,

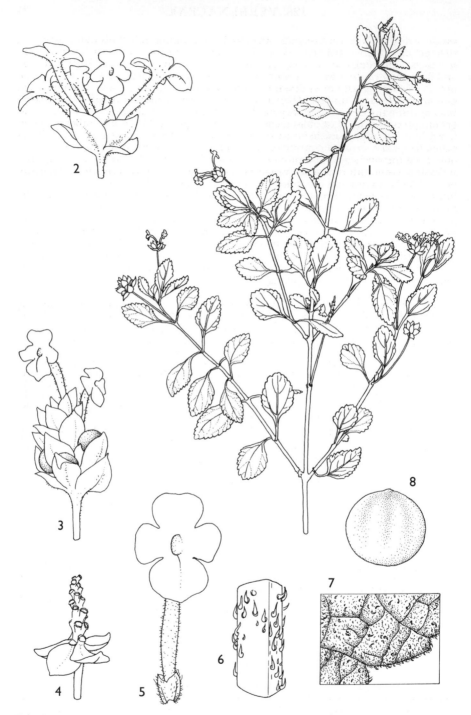

Tab. 4. LANTANA MOLDENKEI. 1, part of flowering branch (× ²/₃); 2, inflorescence (× 3); 3, inflorescence with fruit (× 3), 1–3 from *Wild* 6077; 4, remains of inflorescence after fruit fall (× 3), from *Eyles* 3413; 5, flower (× 5); 6, detail of stem showing hairs on twig (× 20); 7, detail of leaf, upper surface (× 12), 5–7 from *Chase* 6099; 8, fruit (× 5), from *Eyles* 3413. Drawn by Deborah Lambkin.

with sessile glands intermixed. Leaves usually opposite, rarely 3-whorled, petiolate; (0.6)1.2–2.2(2.8) × (0.5)1–1.5(1.8–2) cm, ovate, broadly ovate to subrhombic, somewhat subcircular in the smallest leaves, ± acute to rounded at the apex, rounded or subtruncate at the base with a narrow central portion cuneate-attenuate into the petiole, closely crenate or crenulate on the margins with 3–4 shallow teeth on each side in small leaves and 8–9(10) in the larger leaves, membranous to somewhat coriaceous, ± discolorous and drying dark brownish-green above and pale green to greyish-green beneath, shortly hispidulous to strigose with whitish tubercle-based hairs on the upper surface, becoming ± scabridulous with the tubercle-bases persisting in older leaves, ± tomentose beneath with whitish bristle-like hairs ± obscuring the surface, with minute sessile reddish glands between the hairs on both surfaces, rugose with impressed venation above, ± strongly raised-reticulate beneath; petiole 3–11 mm long. Peduncles solitary in leaf axils, ascending-erect to ± spreading, 10–28 mm long and filiform at anthesis, shorter than to longer than the subtending leaf, increasing to 35(45) mm long in fruit. Flowering spikes up to 1 × 0.7 cm, subspherical, elongating slightly in fruit with drupes spaced along the axis; lower bracts of flowering spikes green, membranous, 4.5–7 × 2.5–4 mm, broadly elliptical or oblong-elliptical, shortly acuminate, sometimes leaf-like with prominent nerves and 1–2 teeth toward the apex, appressed pilose on both surfaces but more densely so and with sessile glands on the outer surface; upper bracts successively smaller; spike-axes up to 13 mm long, with scattered scars. Calyx thinly membranous. Corollas exceeding the bracts, usually mauve, pink, lavender or pale violet, sometimes nearly white; tube 5–7.5 mm long, ± swollen about the middle, densely shortly puberulous; lower lip 4.5–5 mm broad, 3-lobed, the lobes rounded, the median the largest; upper-lip 2.75–3.25 mm broad. Anthers 0.3–0.5 mm long, ovate. Drupes 2–3 × 3–3.75 mm, broader than long, subrhombic in outline.

Zimbabwe. N: Bindura Distr., near Mutepatepa, 14.iii.1963, *Wild* 6077 (BR; K; LISC; SRGH). W: Matobo Distr., Farm Shimbashaba, c. 1478 m, ii.1954, *O.B. Miller* 2151 (K; LISC; SRGH). C: Harare Distr., Chindamora in Makowe "crater", c. 1440 m, 15.iv.1922, *Eyles* 3413 (K; LD; SRGH). E: Mutare Distr., Zimunya's Reserve, c. 1216 m, 6.v.1956, *Chase* 6099 (BM; BR; K; LISC; SRGH).
Known only from Zimbabwe. On or beside granite outcrops, in woodland or grassland, among rocks and in sandy soil; 1150–1480 m.
This species may be distinguished from *L. rugosa* mainly by the shorter, relatively broader leaves (1–1.5 times longer than broad in *L. moldenkei* and usually 2–2.5 times longer than broad in *L. rugosa*), usually obtuse at the apex, and with 3–8(10) teeth on each side in *L. moldenkei* and 9–14(18–21) in *L. rugosa*. *L. moldenkei* has relatively longer petioles, and shorter peduncles and has fruiting spike axes up to 8(13) mm long, whereas the axes are 13–30 mm long in *L. rugosa*.
L. petitiana A. Rich. (from Ethiopia) also with small leaves, has lanceolate or elliptic-lanceolate leaves which are obtuse at the apex (more acute in *L. moldenkei*) with fewer teeth on each margin. The petioles are up to 6 mm long in *L. petitiana* and 11 mm in *L. moldenkei* where they are longer relative to the lamina.

5. **Lantana trifolia** L., Sp. Pl.: 626 (1753). —Miller, Gard. Dict., ed. 8, **4**: 333 et 336 (1768, trad. fr. 1785). —Willdenow., Sp. Pl. **3**: 315 (1800). —Sims in Bot. Mag. **35**: t. 1449 (1812). — Walpers, Repert. Bot. Syst. **4**: 61 (1845). —Schauer in A. de Candolle, Prodr. **11**: 606 (1847). —Briquet in Engler & Prantl, Nat. Pflanzenfam. **4**, 3a: 151 (1897). —Moldenke in Lilloa **4**: 287 (1939). —Meikle in Mem. New York Bot. Gard. **9**: 35 (1954). —Moldenke in Fl. Madag., Verbenaceae: 11 (1956). —Cufodontis in Bull. Jard. Bot. État **32**, Suppl.: 789 (1962). —White, For. Fl. N. Rhod.: 368 (1962). —Binns, First Check List Herb. Fl. Malawi: 103 (1968). —Adams, Fl. Pl. Jamaica: 628 (1972). —Moldenke in Ann. Missouri Bot. Gard. **60**: 48 (1973). —J. Williamson, Useful Pl. Malawi: 147, fig. 146 (1975). —R. Fernandes in Bol. Soc. Brot., sér. 2, **61**: 180, t. 15 & 16 (1989). —Verdcourt in F.T.E.A., Verbenaceae: 46 (1992). Type from West Indies.
Camara trifolia (L.) Kuntze, Revis. Gen. Pl., part 2: 504 (1891).

A weak-stemmed shrub 0.9–3 m tall, or sometimes a woody subshrub. Branches somewhat 4-angled, often purplish, with internodes 2.5–10 cm long, ± weakly strigose with short ± tubercle-based hairs usually curving or ± appressed antrorse, or somewhat hispid with stouter spreading bristle-like hairs, usually with sessile glands intermixed. Leaves aromatic, mostly 3-whorled with some leaves opposite, petiolate; lamina (2.5)5–14 × (1)2–6 cm, mostly 6–8 × 3–4.5 cm, usually 2.5–3 times longer than wide, ovate-lanceolate to lanceolate, or ovate to ovate-oblong or elliptic, ± attenuate towards the apex and ± acute, ± cuneate at the base and somewhat decurrent into the petiole or sometimes rounded and slightly contracted at the base, closely crenate or crenate-serrate on the margins and 24–34(39) shallowly toothed on each side, membranous to somewhat coriaceous becoming somewhat rigid on drying, ± concolorous, shortly and sparsely hispidulous to strigose on upper surface, becoming ± scabridulous with the tubercle-bases persisting in older leaves, puberulous or shortly tomentose beneath, rarely pilose; venation somewhat impressed above but lamina not bullate ± raised or obscure beneath; petiole slender. Peduncles usually 1 per leaf axil or sometimes 2 (up to 6 per node), somewhat rigid, straight, slender, 1.5–9(10) cm long at anthesis, usually shorter than the subtending leaf (petiole included), increasing to 11 cm long in fruit. Flowering spikes 0.8–1.0 × 1.3–1.7 cm, hemispheric, up to 1.7–4.5(5) cm and ± cylindrical in fruit; lower bracts 8–12(14) × 3–5(6.5) mm and ovate-elliptic tapering into an attenuate-subulate apex, or 7.5–9.5 × 4–4.5 and ovate with a tapering-acute apex; upper bracts successively smaller; spike-axes up to 5 cm long, with scattered scars. Calyx thinly membranous. Corolla usually white (in plants of the Flora Zambesiaca area), sometimes mauvish-white, very light pink, rarely pink or purple with yellow throat; tube 5–6(7) mm long; lower-lip 5–5.5 mm broad. Anthers c. 0.5 mm long. Drupes 3–3.5 × 2.5–3 mm, mauve to purple.

This species is widespread and very polymorphic in tropical regions. In the Flora Zambesiaca area at least, the variation exhibited is of little taxonomic value since intermediates occur between most characters. However, some plants from southern Malawi (*Buchanan* 326 and 784 from Shire Highlands, *Best* 180 from Zomba and *Exell, Mendonça & Wild* 894 from Namwera Escarpment) and Niassa Province in Mozambique (*Gomes e Sousa* 1274 and 1305 from Massangulo) may be distinguished by the combination of their longer peduncles and broader bracts, and by the leaf bases less attenuate-decurrent on the petiole. More material needs to be seen before these plants can be recognized as representing a separate variety. In the meantime three forms are recognized in the Flora Zambesiaca area.

1. Peduncles 2 in the axils of subtending leaves, at least in some leaves (up to 6 per node); corollas white · ii) forma *congolensis*
–   Peduncles solitary in leaf axils; corollas white or pale rose-pink in colour · · · · · · · · · · · 2
2. Hairs on the branches and peduncles slender, short and appressed to somewhat spreading · i) forma *trifolia*
–   Hairs on the branches and peduncles bristle-like, long, ± erect and somewhat sparse · iii) forma *hirsuta*

i) Forma **trifolia** —R. Fern. in Bol. Soc. Brot., sér. 2, **61**: 188 (1989).
      Lantana salviifolia sensu J.G. Baker in F.T.A. **5**: 276 (1900), as "*salvifolia*" pro parte quoad saltem specimina *Buchanan* 88, 326 et *Scott-Elliot* 8568 in Shire Highlands lecta, *Kirk* s.n. in Mambane lectum, *Kirk* s.n. et *Meller* s.n. in Manganja Hills lecta, *Whyte* s.n. in Mt. Chiradzulu lectum, etc. —sensu Hiern, Cat. Afr. Pl. Welw. **1**, 4: 827 (1900) quoad specim. *Welwitsch* 5666. —sensu Brenan, Check-list For. Trees Shrubs Tang. Terr.: 639 (1949) pro parte, non Jacq. (1798).
      Lippia schliebenii Moldenke in Phytologia **2**: 316 (1947); op. cit. **12**: 483 (1966) quoad saltem *Quarré* 3448. —R. Fernandes in Bol. Soc. Brot., sér. 2, **59**: 263 (1986); op. cit. **61**:184 (1989) in adnot. Type from Tanzania.
      Lantana trifolia forma *oppositifolia* Moldenke in Phytologia **4**: 179 (1953); in Ann. Missouri Bot. Gard. **60**: 49 (1973). —R. Fernandes in Bol. Soc. Brot., sér. 2, **61**: 187 (1989) in adnot. Type from Java.
      Lantana mearnsii sensu Moldenke, Fifth Summ. Verbenaceae: 230 pro parte; 237 pro parte; 246 pro parte; 249 pro parte et 252 pro parte (1971), non Moldenke (1940).
      Lantana mearnsii var. *congolensis* sensu Moldenke, Fifth Summ. Verbenaceae: 230 pro parte; 252 (1971), non Moldenke (1947).
      Lantana trifolia forma *albiflora* Moldenke in Phytologia **6**: 327 (1958). —H. Moldenke & A. Moldenke in Rev. Fl. Ceylon **4**: 215 (1983). —R. Fernandes in Bol. Soc. Brot., sér. 2, **61**: 187 (1989) in adnot. Type from Ecuador.
      Lantana mearnsii var. *punctata* Moldenke in Phytologia **12**: 428 (1965). Type: Malawi, Thyolo (Cholo) Mt., c. 1200 m, *Brass* 17719 (NY, holotype; K; US).
      Lantana sp. (*Chapman* 351), Chapman, Veg. Mlanje Mt., Nyasal.: 34 (1962).

Moldenke (Fifth Summ. Verbenaceae: 535–541, 1971), has listed many other synonyms, but the type specimens of these have not been verified by me.

Zambia. N: Mporokoso, 1280 m, 29.iv.1957, *Vesey-FitzGerald* 1230 (SRGH). W: Ndola, 30.i.1954, *Fanshawe* 747 (BR; K). C: Kabwe (Broken Hill), vi.1920, *Rogers* 26079 (Z). S: Monze

Distr., Pemba, 30.xii.1911, *Rogers* 8920 (Z). **Zimbabwe.** C: Mutasa Distr., St. Triashill (St. Trias Hill) Mission, ii.1917, *Mundy* s.n. (K). E: Mutare (Umtali) Heights, c. 1280 m, 29.iii.1959, *Corner* s.n. (E). **Malawi.** N: Mzimba Distr., South Viphya Plateau, 11 km south of Chikangawa on Viphya Link Road to Luwawa, 1740 m, 8.v.1970, *Brummitt* 10455 (K; SRGH). C: Dedza Mountain, 5.iv.1978, 1750 m, *Pawek* 14229 (K). S: Mt. Mulanje (Mlange), above N.A. Nkanda's H.Q., 10.iv.1957, *Chapman* 351 (BM; BR; FHO; K; PRE). **Mozambique.** N: Mandimba, 816 m, 23.iii.1942, *M.G. Hornby* 3724 (PRE). Z: Milange Distr., Serra de Tumbine, 12.xi.1942, *Mendonça* 1415 (COI; LISC; M).
Native of tropical America, it is now widely spread throughout nearly all tropical regions. In Africa it is found in Angola, Dem. Rep. Congo, S Nigeria, Ethiopia, Uganda, Kenya and Tanzania. Miombo woodland often on termitaria, riverine vegetation and dambo margins; also in montane grassland and evergreen forest margins; 816–1750 m.

Some specimens which appear to be intermediate between *L. trifolia* and *L. rhodesiensis* may be hybrids. They approach *L. rhodesiensis* in their habit – the stems usually simple with long internodes and arising from a woody rhizome, and in their leaves – in shape, crenations, colour and indumentum, and the usually short peduncles. They approach *L. trifolia* in their thin lamina and smaller spikes. Their lower bracts are narrower than in *L. rhodesiensis* and not as attenuate-cuspidate as in *L. trifolia*, and the plants therefore appear intermediate between both. The following specimens are probable hybrids between *L. trifolia* and *L. rhodesiensis*: **Zambia** N: Sunzu, Kalambo Farm, *Richards* 3939 (K). W: Mufulira, *Cruse* 126 (K) and 241 (BR; K). S: Mazabuka, *Rogers* 26104 (Z). **Malawi** S: Chikwawa Escarpment near Blantyre, *Benson* s.n. (PRE 28121); Blantyre, *Jackson* 1803 (BR; K). **Mozambique** N: Mutuáli, Malema Road, *Gomes e Sousa* 4198 (COI; LMA; SRGH); Valleys of Mandimba, *Hornby* 3497 (K; PRE); entre Nampula e Murrupula, *Torre* 675 (COI; LISC); Nampula, *Torre* 1260 (COI; LISC) and 1258 (COI); *Torre & Paiva* 9749 (COI; LISC; P; SRGH). See R. Fernandes in Bol. Soc. Brot., sér. 2, **61**: 194–195 (1989) for further discussion.

Some specimens with opposite leaves and appearing intermediate between *L. trifolia* and *L. viburnoides* subsp. *richardii* may represent another probable hybrid. Such specimens include: **Malawi** N: Mzimba, *Whyte* 2504 (FHO; K). **Mozambique** N: Massangulo, *Pedro & Pedrógão* 3515 (LMA; LMU).

ii) Forma **congolensis** (Moldenke) R. Fern. in Bol. Soc. Brot., sér. 2, **59**: 254 (1987); in op. cit. **61**: 190 (1989). Type from Dem. Rep. Congo.
    Lantana salviifolia sensu J.G. Baker in F.T.A. **5**: 277 (1900) pro parte quoad specim. *Welwitsch* 5691 et 5742. —sensu Hiern, Cat. Afr. Pl. Welw. **1**, 4: 827 (1900) pro parte quoad specim. *Welwitsch* 5643, 5691, 5742 et 5743, non Jacq. (1798).
    Lippia burtonii Baker in F.T.A. **5**: 281 (1900). Type from Dem. Rep. Congo.
    Lantana mearnsii Moldenke var. *congolensis* Moldenke in Phytologia **2**: 313 (1947). Type from Dem. Rep. Congo.
    Lantana trifolia sensu Richards & Morony, Check List Fl. Mbala: 238 (1969) non L. (1753).

Peduncles 2 per leaf axil, at least in some leaves, giving a maximum 6 peduncles per node. Indumentum on the stem, branches and peduncles consisting of short ± tubercle-based hairs usually curving or ± appressed antrorse. Corollas white.

Zambia. N: Mbala Distr., Dhul'miti Kloof, 12.v.1952, *Richards* 1718 (BR; K). W: Mwinilunga Distr., Matonchi Farm, 1936, *K.R. Paterson* 12 (K). **Malawi.** C: Dedza Distr., 11 km NW of Dedza, c. 1500 m, 1.iv.1970, *Brummitt* 9568 (K; SRGH). S: Blantyre Distr., Soche Mt., NE side, c. 1230 m, 13.iii.1970, *Brummitt & Moriarty* 9059 (K). **Mozambique.** T: Angónia, arredores de Vila Velha prox. N'dzeulé, 11.ii.1980, *Macuácua, Stefanesco & Mateus* 960 (LMA).
Also in Angola and Dem. Rep. Congo. Habitat as for *L. trifolia* forma *trifolia*; 1230–1600 m.

iii) Forma **hirsuta** Moldenke in Phytologia **3**: 113 (1949); in Ann. Missouri Bot. Gard. **60**: 49 (1973). —H. Moldenke & A. Moldenke in Rev. Fl. Ceylon **4**: 213 (1983). —R. Fernandes in Bol. Soc. Brot., sér. 2, **61**: 191 (1989). Type from Colombia.
    Lantana trifolia sensu J.G. Baker in F.T.A. **5**: 278 (1900) quoad specim. *Rowland* s.n. — sensu Hutchinson & Dalziel, F.W.T.A. **2**: 269 (1931). —sensu Meikle in F.W.T.A., ed. 2, **2**: 435, 1963; in Dale & Greenway, Kenya Trees & Shrubs: 587 (1961), non L. (1753).
    Lantana salviifolia sensu Hiern, Cat. Afr. Pl. Welw. **1**, 4: 827 (1900) pro parte quoad specim. *Welwitsch* 5727 (1900), non Jacq. (1798).
    Lantana mearnsii Moldenke in Phytologia **1**: 421 (1940); in Phytologia **2**: 313 (1947). Type from Kenya.

Zambia. W: Solwezi Distr., near Kifubwa R., 5 km east of Solwezi, 17.iii.1961, *Drummond & Rutherford-Smith* 6968 (K; LISC; SRGH). **Malawi.** N: Chitipa Distr., Misuku Hills, Mugesse (Mughesse) Forest, Songwe View, c. 1728 m, 25.iv.1972, *Pawek* 5230 (K). **Mozambique.** N: Mueda Distr., 12 km from Mueda to Diaca, c. 700 m, 15.iv.1964, *Torre & Paiva* 11976 (LISC).

Widespread throughout tropical America, Hawaiian Islands, Indonesia, Burma, India and Sri Lanka. Also in Ethiopia, Uganda, Kenya, Tanzania, Nigeria, Dem. Rep. Congo, Angola and probably elsewhere. Habitat as for *L. trifolia* forma *trifolia*; 700–1728 m.

Moldenke described *L. mearnsii* for a plant frequently found in east Africa with 3-whorled leaves and pink corollas. However, many intermediates occur and it cannot be kept separate from *L. trifolia*.

Specimens of forma *hirsuta* from the Flora Zambesiaca area mostly have opposite relatively broad leaves and white corollas, and in these characters are similar to plants from Tanzania which are probably a hybrid between forma *hirsuta* and *L. viburnoides* var. *richardii* (cf. R. Fernandes in Bol. Soc. Brot., sér. 2, **61**: 195 (1989)). However, the Flora Zambesiaca area plants have shorter peduncles and narrower bracts more attenuate-cuspidate towards the apex and are therefore distinct from the hybrid.

Gilges 358 (K; SRGH), from Barotseland in Zambia, was determined by Meikle as *L. subtracta* Hiern, a species known only from Angola. However, it differs from the Angolan species in many characters: stem erect (not prostrate as in *L. subtracta*), with longer internodes, covered with a dense indumentum of longer hairs; leaves somewhat attenuate towards the apex, subtomentose on the lower surface (whereas they are sparsely hairy with the hairs short and appressed in *L. subtracta*), with smaller marginal crenae; bracts longer and a little broader, etc. It is possible that the specimen *Gilges* 358, is a hybrid between *L. trifolia* L. and *L. machadoi* R. Fern. The latter species was described from the Angolan province of Moxico which is adjacent to Barotseland (cf. R. Fernandes in Bol. Soc. Brot., sér. 2, **61**: 142 (1989)).

6. **Lantana swynnertonii** Moldenke in Phytologia **3**, 5: 270 (1950). —R. Fernandes in Bol. Soc. Brot., sér. 2, **61**: 174, t. 13 & 14 (1989). Type: Zimbabwe, Chirinda, 1906, *Swynnerton* 259 (K, holotype).

    Lantana mearnsii sensu Moldenke, Fifth Summ. Verbenaceae: 248 et 252 (1971) quoad distrib. in Zimbabwe et Mossambique pro parte, non Moldenke (1940).

    Lantana mearnsii var. *congolensis* sensu Moldenke, Fifth Summ. Verbenaceae: 252 (1971) quoad distrib. in Mossambique pro parte, non Moldenke (1947).

    Lantana rugosa sensu Moldenke, Fifth Summ. Verbenaceae: 252 (1971) quoad distrib. in Mossambique pro parte, non Thunb. (1800).

    Lantana rugosa var. *tomentosa* sensu Moldenke, Fifth Summ. Verbenaceae: 252 (1971) quoad distrib. in Mossambique pro parte, non Moldenke (1948).

Shrub up to 2.5 m tall, or bushy subshrub to 1.5 m. Branches usually simple with internodes 3–10 cm long; hispid with long white ± patent tubercle-based hairs, or ± weakly strigose with ± short tubercle-based hairs usually curving or ± appressed antrorse, usually with sessile glands intermixed. Leaves usually all 3-whorled, less often also opposite or 4-whorled in the same plant, more rarely all opposite, petiolate; lamina (2)3–5.5(7.5) × (1)2–3.5(5.5) cm, ovate to broadly ovate or lanceolate, acute at the apex, rounded at the base and sometimes the middle part of the rounded base cuneate and somewhat decurrent into the petiole, closely crenate or crenate-serrate on the margins and 16–24 shallowly toothed on each side, somewhat coriaceous when mature becoming rigid when dry, deep green, concolorous or somewhat darker above, hispid on upper surface with ± scattered tubercle-based bristles, becoming ± scabrid with the tubercle-bases persisting in older leaves, hispidulous-pubescent to somewhat tomentose beneath with bristles mainly on the raised venation, indumentum intermixed with sessile glands; venation somewhat impressed above and raised beneath; petiole 0.5–0.7 cm long. Peduncles 1 per leaf axil, slender, ascending, hispid-pubescent with bristles usually spreading, (4)6–12.5(14.5) cm long at anthesis, usually 1.5–2.5 times longer than the subtending leaf, rarely shorter than the leaf, increasing to 14.5 cm long in fruit. Flowering spikes 1–2(2.5) cm in diameter, hemispheric, elongating and ± cylindrical in fruit; lower bracts 5–8(10) × 3–5(6) mm, ovate or lanceolate to elliptic, acute to acuminate-cuspidate at apex, flat on the margins, appressed pubescent on both surfaces; upper bracts successively smaller and ± lanceolate, acuminate-cuspidate at apex; spike-axes 1–3.3(5) cm long and slender in fruit with closely-spaced scars. Calyx c. 1.5 mm long, cylindric, truncate, thinly membranous and appressed puberulous. Corolla pinkish-mauve to purplish, greatly exceeding the bracts especially in the upper flowers; tube 5–7.5(8) mm long, ± ventricose; lower lip 4–5.5(6) mm broad. Drupes 2–3 mm long, bright purple.

Zimbabwe. E: Mutare Distr., Inyamatshira Mt. Range, commonage in foothills, c. 1248 m, 8.x.1953, *Chase* 5111 (BM; K; LISC; SRGH). **Mozambique**. Z: Serra de Morrumbala, 1.v.1943,

Torre 5255 (BM; LISC). MS: Mossurize Distr., Mts. de Espungabera, c. 1000 m, 14.ii.1966, *Pereira, Sarmento & Marques* 1339 (BR; LMU).

Recorded mostly from the lower, drier slopes and foothills of the eastern border mountains of Zimbabwe, it would be expected to occur in adjacent localities in Mozambique; it has also been collected on Morrumbala Mt. In thickets and open scrub on lower mountain slopes, in riverine scrub and *Brachystegia* woodland bordering evergreen forest; from 750 to c. 1850 m.

L. rugosa may be distinguished from this species in having: opposite leaves and longer petioles, shorter flowering peduncles, bracts narrower at the base and revolute in the upper half, and corollas smaller (tube up to 4 mm long).

Moldenke, in Phytologia **3**, 5: 271 (1950) considered that *L. swynnertonii* belonged in the *L. camara* L. group, although it "differs conspicuously in its broad bractlets". However, this relationship is not accepted in this account, because, in addition to the relatively broader bracts, *L. swynnertonii* usually has 3-whorled leaves (never opposite as in *L. camara*) and the spike axis becomes elongate in fruit (even when only the lower drupes are developed).

7. **Lantana viburnoides** (Forssk.) Vahl, Symb. Bot. **1**: 45 (1790). —J.G. Baker in F.T.A. **5**: 276 (1900) pro parte. —Schwartz in Mitt. Inst. Bot. Hamb. **10**: 215 (1939). —R. Fernandes in Bol. Soc. Brot., sér. 2, **61**: 196 (1989). —Verdcourt in F.T.E.A., Verbenaceae: 40 (1992). Type from Yemen.

    Charachera viburnoides Forssk., Fl. Aegypt.-Arab.: CXV, 116, 379 (1775).

Subsp. **richardii** R. Fern. in Bol. Soc. Brot., sér. 2, **61**: 200, t. 19 (1989). Type from Ethiopia.
    Lantana viburnoides var. *richardii* R. Fern. in Bol. Soc. Brot., sér. 2: **61**: 201, tab. 19 & 20 (1989).
    Lantana viburnoides sensu A. Rich., Tent. Fl. Abyss. **2**: 168 (1850) excl. specim. *Schimper* 257. —sensu J.G. Baker in F.T.A. **5**: 276 (1900) quoad specim. *Schimper* 455. —sensu Meikle in Dale & Greenway, Kenya Trees & Shrubs: 587 (1961). —sensu Cufodontis in Bull. Jard. Bot. État **32**, Suppl.: 790 (1962) pro parte. —sensu White, For. Fl. N. Rhod.: 370 (1962). —sensu Richards & Morony, Check List Fl. Mbala: 238 (1969). —sensu Moldenke, Fifth Summ. Verbenaceae: 212 (pro parte?), 237 (pro parte?), 241 (pro parte?), 248 pro parte, 250 pro parte (1971), non (Forssk.) Vahl (1790).
    Lantana salviifolia sensu J.G. Baker in F.T.A. **5**: 277 (1900), as "*salvifolia*", pro parte quoad saltem specim. abyss. *Roth* 488, *Schimper* 377, 488 et 2193, *Steudner* 1303, keniens. *Hildebrandt* 1988 et *Wakefield* s.n., zanzibarens. *Bojer* s.n. et angolens. *Welwitsch* 5720. —sensu Hiern, Cat. Afr. Pl. Welw. **1**, 4: 827 (1900) quoad. specim. *Welwitsch* 5720. —sensu R. Good in J. Bot. **68**, Suppl. 2 (Gamopet.): 139 (1930) pro parte quoad specim. *Gossweiler* 1517, non Jacq. (1798).
    Lantana rugosa sensu Moldenke, Fifth Summ. Verbenaceae: 241 (1971) pro parte, non Thunb. (1800).

Shrub 1–2 m tall or a subshrub 30–50 cm tall. Branches usually simple with internodes up to 10 cm long or sometimes internodes c. 5 mm long or less, scars of leaf bases distinctly raised; young branches hispid-pubescent with short whitish tubercle-based hairs usually curving and upwardly directed or ± appressed antrorse, usually with sessile glands intermixed; older branches somewhat woody, ± glabrescent and scabridulous. Leaves opposite, petiolate; lamina mostly 3.8–7.5(12) × 1.5–4(5) cm, ovate or ovate-lanceolate to elliptic, somewhat tapering to an acute to subobtuse apex, rounded at the base and sometimes the middle part of the rounded base cuneate and somewhat decurrent into the petiole, crenate on the margins with broad shallow teeth, submembranous when mature becoming ± rigid when dry, concolorous to somewhat discolorous drying dark olive-green or dull green above and pale green beneath, sparsely hispidulous on upper surface with ± scattered short ± appressed tubercle-based bristles becoming ± scabridulous with the tubercle-bases persisting in older leaves, ± densely hispidulous-pubescent beneath with hairs mainly on the venation and not tubercle-based, indumentum intermixed with sessile glands; venation not or hardly impressed above or raised beneath; petiole up to c. 1 cm long, slender, merging into the decurrent base of the lamina. Peduncles 1 per leaf axil, usually erect or ascending, (1.5)2.5–13.5 cm long, hispid-pubescent with short whitish tubercle-based upwardly directed hairs. Flowering spikes 0.7–1.5 × 1.1–2.3 cm, usually shorter than broad, sub-hemispheric, up to 2.4 cm long and ± cylindrical in fruit; lower bracts 6–14(15) × 5–8 mm, ovate or broadly lanceolate, acuminate-cuspidate at apex, sometimes 9–11 × 3–4 mm ± oblong; upper bracts successively smaller and ± lanceolate, acuminate-cuspidate at apex, equalling or exceeding the upper corollas at anthesis, shorter than the spikes in fruit; spike-axes 1–1.7 cm long in fruit, with ± spaced scars. Corolla white; tube c. 5.5 mm long, ± ventricose; lower lip c. 3.5 mm broad. Drupes c. 2.5–3.5 × 3–3.5 mm, truncate at the top, purple-lilac.

Zambia. N: Mbala Distr., Chilongowelo, Taskers Deviation, 2.iii.1952, *Richards* 858 (K). **Zimbabwe**. E: Chimanimani Distr., Chikwizi R., Mutare to Birchenough Bridge road, Muwushu Reserve, 15.xii.1963, *Chase* 8081 (K; LISC; SRGH). **Malawi**. N: Mzimba Distr., Mzuzu, Marymount, c. 1370 m, 5.vi.1969, *Pawek* 2459 (K). **Mozambique**. N: Ngauma Distr., Massangulo, iii.1933, *Gomes e Sousa* 1310 (COI; K). T: Macanga Distr., Mt. Furancungo, 1265–1380 m, 15.iii.1966, *Pereira, Sarmento & Marques* 1786 (BR; LMU).

Also in Ethiopia, Uganda, Kenya, Tanzania and Angola. Miombo and mixed deciduous woodland and riverine forest, sometimes on termite mounds; 600–1500 m.

Richards 118 & 824 (both in K), from Zambia (Chilongowelo near Mbala), differ from typical subsp. *richardii* in having narrower bracts and shorter peduncles, and perhaps represent a hybrid between *L. trifolia* and *L. viburnoides* subsp. *richardii*.

Subsp. *viburnoides* occurs in Arabia, and subsp. *richardii* differs from it mainly in the following characters: leaves usually broader and not as narrowly tapering to the apex; apex less acute; margins with fewer, broader teeth; lamina thinner in texture and not as stiff on drying, ± smooth and not rugose on the upper surface, and not raised reticulate beneath; peduncles usually longer than the leaf (± half as long as the leaf in the typical subsp.); bracts relatively broader (ovate-oblong or oblong-lanceolate and only 1.5–3.5 mm broad in the typical subsp.); spikes broader, and more than 1 cm in diameter.

This account differs from the taxonomy of Verdcourt, in F.T.E.A., Verbenaceae: 40 (1992), at infraspecific level.

8. **Lantana rhodesiensis** Moldenke in Phytologia **3**: 269 (1950). —F.W. Andrews, Fl. Pl. Anglo-
Egypt. Sudan **3**: 196 (1956). —White, For. Fl. N. Rhod.: 370 (1962). —Richards & Morony,
Check List Fl. Mbala: 238 (1969). —R. Fernandes in Bol. Soc. Brot., sér. 2, **61**: 156, t. 9–11
(1989). —Verdcourt in F.T.E.A., Verbenaceae: 45 (1992) in adnot. Type: Zambia, ncar
Mumbwa, 15°S, 28°E, 1911, *Macaulay* 735 (K, holotype).

Lantana milne-redheadii Moldenke in Phytologia **3**: 268 (1950). —White, For. Fl. N.
Rhod.: 370 (1962) in adnot. Type: Zambia, Mwinilunga Distr., west of Matonchi Farm,
7.xii.1937, *Milne-Redhead* 3542 (K, holotype; BR).

Lantana viburnoides var. *velutina* Moldenke in Phytologia **3**, 3: 120 (1949). Type from Sudan.

Lantana ukambensis sensu Verdcourt in F.T.E.A., Verbenaceae: 43 (1992) pro parte (incl.
fig. 6) non (Vatke) Verdc.

Subshrub 0.30–0.90(1.20) m tall, bushy erect or with sprawling procumbent stems, rarely a shrub. Stems several to many from a woody rootstock, annual or becoming ± woody below, somewhat angular, simple or sometimes branched; internodes (6)8–10.5(11–14) cm long, usually longer than the leaves; indumentum hispid-pubescent with long whitish tubercle-based hairs usually curving and upwardly directed or ± spreading, intermixed with very short patent bristles and usually with scattered sessile glands; branches 1–3, slender, 9–30 cm long. Leaves 3-whorled or sometimes opposite, both often in the same plant, all opposite on some branches, usually shortly petiolate; lamina mostly (4.2–4.5)5–8(8.5–10) × (2)2.5–4(5) cm, smaller on the branches and lower part of stem, ovate or narrowly ovate, tapering-attenuate to an acute apex, ± abruptly rounded at the base with the middle part of the rounded base cuneate and somewhat decurrent into the petiole, serrate-crenate or crenulate on the margins with 14–23 shallow teeth along almost the entire length of each side, somewhat membranous or chartaceous, light green and velvety-hispidulous above with spreading tubercle-based bristles, becoming ± scabrous with the tubercle-bases persisting in older leaves, paler and ± densely hispidulous-pubescent to tomentose beneath with hairs mainly on the venation and not tubercle-based, indumentum intermixed with sessile glands; venation slightly impressed on leaf upper surface and slightly raised beneath; petiole 3–8 mm long, slender. Peduncles 1 per leaf axil with 2–3 per node, or sometimes 2 per axil, erect or ascending, 0.3–2.5(3–5) cm long, usually no longer than half the length of the subtending leaf, (0.5)1–4(5.5) cm long in fruit. Flowering spikes somewhat clustered, (10)13–18 × (4)9–14 mm, ovoid or subspherical in flower, elongating to 3.5 cm long in fruit and becoming ovoid-oblong to subcylindric; lower bracts leaf-like, as long as or exceeding the flowering spike and obscuring the upper corollas at anthesis, (8.5)11–17(20) × (5)7–11(13) mm, ovate to broadly ovate, rarely ovate-lanceolate, acute to acuminate-cuspidate at apex, truncate or rounded at the base, with 5–7 ± prominent nerves from the base, appressed hispidulous-puberulous mainly on the nerves, often purple or brownish-purple at least in the upper bracts; fruiting spike-axes 2.5 cm long, with ± spaced prominent scars often within the remains of persistent calyces, sometimes with pedicels up to c. 3 mm long. Corolla usually purple, sometimes magenta or mauvish-pink, with or without white or yellow centre; tube (2)3–4(5) mm long; limb (1)1.5–3 mm wide. Drupes 2–3.5 × 2–4 mm, subglobose, shiny, mauve, magenta to metallic-purple when mature.

Zambia. N: Mpika, near the airfield, 13.i.1975, *Brummitt & Polhill* 13776 (K). W: Kitwe Distr., 20.iii.1955, *Fanshawe* 2163 (K; SRGH). C: Serenje Distr., Kundalila Falls, south of Kanona, c. 1350 m, 13.iii.1975, *Hooper & Townsend* 717 (K). E: Katete, 10.ii.1957, *Angus* 1501 (BM; BR; FHO; K). S: Machili, 19.x.1960, *Fanshawe* 5850 (FHO). **Zimbabwe**. N: Mount Darwin Distr., Kandeya C.L. (Native Reserve), c. 1024 m, 17.i.1960, *Phipps* 2305 (K; SRGH). **Malawi**. N: Mzimba Distr., Viphya Plateau, c. 59.5 km SW of Mzuzu, Vernal Pool, c. 1760 m, 23.ii.1975, *Pawek* 9099 (BR; K). C: Lilongwe Distr., Chitedze, 1150 m, 22.iii.1955, *Exell, Mendonça & Wild* 1122 (BM; LISC; SRGH). **Mozambique**. N: NW of Mandimba, 6.i.1942, *M.G. Hornby* 3546 (K).

Also in Uganda, Sudan and Tanzania, perhaps elsewhere. In miombo woodland, plateau woodland and mopane, submontane grassland and grassland on hillsides, dambo margins and sometimes in shallow soil overlying rock; 760–1670 m.

L. rhodesiensis is extremely polymorphic and has been confused, in various herbaria, with *L. mearnsii*, *L. rugosa* (as *L. salviifolia*), *L. trifolia*, *L. viburnoides* and even with *Lippia baumii*, *Lippia radula* and *Lippia wilmsii* [See R. Fernandes in Bol. Soc. Brot., sér. 2, **61**: 156–162 (1989), and discussion by Verdcourt, F.T.E.A., Verbenaceae: 45 (1992)]. Specimens intermediate between this species and *L. angolensis*, and between it and *L. trifolia*, are discussed under those species in this treatment.

Specimens from west tropical Africa were included in *Lantana rhodesiensis* by Meikle in F.W.T.A., ed. 2, **2**: 435 (1963). However, those specimens seen by myself (from Guinea Bissau, Cameroon, and N and S Nigeria) differ from typical *L. rhodesiensis* mainly in the smaller bracts, the lower ones of which do not reach the top of the spike, and the upper ones of which are nearly filiform at the apex and narrower than in *L. rhodesiensis*. These west tropical African specimens also have a shrubby habit and larger relatively broader leaves. Perhaps these plants represent a separate subspecies of *L. rhodesiensis*. Meikle (loc. cit.) also considered *L. mearnsii* var. *latibracteolata* Moldenke to be a synonym of *L. rhodesiensis*. However, the type of this variety, a specimen from Dem. Rep. Congo, differs not only from the west tropical African specimens but also from typical *L. rhodesiensis* in having retrorse hairs on the stem, obtuse leaves, narrower spikes, smaller bracts and smaller drupes.

Verdcourt, in F.T.E.A., Verbenaceae: 45 (1992), sank *L. rhodesiensis* into *L. ukambensis* (Vatke) Verdc., a species based on *Lippia ukambensis* Vatke, the type specimen of which (*Hildebrandt* 2739; B†, holotype; K, isotype + fragment of holotype) is a small branch of a shrub with short internodes, opposite leaves and oblong floral bracts; whereas *L. rhodesiensis* is usually a many-stemmed perennial subshrub with unbranched stems, very long internodes, and larger spikes with larger ovate to broadly ovate floral bracts. In some respects the isotype of *Lantana ukambensis* approaches *Lantana angolensis* Moldenke more closely than *L. rhodesiensis*. On the other hand, the type specimen of *L. ukambensis* may be a hybrid between *L. rhodesiensis* and another species.

With regard to the difficulty in distinguishing taxa Verdcourt (loc. cit.) pointed out that the situation is confused by the almost certain existence of hybrids between the '*L. ukambensis* - *L. rhodesiensis* - *L. angolensis*' complex and both *L. trifolia* and *L. viburnoides*. He suggested that a possible solution may be to recognise a broad taxon with three subspecies based on *L. ukambensis*, *L. rhodesiensis* and *L. angolensis*.

9. **Lantana angolensis** Moldenke in Phytologia **3**: 37 (1948). —White, For. Fl. N. Rhod.: 370 (1962). —Friedrich-Holzhammer et al. in Merxmüller, Prodr. Fl. SW. Afrika, fam. 122: 7 (1967). —R. Fernandes in Bol. Soc. Brot., sér. 2, **61**: 128, t. 1 & 2 (1989). Type from Angola.
  Lantana salviifolia sensu Gürke in Engler, Pflanzenw. Ost-Afrikas C: 337 (1895) pro parte; in Warburg, Kunene-Samb.-Exped. Baum: 349 (1903). —sensu J.G. Baker in F.T.A. **5**: 276–277 (1900), as "*salvifolia*", quoad saltem specimina *Baines* s.n., *E. Holub* s.n., *Schinz* 460, *Welwitsch* 5755 et 5761. —sensu Hiern, Cat. Afr. Pl. Welw. **1**, 4: 827 (1900), as "*salvifolia*", pro parte quoad specim. *Welwitsch* 5755 et 5761. —sensu H. Pearson in F.C. **5**, 1: 190 (1901) pro parte, non L. (1759) neque Jacq. (1798).
  Lantana milne-redheadii sensu Moldenke, Fifth Summ. Verbenaceae: 244 (1971) pro parte quoad distrib. in Huila (Angola), non Moldenke (1950).
  Lantana kisi sensu Moldenke, Fifth Summ. Verbenaceae: 246 (1971) quoad distr. in Zambia, non A. Rich. (1850).
  Lantana mearnsii var. *latibracteolata* sensu Moldenke, op. cit.: 246 et 248 (1971) quoad distrib. in Zambia et Zimbabwe, non Moldenke (1947).
  Lantana viburnoides var. *velutina* sensu Moldenke, loc. cit. (1971) quoad distr. in Zambia pro parte, non Moldenke (1949).

Lantana mearnsii var. *congolensis* sensu Moldenke, op. cit.: 248 (1971) quoad distr. in Zimbabwe, non Moldenke (1947).
Lantana rugosa sensu Moldenke, loc. cit. (1971) quoad distrib. in Zimbabwe pro parte, non Thunb. (1800).
Lantana rugosa var. *tomentosa* sensu Moldenke, op. cit.: 248 & 254 (1971) quoad distrib. in Zimbabwe et in Namibia, non Moldenke (1948).
Lantana viburnoides sensu Moldenke, op. cit.: 248 (1971), quoad distr. in Zimbabwe pro parte, non (Forssk.) Vahl (1790).

Shrub up to 1.5(2) m tall or a ± bushy perennial herb 20–70 cm tall. Stems several to many from a woody rootstock, becoming ± woody, somewhat 4-angular, simple or divaricately-branched; branches stout, leafy; branchlets often short with ± crowded leaves; indumentum hispid-pubescent with long whitish tubercle-based hairs usually curving and upwardly directed, twigs sparsely to ± densely hispid-tomentose with long spreading bristles especially on young growth, usually with scattered sessile glands, older branches and stems glabrescent. Leaves usually opposite, rarely 3-whorled, distinctly petiolate; lamina (1)3–6(9) × (0.7)2–3.5(5) cm, ovate to broadly ovate or sometimes oblong-ovate, mostly ± tapering to an acute or subobtuse apex, ± abruptly rounded to subcordate at the base with the middle part of the rounded base cuneate and somewhat decurrent into the petiole, crenate or crenate-serrate on the margins with 16–28 teeth along almost the entire length of each side, somewhat membranous or chartaceous, light green and sparsely appressed hispidulous or sometimes velvety-hispidulous above with short spreading bristles, becoming ± scabridulous with the tubercle-bases persisting in older leaves, paler and ± densely hispidulous-pubescent to tomentose beneath with hairs mainly on the venation and not tubercle-based, indumentum intermixed with sessile glands; venation impressed on leaf upper surface and raised beneath; petiole 0.4–1.5 cm long. Peduncles 1 per leaf axil with 2–3 per node, erect or ascending, 0.2–1.5(2.5) cm long, usually no longer than half the length of the subtending leaf and often shorter than the petioles, elongating in fruit. Flowering spikes 0.7–1.5 cm long, subspherical to ovoid in flower, elongating in fruit to c. 2(4) cm long and becoming ovoid-oblong to subcylindric; lower floral bracts leaf-like, shorter than the flowering spike, 5.5–10 × 3.5–6 mm, ovate or oblong-ovate, ± acuminate-cuspidate to acute at the apex, truncate or rounded at the base, flat or with ± revolute margins towards the apex, ± inconspicuously nerved, appressed hispidulous-puberulous, only slightly enlarged in fruiting spikes, usually soon caducous in fruit; fruiting spike-axes up to 27 mm long, with ± spaced prominent scars often within the remains of persistent calyces, white hispidulous-pubescent to strigose. Corolla lilac or mauve to pink or purple, with a yellow throat; tube 5–6 mm long; upper lip 2.5–3 mm wide, slightly emarginate; lower lip 4–6 mm wide with the 3 lobes entire, rounded. Anthers c. 0.5 mm long, oblong. Drupes 3–3.5 × 3 mm, lilac to purple or dark purple when mature.

Caprivi Strip. c. 63 km from Katima Mulilo on road to Linyanti, c. 914 m, 27.xii.1958, *Killick & Leistner* 3131 (K; SRGH). **Botswana**. N: Ngamiland Distr., on road to Narragha Valley, c. 18 km from Maun, 6.v.1967, *Lambrecht* 177 (K; LISC; SRGH). SW: Ghanzi Distr., Farm 70, D'kar, 976 m, 18.ii.1970, *R.C. Brown* 8676 (PRE; SRGH). SE: Mochudi, 1.iv.1914, *Harbor* in *Rogers* 6547 (PRE; Z). **Zambia**. B: above Senanga, 3.v.1925, *Pocock* 183 (PRE). W: Mufulira, 8.v.1948, *Cruse* 338 (K; S). C: Kabwe Distr., Golden Valley Farm, 17.iii.1933, *Michelmore* 687 (K). E: Chadiza, 850 m, 27.xi.1958, *Robson* 734 (BM; BR; K; LISC; SRGH). S: Mazabuka Distr., Kaleya Estates, 20.iii.1963, *van Rensburg* 1868 (K; SRGH). **Zimbabwe**. N: Murehwa–Mutoko (Mrewa–Mtoko) main road, c. 1500 m, v.1956, *Davies* 1934 (SRGH). W: Victoria fälle der Sambesi, 1000 m, 1.xii.1913, *A. Peter* 51033 (B). C: Kwekwe Distr., Erin Farm, c. 23 km from Kwekwe (Que Que), c. 1312 m, 1.iv.1966, *Biegel* 1053 (SRGH). **Malawi**. C: Kasungu Distr., Kasungu National Park, c. 1120 m, 22.xii.1970, *Hall-Martin* 1298 (K; SRGH). **Mozambique**. T: Cahora Bassa Distr., between the Songo Plateau and Zambezi R., 23.ii.1972, *Macêdo* 4893 (LISC; LMA).

Also in Angola, Namibia and South Africa (Transvaal). In *Brachystegia* woodlands often on rocky outcrops and escarpments, and in scrub on Kalahari Sands, also in *Acacia* woodlands, thickets, wooded grassland and grassland on floodplains and in riverine vegetation, sometimes in overgrazed areas; 850–1500 m.

Some specimens, mainly those from Botswana, have an indumentum of ± long spreading hairs on the branches, branchlets, petioles and peduncles, and appear to differ from typical *L. angolensis* where the hairs are shorter, appressed and antrorse. However, transitional forms between these two indumentum types are also found, and the recognition of the Botswana material as a separate variety (based on the indumentum) is not justified.

Specimens belonging to *L. angolensis* have, in the past, been confused with a number of other taxa (see R. Fernandes in Bol. Soc. Brot., sér. 2, **61**: 129–130 (1989)), and in particular with *L. rugosa* Thunb. (as *L. salviifolia* Jacq.) where intermediates between it and *L. angolensis* occur. *L. angolensis* may be distinguished from *L. rugosa* by its more robust branches usually bearing short densely leafy branchlets, and by its leaves mostly larger and relatively broader. Its flowering peduncles are shorter, usually less than half the length of the leaf, whereas in *L. rugosa* they equal or exceed the middle of the subtending leaf, and are sometimes up to 2 or more times as long as it. It is also distinguished from *L. rugosa* by its bracts which are not or hardly abruptly contracted and acuminate above, with flat or only slightly revolute margins; and by its denser indumentum of longer hairs, mainly on the lower surface of the leaf lamina.

Other specimens possess characters intermediate between *L. angolensis* and *L. rhodesiensis*. These specimens usually have bracts a little broader than those of *L. angolensis*, but not as large as in *L. rhodesiensis*, and the lower bracts do not reach the top of the flowering spike which is narrower than in *L. rhodesiensis*. In addition these specimens usually have opposite leaves, which is the normal arrangement in *L. angolensis* (although the leaves here are sometimes also 3-whorled), whereas in *L. rhodesiensis* the leaves are normally 3-whorled. The corollas are smaller than in *L. angolensis*, and approach *L. rhodesiensis* in this character. These specimens, some of which have been determined in various herbaria as *L. mearnsii* var. *latibracteolata*, may represent hybrids between these two species. Examples of specimens intermediate between *L. angolensis* and *L. rhodesiensis* are: **Zambia**: *Angus* 1845 (BR; K; LISC); *Bock* 169 (K; PRE). **Zimbabwe**: *Basera* 188 (SRGH); *Biegel* 1532 (K; SRGH); *Miss S. Davies* s.n. (BM); *Davison* s.n. (PRE); *Eyles* 7306 (SRGH); *Fries, Norlindh & Weimarck* 3720 (K; LD); *Godman* 199 (BM); *Jacobsen* 2059 (PRE; K); *Philcox, Leppard & Dini* 8690 (K); *Phipps* 100 (SRGH); *Rogers* 5795 (SRGH); *Taylor* 98 (SRGH); *West* 3204 (SRGH). **Malawi**: *Pawek* 12314 B (K).

DOUBTFUL TAXON

Mendonça 1038 (BM; LISC), from Mozambique, Cabo Delgado Province (Palma, on the road to the Lighthouse, 21.x.1942) was originally determined as *L. salviifolia* Jacq. and later redetermined by Moldenke as *L. viburnoides* (Forssk.) Vahl. Although in some features it approaches material belonging to *L. rugosa* Thunb., or in others *L. viburnoides* subsp. *richardii*, or material from east Africa, it appears to be distinct. It is an undershrub of the open bush on coral rocks, with long lateral branches, short internodes and raised nodes. Its leaves are opposite, small, up to 2.5 × 1.4 cm, and its peduncles are up to 3.7 cm long, slender, and longer than the subtending leaf. The flowers are apparently white. More material from Cabo Delgado and from Tanzania and Kenya is needed in order to determine the taxonomic status of this plant.

3. LIPPIA L.

Lippia L., Sp. Pl.: 633 (1753); Gen. Pl., ed. 5: 282 (1754). —Moldenke in Phytologia **12**, **13** (1965–1966) (numerous contributions). —R. Fernandes in Bol. Soc. Brot., sér. 2, **59**: 245–272 (1987). —Verdcourt in F.T.E.A., Verbenaceae: 27–37 (1992). —Atkins in Kadereit (ed.), Fam. Gen. Vasc. Pl. (Kubitzki, ed. in chief) **VII**: 462 (2004).

Erect shrubs, undershrubs or perennial herbs, mostly pyrophytes from large woody rootstocks, often aromatic, pubescent, hairy or setose. Leaves opposite or in whorls of 3 or 4, rarely alternate, entire, crenate, serrate, dentate or lobed. Inflorescences spicate; spikes densely flowered, usually short and head-like in flower, often somewhat elongate and cylindric in fruit, 1–several per axil and sometimes forming terminal corymbs or panicles. Flowers small, sessile, solitary in the axils of bracts, usually 4-ranked. Calyx campanulate or compressed and laterally 2-keeled or 2-winged, undulate, truncate, 2–4-fid or 4-dentate at the apex. Corolla white, greenish or creamy-yellow, often darker at the throat, rarely magenta; corolla tube cylindric or funnel-shaped, straight or incurved; corolla limb weakly 2-lipped, oblique, with the anterior lip larger than the posterior one. Stamens 4, didynamous, inserted in the corolla tube, included or slightly exserted; anthers ovate, with parallel thecae. Ovary 2-locular, each locule 1-ovulate; style usually short, with an oblique or recurved stigma. Fruit a dry schizocarp surrounded by the appressed calyx; mericarps 2, separating at maturity; pericarp papery or hard. Seeds without endosperm.

A genus of about 207 species of warmer regions, mainly of South and Central America but many also in Africa.

1. Pyrophyte 1–8(18) cm tall, flowering before the leaves appear; the first inflorescences borne at the lowermost node · 7. *praecox*
  − Subshrubs, shrubs or perennial herbs 0.3–3 m tall, flowering after the leaves appear; inflorescences borne at the upper nodes · 2
2. Inflorescences essentially terminal on stems and upper branches, with additional elements (usually 1–2 long-peduncled dense head-like spikes) in the upper leaf axils; bracts ± patent, at least after anthesis, the lower-ones usually ± reflexed; pyrenes 1–1.2 × 0.6 mm · · · 1. *plicata*
  − Inflorescences essentially axillary, often in many axils ± scattered along the stem and branches, the uppermost sometimes overtopping the shoot apex and appearing terminal; bracts not reflexed; pyrenes larger · 3
3. Leaves 3-whorled; lower bracts 6–7 × 5–5.5 mm, broadly ovate to suborbicular, rounded or very shortly acuminate at the apex, longer than the flowers, densely puberulous-tomentellous to silky on the outside · 2. *oatesii*
  − Leaves usually opposite; lower bract characters not combined as above · · · · · · · · · · · · 4
4. Peduncles 1–4 per axil; bracts 2–3 × 1.75–2 mm (up to 3.5 mm long in fruit), shortly to obsoletely apiculate or attenuate into an acute point; a much branched perennial herb or shrub 1–2(3) m high · 3. *javanica*
  − Peduncles usually 1 (rarely 2) per axil; bracts larger, ± acuminate at the apex; smaller many-stemmed perennial herbs or subshrubs · 5
5. Peduncles up to 18.5 cm long, 2–3 times longer than the subtending leaf, spreading-patent · 4. *baumii*
  − Peduncles up to 6.5 cm long, shorter to slightly longer than the subtending leaf · · · · · · 6
6. Bracts 5–9.5 × 1.5–2.5 mm, ovate-lanceolate to lanceolate, attenuate subulate; calyx truncate or undulate at the apex, densely long white setose on the sides and subglabrous on the anterior and posterior faces between · 6. *woodii*
  − Bracts up to 12 × 6 mm, broadly ovate, shortly acuminate; calyx uniformly hispidulous · 5. *scaberrima*

1. **Lippia plicata** Baker in F.T.A. **5**: 281 (1900). —R.E. Fries, Wiss. Ergebn. Schwed. Rhod.-Kongo-Exped.: 273 (1916). —Meikle in Mem. New York Bot. Gard. **9**: 36 (1954). —Moldenke, Résumé Verbenaceae: 142, 145 et 148 (1958); in Phytologia **12**: 350 (1965); op. cit. **39**: 394 (1978); op. cit. **48**: 256 (1981). —White, For. Fl. N. Rhod.: 370 (1962). —Richards & Morony, Check List Fl. Mbala: 238 (1969). —Gilli in Ann. Naturhist. Mus. Wien **77**: 30 (1973). —Verdcourt in F.T.E.A., Verbenaceae: 34, fig. 5 (1992). TAB. **5**. Lectotype: Zambia N: Fwambo, ix.1893, *Carson* 81 (K), chosen by Verdcourt.
    Lippia adoensis sensu J.G. Baker, op. cit.: 280–281 quoad specim. angolens. pro parte, non Hochst. (1841).
    Lippia adoensis Hochst. var. *multicaulis* Hiern, Cat. Afr. Pl. Welw. **1**, 4: 829–830 (1900). —R. Good in J. Bot. **68**, Suppl. 2 (Gamopet.): 139 (1930). Syntypes from Angola.
    Lippia strobiliformis Moldenke in Phytologia **2**: 317 (1947); Résumé Verbenaceae: 463 (Taxon. Index, 1958). Type from Tanzania.
    Lippia strobiliformis var. *acuminata* Moldenke in Phytologia **2**: 317 (1947); loc. cit. (1958). Type from Dem. Rep. Congo.
    Lippia strobiliformis var. *parvifolia* Moldenke in Phytologia **2**: 318 (1947); loc. cit. (1958). Type from Dem. Rep. Congo.
    Lippia oatesii sensu Moldenke, Résumé Verbenaceae: 147 (1958); in Phytologia **12**: 307 (1965) quoad specim. angolens. *Gossweiler* 9364; Fifth Summ. Verbenaceae: 244 (1971), non Rolfe (1889).
    Lippia plicata var. *acuminata* (Moldenke) Moldenke in Phytologia **12**: 352–353 (1965); op. cit. **13**: 365 (1966); op. cit. **39**: 395 (1978). Type as for *L. strobiliformis* var. *acuminata*.
    Lippia plicata var. *parvifolia* (Moldenke) Moldenke, tom. cit.: 353 (1965); loc. cit. (1971). Type as for *L. strobiliformis* var. *parvifolia*.
    Lantana salviifolia sensu Richards & Morony, Check List Fl. Mbala: 238 (1969) non Jacq. (1798).

  A many-stemmed perennial herb 0.5–1.5 m high from a woody rootstock, or shrub 1–3 m high, leaves aromatic when crushed. Stems erect, 4-angular ± striate and sulcate with internodes up to 11 cm long, strigose with bulbous-based whitish hairs, the indumentum similar on branches, petioles and peduncles; branches suberect, straight. Leaves usually opposite, occasionally 3-whorled, up to 12.5 × 4.5(5.8) cm, smaller on the branches, ovate or ovate-lanceolate to elliptic, acute at the apex, shortly cuneate or rounded at the base, serrate-crenate or minutely crenate, except at the base, chartaceous, closely rugose-bullate, reticulate, scabrid above with whitish tubercle-based hairs, densely pubescent and less scabrid beneath, the hairs longer

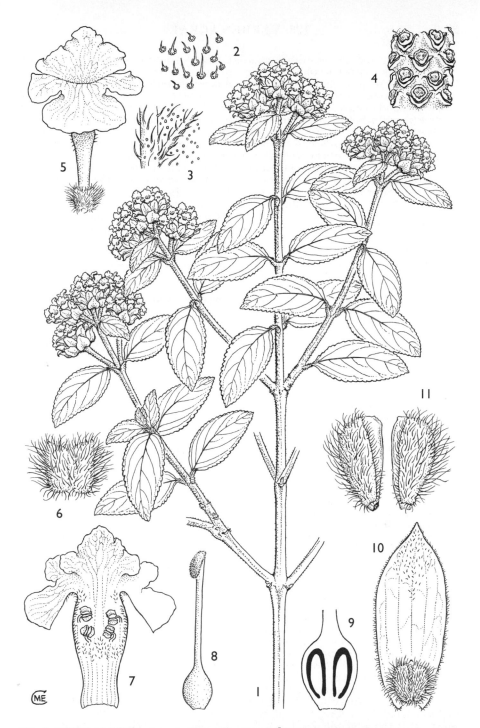

Tab. 5. LIPPIA PLICATA. 1, part of flowering stem (× ²/₃); 2, detail of leaf upper surface hairs (× 10); 3, detail of lower surface of leaf (× 10), 1–3 from *Bullock* 1945; 4, part of floral axis, showing scars of bracts and flowers (× 8), from *Newbould & Harley* 4251; 5, flower (× 6); 6, calyx (× 14); 7, corolla, opened out (× 6); 8, ovary and style (× 14); 9, longitudinal section of ovary (× 24), 5–9 from *Bullock* 1945; 10, bract and fruit (× 6); 11, 2 nutlets (× 16), 10 & 11 from *Newbould & Harley* 4251. Drawn by M.E. Church. From F.T.E.A.

and denser on the nerves, with numerous sessile glands, sometimes glabrescent; midrib prominent below with 5–8 pairs of main nerves, ascending, ± arcuate, impressed above, paler and somewhat raised beneath, reticulation impressed above and prominent beneath; petiole 2–7(10) mm long. Spikes 1–4 per axil and clustered at the stem and branch apices and often also in the axils of the upper leaves, densely many-flowered, 1.5–3.8 × 1–1.5 cm, somewhat hemispherical in flower, ovoid-oblong in fruit; peduncles 1–7 cm long, slender, suberect, shortest in the terminal inflorescences; lower bracts 7–10 × 2.5–5 mm and longer than the flowers, progressively smaller upwards, ovate, ovate-oblong or ovate-lanceolate, abruptly acute to ± attenuate apically, membranous, green or pale green, ± densely appressed-pubescent and minutely glandular outside, glabrous or sparsely pubescent towards the apex on the inside, spreading or reflexed at least after anthesis, or the lower ones completely deflexed, easily caducous. Calyx 2-lobed, c. 1.5 mm long, uniformly white-hispid. Corolla slightly to strongly scented, white with yellow throat, 2-lipped; tube c. 5 mm long, cylindric, suddenly dilated into the limb, densely and shortly pubescent and also minutely glandular towards the top; limb upper lip c. 3.5 mm wide, subtriangular, limb lower lip 4–5.5 mm wide, 3-lobed, the median lobe the largest up to 1.5 × 2 mm, the lateral lobes suboblong. Mericarps plano-convex, 1–1.2 × 0.6 mm on the commisural face.

Zambia. N: Mbala (Abercorn), foot of Nchalanga Hill, Kawimbe, c. 1856 m, 13.iv.1961, *Richards* 15039 (K; SRGH). W: Mwinilunga, 22.v.1969, *Mutimushi* 3433 (K). **Malawi**. N: Chisenga, foot of Mafinga Hills, 1850 m, 13.xi.1958, *Robson & Fanshawe* 610 (BM; BR; K; LISC; SRGH). C: Dedza Distr., Chongoni Forest, 16.viii.1960, *Chapman* 865 (K; SRGH). S: Mangochi Distr., east of Escarpment, 1090 m, 2.v.1960, *Leach* 9897 (K; LISC; SRGH). **Mozambique**. N: Maniamba, 22.v.1948, *Pedro & Pedrógão* 3824 (LMA; LMU). T: Dómuè Mt., c. 1450 m, 9.iii.1964, *Torre & Paiva* 11072 (LISC).
Also in Tanzania, Dem. Rep. Congo and Angola. On damp ground amongst tall grass, often in *Brachystegia* woodland, also in rupideserta; c. 830–2350 m.
Richards 9603 from Zambia (Kawimbe (BR; K)), with relatively longer leaves and shorter peduncles and with somewhat narrower spikes with smaller more appressed bracts, seems intermediate between *L. plicata* and *L. javanica*, being perhaps a hybrid between the two species.
Var. *acuminata* Moldenke is perhaps a variant without taxonomic value. Intermediate forms between typical *L. plicata* (with bracts ovate or ovate-oblong and abruptly acute at the apex) and the type of var. *acuminata* (with bracts ovate-lanceolate, attenuate and more acute at the apex) are sometimes found (e.g. *Fanshawe* 3675, *Mutimushi* 3433 from Zambia; *Monteiro* 35, *Torre & Paiva* 11072 from Mozambique).

2. **Lippia oatesii** Rolfe in Oates, Matabeleland Victoria Falls, ed. 2, appendix V: 407 (1889). — J.G. Baker in F.T.A. **5**: 279 (1900). —Moldenke, Résumé Verbenaceae: 149 (1958); in Phytologia **12**: 306–307 (1965) excl. specim. angol. *Gossweiler* 9364; op. cit. **39**: 259 (1978). —Wild in Kirkia **5**: 64 (1965); op. cit. **7**, 1: 57 (1968–69). —Martineau, Rhodesia Wild Fl.: 65 (1953). Type: Zimbabwe, Matabeleland, *Oates* s.n. (K, holotype?).

Perennial herb or subshrub up to 1.3 m high from a woody rootstock, with leaves pleasantly aromatic when crushed, densely and shortly whitish puberulous on stem, branches, leaves and peduncles with numerous minute pale yellowish spherical sessile glands intermixed with the hairs; branchlets spreading, slender, with the inflorescences borne in the axils of the upper 1–10 leaves. Leaves 3-whorled, sessile or shortly petiolate; lamina up to 10 × 2.8 cm, usually smaller on the branchlets, narrowly elliptic to elliptic, acute or nearly so at the apex, cuneate at base, minutely serrate to slightly or distinctly crenate-serrate at the margins towards the apex, chartaceous, grey-green, often folded along the midrib; midrib and main lateral nerves somewhat impressed above, ± raised beneath, reticulum slightly prominent beneath. Spikes solitary in each axil, 0.8–3 × 0.8–1.4 cm, subspheric to subcylindric; peduncles 0.8–5 cm long, ascending or becoming somewhat deflexed in fruit, shorter than the subtending leaf, slender but somewhat thickened towards the apex; bracts yellow-green, imbricate, persistent, accrescent, light green or yellow-green, ± concave, with ciliate somewhat revolute margins, densely puberulous-tomentellous or silky outside, sparsely pubescent at the apex and glabrous on the inside, the lower bracts 6–7 × 5–5.5 mm, orbicular to broadly obovate, usually rounded and blunt at the apex, progressively smaller and relatively narrower upwards, usually ± clearly longitudinally nerved. Calyx 2-lobed, 1.5–2 mm long, the lobes c. $^1/_4$–$^1/_3$ as long as the tube, densely hairy everywhere on the outside, the hairs rather long, white. Corolla slightly to strongly scented, white, 3–5 mm long, pubescent on the upper half except for the lobes which are mainly glandular. Mericarps plano-convex, c. 1.5 × 1.2 mm, oblong.

Zimbabwe. W: c. 13 km SE of Bulawayo, c. 1500 m, v.1961, *O.B. Miller* 7947 (K; LISC; SRGH). C: Gweru Distr., c. 41 km north of Gweru (Gwelo) on Lower Gweru Road at Hunters Road, c. 24 km from turn-off, 22.ii.1968, *Biegel* 2556 (K; SRGH); Zhombe Distr., Empress Mine, 7.iv.1981, *G.V. Pope* 1941 (K; SRGH). S: Masvingo Distr., Dindingwe Farm, c. 19 km from Masvingo (Fort Victoria), on road to Great Zimbabwe, 15.v.1965, *Ngoni* 1 (K; LISC; SRGH).

Known only from Zimbabwe. Stony hillsides, and the serpentine soils and lateritic areas of the Great Dyke, in short grassland and mixed deciduous woodlands with sparse ground cover; 750–1500 m.

Moldenke, in Phytologia **12**: 307 (1965); op. cit.: 147 (1958); op. cit.: 244 (1971), has erroneously recorded this species from Botswana.

3. **Lippia javanica** (Burm.f.) Spreng., Linnaeus Syst. Nat., ed. 16, **2**: 752 (1825). —Walpers, Repert. Bot. Syst. **4**: 50 (1845). —A. Meeuse in Blumea **5**: 68 (1942) excl. syn. pro parte. —Meikle in Mem. New York Bot. Gard. **9**: 36 (1954) excl. distr. Angola. —Chapman, Veg. Mlanje Mt. Nyasal.: 47 et 54 (1962). —White, For. Fl. N. Rhod.: 370 (1962), excl. syn. *L. asperifolia.* —Moldenke in Phytologia **12**: 225 (1965); op. cit. **13**: 358 (1966); op. cit.: **14**: 408–409 (1967); op. cit.: **39**: 99 et 437 (1978); op. cit.: **48**: 176 (1981); Résumé Verbenaceae: 461 (Taxon. Index, 1958). —Wild in Kirkia **5**: 64 (1965). —Richards & Morony, Check List Fl. Mbala: 238 (1969). —J.H. Ross, Fl. Natal: 299 (1972), excl. syn. *L. asperifolia.* —Gilli in Ann. Naturhist. Mus. Wien **77**: 30 (1973). —Jacobsen in Kirkia **9**: 172 (1973). —Verdcourt in F.T.E.A., Verbenaceae: 30 (1992). —M. Coates Palgrave, ed. 3 of K. Coates Palgrave, Trees South. Africa: 976 (2002). Type a cultivated plant.

Verbena javanica Burm.f., Fl. Ind.: 12, t. 6 fig. 2 (1768). —Willdenow, Sp. Pl., ed. 4, **1**: 117 (1797).

Blairia javanica (Burm.f.) Gaertn., Fruct. Sem. Pl. **1**: 265, t. 56 (1788).

Zapania javanica (Burm.f.) Lam., Tabl. Encycl. [Illustr. Gen.] **1**: 59, t. 17 fig. 2 (1791). —Poiret, Encycl. Méth. Bot. **8**: 840 (1808).

Verbena capensis Thunb., Prodr. Pl. Cap. part 2: 96 (1800). Type from South Africa.

Lippia capensis (Thunb.) Spreng., Linnaeus Syst. Nat., ed. 16, **2**: 751 (1825). —Walpers, Repert. Bot. Syst. **4**: 43 (1845).

Lantana lavandulacea Willd., Sp. Pl., ed. 4, **3**: 319 (1800). —Jacquin, Pl. Hort. Schoenbr. **3**: 59, t. 361 (1798). —Persoon, Syn. Pl. **2**: 141 (1806). —Walpers, Repert. Bot. Syst. **4**: 64 (1845). Type a cultivated plant.

Lippia asperifolia sensu Schauer in A. de Candolle, Prodr. **11**: 583 (1847) quoad syn. pro parte et distrib. Afr. —sensu Klotzsch in Peters, Naturw. Reise Mossambique **6**, part 1: 256 (1861). —sensu Gürke in Engler, Pflanzenw. Ost-Afrikas C: 338 (1895). —sensu Briquet in Engler & Prantl, Nat. Pflanzenfam. **4**, 3a: 152 (1897) quoad fig. 58 C, et distr. Afr. —sensu J.G. Baker in F.T.A. **5**: 280 (1900) excl. saltem *Holst* 8893 et *Scott-Elliot* 6331. —sensu H. Pearson in F.C. **5**, 1: 195 (1901) quoad syn. pro min. parte et pl. afr. —sensu Hutchinson, Botanist South. Africa: 672 (1946). —sensu Brenan, Check-list For. Trees Shrubs Tang. Terr.: 640 (1949). —sensu Martineau, Rhodesia Wild Fl.: 65 (1953), non Marthe (1801).

Lippia galpiniana H. Pearson in F.C. **5**, 1: 189 (1901). Type from South Africa.

Phyla javanica (Burm.f.) Moldenke in Herter, Rev. Sudan. Bot. **5**: 2 (1933) quoad basion?

Lippia whytei Moldenke in Phytologia **1**: 428 (1940); op. cit. **2**: 340–341 (1947); op. cit. **13**: 170 (1966); op. cit. **40**: 202 (1978); Résumé Verbenaceae: 463 (Taxon. Index, 1958). Type: Malawi, s.l. (?), *Whyte* s.n. (NY, holotype; K?).

Perennial herb or shrub 1–2 m (3 m or more?) high, much branched often from the base or many-stemmed, leaves strongly aromatic when crushed. Stems usually erect, subterete or somewhat angular, green, soon turning brown, striate, the stems and branches rather leafy with inflorescences in nearly all axils, ± densely hispidulous with appressed or spreading short stiff tubercle-based whitish hairs and small glands. Leaves opposite-decussate, shortly petiolate; lamina 2–9.5(13.5) × 0.8–3.2(5) cm, lanceolate, oblong-lanceolate or elliptic or sometimes obovate, subacute or obtuse at the apex, rounded then cuneate at the base, crenate-serrate or closely serrulate on the margins except near the leaf base, papyraceous, the upper surface dull green, strongly rugose and scabrid above with tubercle-based hairs, the lower surface pale or grey-green, softly and densely often ± velvety appressed pubescent with small glands, the nerves ± prominent, lateral nerves 4–7(10) on each side, ascending, impressed above and prominent beneath. Spikes 1–3(4) per axil, shorter than the subtending leaf, 3–8.5(10) × 3–4 mm, subspherical or ovoid, sometimes subcylindric in flower, up to 19 × 7 mm and ovoid-oblong to cylindric in fruit, purple or dull-reddish, ± dark brownish on drying; peduncles 0.6–3 cm long, equal or unequal in length, usually erect, slender, with an indumentum similar to that of the branches; bracts imbricate, sparsely to densely spreading white pubescent or silky, glandular, 2–3 × 1.75–2 mm, broadly ovate or obovate, accrescent and up to c. 3.75 mm broad in fruit, abruptly and shortly apiculate to long acuminate, the lower bracts deeply concave and truncate to shortly cuneate at the base, progressively smaller and less concave upwards, appressed not spreading or reflexed. Calyx 2-lobed, half as long as the corolla, 1–1.25 mm broad, ovate-campanulate, compressed in anterior-posterior plane, slender, membranaceous, pale yellow-green, usually densely spreading pubescent to silky outside, somewhat accrescent in fruit, persistent. Corolla white or creamy-white, yellow in the throat, glandular and pubescent outside in the upper half, 2–3 mm long, caducous; tube narrowly funnel-shaped from a very narrow base, 1–1.5 mm in diameter at the throat; limb ± 2-lipped with the posterior lip entire and the anterior one 3-lobed, the median lobe larger than the lateral ones. Anthers sulphur-yellow. Mericarps brown on the convex face, white on the flat commissural face, 1.5–1.75 × 1–1.25 mm, half-ovoid or sub-hemispheric, glabrous.

Botswana. SE: South East Distr., 13 km NE of Gaborone, 24°31'S, 25°58'E, 1050 m, 10.xi.1978, *O.J. Hansen* 3534 (BM; GAB; K; PRE; S; SRGH). **Zambia**. B: Sesheke Distr., Masese, l9.vi.1960, *Fanshawe* 5747 (BR; FHO; K; LISC). N: Mbala Distr., Kawimbe Marsh, c. 1600 m, 20.v.1955, *Richards* 5777 (BR; K; LISC). W: Ndola, 26.v.1953, *Fanshawe* 34 (BR; K; SRGH). C: Lusaka Distr., Mt. Makulu, 30.vi.1956, *P. Simwanda* 1 (K; SRGH). E: Chipata Distr., Lumvi, Munkanya, Luangwa Valley, 24.iv.1968, *R. Phiri* 199 (K; SRGH). S: Kalomo Distr., Mochipapa near Choma, 9.iii.1962, *Astle* 1466 (K; SRGH). **Zimbabwe**. N: Hurungwe Distr., Mwami (Miami), Experimental Farm, 5.iii.1947, *Wild* 1829 (K; SRGH). W: Matobo Distr., Farm Shumbashaba, 1536 m, i.1957, *O.B. Miller* 4048 (BR; K; SRGH). C: Marondera (Marandellas); 13.ii.1941, *Dehn* 53 (M; SRGH). E: ad pagum Nyanga (Inyanga), 1700 m, 13.i.1931, *Norlindh & Weimarck* 4245 (K; LD; LISC; M). S: Mwenezi Distr., Bubye R., near Bubye Ranch homestead, 576 m, 8.v.1958, *Drummond* 5700 (BR; K; LISC; SRGH). **Malawi**. N: Mzimba, 1440 m, 30.vii.1960, *Leach & Brunton* 10370 (K; LISC; SRGH). C: Lilongwe Distr., Namitete R. below bridge on Lilongwe–Chipata (Fort Jameson) road, 1150 m, 5.ii.1959, *Robson* 1463 (BM; BR; K; LISC). S: Zomba Distr., Mingoli Estate, 14.viii.1962, *Adlard* 477 (K; LISC; SRGH). **Mozambique**. N: Mueda, 16.ix.1948, *Pedro & Pedrógão* 5237 (LMA; LMU). T: Tete, from Marueira to Songo, at c. 6 km from Marueira, 5.ii.1972, *Macêdo* 4789 (LISC; LMA; LMU). Z: Morrumbala Distr., at the summit of Morrumbala Mt., 1000 m, l.v.1943, *Torre* 5266 (K; LISC; M; P). MS: Manica Distr., south of Bandula, 23.iv.1948, *Andrada* 1147 (LISC; M; SRGH). GI: Chokwè Distr., Guijá, right bank of Limpopo R., 9.vi.1947, *Pedrógão* 264 (COI; K; LISJC; LMA; SRGH). M: Namaacha, at km 10 to Matianine, 14.v.1974, *A. Marques* 2447 (K; LISC; LMU; M; SRGH).

Also in Ethiopia, Uganda, Kenya, Tanzania and South Africa (KwaZulu-Natal, Transvaal and Cape Provs.). Widespread, in low to high altitude woodlands and wooded grasslands, in riverine vegetation and on margins of dambos and swampy ground, sometimes on anthills, in montane grassland and on evergreen forest margins; also in disturbed ground beside roads and in cultivated land; 10–2240 m.

As recognized here this species exhibits great variation in the number of spikes per axil and their peduncle length. An attempt was made by this author to distinguish two varieties based mainly on peduncle and spike length – var. *javanica* with peduncles usually slender and mostly more than 1 cm (up to 3 cm) long, longer than the spikes which are usually small even in fruit, and var. *whytei* (corresponding to *Lippia whytei* Moldenke), with peduncles mostly less than 1 cm long (or spikes nearly sessile), usually shorter than the spikes which are longer, thicker and, in fruit, darker purplish in colour. In typical var. *javanica* the spikes are borne in lax axillary groups on peduncles of unequal length. While in characteristic var. *whytei* the spikes are borne in ± compact groups, on short ± equal peduncles, or some spikes subsessile.

However, variation between the two varieties so delimited was found to be ± continuous, and sometimes duplicates of the same gathering would be assigned to different varieties in different herbaria. Further study is necessary to establish to what extent peduncle and spike length increases with the maturity of the plant and/or with the progress of the season. Therefore despite their very different appearance the two varieties are not formally separated here.

It should be pointed out that var. *javanica* is more common in the south of the Flora Zambesiaca area while var. *whytei* is more common in the north. In Botswana almost only var. *javanica* has been recorded, while in Malawi 25 gatherings of var. *whytei* and 12 of var. *javanica* were seen. In Zimbabwe 47 specimens of var. *javanica*, and 22 of var. *whytei* were seen. In northern Mozambique (Niassa and Tete Provinces) var. *whytei* is the more common (15 specimens as opposed to 4 of var. *javanica*), and var. *javanica* is most common in the rest of the country, with var. *whytei* apparently absent from Gaza and Maputo Provinces.

Chapman 298 from Malawi (BR; SRGH), with exceptionally long stout peduncles and large spikes, is probably a hybrid between *L. javanica* and *L. woodii*, a taxon also found in Malawi. The aborted ovaries and empty anthers (if pollen present, then nearly all grains sterile) support this hypothesis.

4. **Lippia baumii** Gürke in Warburg, Kunene-Samb.-Exped. Baum: 350 (1903). —R. Fernandes in Bol. Soc. Brot., sér. 2, **59**: 247 (1987). —Verdcourt in F.T.E.A., Verbenaceae: 32 (1992). Type from Angola.

An erect perennial herb with annual stems from a woody rootstock, leaves aromatic when crushed. Stems 1–many, up to 1 m tall, green, simple or branched above, striate, subterete to somewhat angular in the upper part, stems and branches rather leafy with inflorescences in upper axils, densely sericeous to hispid-strigose with patent or ± appressed tubercle based whitish hairs and small sessile glands. Leaves decussate, subsessile or shortly petiolate; lamina 2–9.5(10) × 1–3 cm, narrowly oblong-elliptic to oblong-lanceolate or 1.5–5 × 1.4–3.5 cm, ovate-lanceolate to ovate or rotund, obtuse to ± acute at the apex, broadly cuneate to rounded or subcordate at the base, crenate to crenulate serrate on the margins in the upper part and entire in the lower one third, venation impressed above, prominent and reticulate beneath, ± densely sericeous to hispid-strigose on both surfaces, sometimes ± scabridulous, with small sessile glands. Spikes head-like, up to 10 × 10 mm and usually broader than long at flowering time, increasing to c. 3 cm long in fruiting spikes; peduncles usually solitary in the upper axils, widely spreading or ± ascending, 2–14 cm long and longer than the subtending leaf, usually stout, with indumentum as in the stems and branches; lowermost bracts ± leaf-like, up to c. 12(25) × 5(10) mm, inner bracts imbricate, longer or shorter than the flowers they subtend, ovate and shortly acuminate to truncate at the apex to ± attenuate, sericeous to hispid, accrescent in fruit. Calyx c. 1.5 mm long, 2-lobed, ± densely appressed white-hairy to puberulous outside. Corolla yellow or white with a yellow centre, or in one variety deep magenta; tube hairy outside except at the base and below the upper lip; limb c. 2-lipped, c. 4 mm wide, the upper lip entire, rounded, the lower one 3-lobed, the lobes entire. Mericarps ovoid, very convex, 2–2.5(3) × 1.25–2 mm, brownish and slightly shining outside, flat and white inside.

Indumentum less dense, of hispid-strigose markedly unequal acicular setae; leaves usually ± scabridulous, oblong to lanceolate or ovate, mostly much longer than wide; inner bracts ovate, usually shortly acuminate or truncate at the apex and over-topped by the subtended flowers · var. *baumii*
Indumentum of densely patent long subequal softly sericeous hairs; leaves relatively shorter and broader, not much longer than wide; inner bracts ± attenuate, exceeding the subtended flowers · var. *nyassensis*

Var. **baumii** —R. Fernandes in Bol. Soc. Brot., sér. 2, **59**: 247 (1987).
    Lippia baumii sensu Moldenke, Résumé Verbenaceae: 461 (Taxon. Index, 1958), excl. saltem distrib. in S.W.A.; in Phytologia **12**: 97 (1965) excl. saltem specim. *Baum* 250; Fifth Summ. Verbenaceae: 890 (Taxon. Index, 1971), excl. saltem distrib. in S.W.A. —sensu Richards & Marony, Check List Fl. Mbala: 238 (1969). Type from Angola.
    Lippia wilmsii H. Pearson var. *villosa* sensu Moldenke in Phytologia **13**: 175–176 (1966) quoad saltem specim. *Milne-Redhead* 1155; Fifth Summ. Verbenaceae: 895 (Taxon. Index, 1971), quoad saltem distrib. in Zambia, non Moldenke (1953) quoad saltem typum var.
    Lippia wilmsii sensu Richards & Morony, Check List Fl. Mbala: 238 (1969).
    Lippia sp. of Richards & Morony, Check List Fl. Mbala: 239 (1969).

An erect perennial herb or subshrub up to 100 cm high, ± scabridulous. Branches spreading. Indumentum of hispid-strigose markedly unequal acicular setae. Cauline median leaves 4.8–10 × 2–3(5) cm, mostly longer than the internodes, oblong to lanceolate or ovate, mostly much longer than wide, crenulate-serrate to coarsely crenate-serrate or obsolete-crenate in the upper $^1/_2$–$^2/_3$, entire at the lower $^1/_2$–$^2/_3$, with 3–6 lateral nerves at each side, sessile or subsessile; lower cauline leaves smaller, shorter than the internodes. Peduncles up to 18.5 cm long, spreading-patent to ascending, slender to somewhat stout. Lower bracts leaf-like, with crenulate margins, and midrib and nerves sunken above and raised beneath; inner bracts ovate, usually shortly acuminate or truncate at the apex and over-topped by the subtended flowers. Corolla (3.5)4–5.5 mm long, pale yellow or white with yellow centre.

Zambia. N: Mpika, 28.i.1955, *Fanshawe* 1875 (BR; K; SRGH); Mbala Distr., pans near Mbala (Abercorn), 20.i.1955, *Richards* 4183 (K).
Also in Tanzania and Angola. Plateau woodlands and high altitude grasslands; 650–1740 m.

Var. **nyassensis** R. Fern. in Bol. Soc. Brot., sér. 2, **59**: 251 (1987). —Verdcourt in F.T.E.A., Verbenaceae: 32 (1992). Type: Malawi Mzimba Distr., Viphya Plateau, 25.iv.1967, *Salubeni* 653 (SRGH, holotype; BR; K; LISC).
    Lippia baumii sensu Binns, First Check List Herb. Fl. Malawi: 103 (1968) pro parte quoad Distr. Mzimba et Nkhata Bay.
    Lantana primulina Moldenke in Phytologia **28**: 402 (1974). Type: Malawi, Mafinga Hills, *E.A. Robinson* 4452 (MO, holotype; BR; K; M).

An erect many-stemmed perennial herb up to 100 cm high, softly hairy. Branches widely spreading. Indumentum denser, of longer and more spreading hairs. Stems simple, more robust and with shorter internodes than in var. *baumii*. Cauline median leaves 1.5–5 × 1.4–3.5 cm, broadly ovate to roundish, usually not much longer than wide, crenulate-serrate to coarsely crenate-serrate or obsolete-crenate in the upper $^1/_2$–$^2/_3$, entire at the lower $^1/_2$–$^1/_3$, sessile or subsessile; lower cauline leaves smaller, shorter than the internodes. Peduncles up to 18.5 cm long, spreading-patent to ascending, slender. Lower bracts leaf-like, with crenulate margins, and midrib and nerves sunken above and raised beneath; inner bracts ± attenuate, exceeding the subtended flowers. Corolla yellowish.

Malawi. N: Nkhata Bay Distr., c. 50 km SW of Mzuzu, Viphya Plateau, c. 1800 m, 27.iii.1976, *Pawek* 10950 (BR; K; SRGH); Chitipa Distr., Mafinga Mts., c. 2135 m, 2.iii.1982, *Brummitt, Polhill & Banda*, 16232 (K; MAL).
Known only from the Flora Zambesiaca area. Montane short grassland; 1740–2135 m.

5. **Lippia scaberrima** Sond. in Linnaea **23**: 87 (1850). —H. Pearson in F.C. **5**, 1: 194 (1901). —Moldenke in Phytologia **12**: 460 (1965); op. cit. **40**: 58 (1978); op. cit. **41**: 150 (1979); op. cit. **48**: 261 (1981). —Watt & Breyer-Brandwijk, Medic. & Pois. Pl. S. & E. Africa: 1053 (1962). —J.H. Ross, Fl. Natal: 299 (1972). Type from South Africa (Free State).
    Camara salviifolia var. *transvaalensis* Kuntze, Revis. Gen. Pl., part 3: 250 (1898). Type from South Africa (Transvaal).
    Lippia transvaalensis (Kuntze) Moldenke in Phytologia **1**: 428 (1940). Type as for *Camara salviifolia* var. *transvaalensis*.

Perennial herb, undershrub or low shrub, from a woody rootstock, leaves strongly aromatic when crushed. Stems 1–several, 30–100 cm tall, erect or spreading, simple or shortly branched above, somewhat 4-angular, striate, stems and branches leafy with inflorescences in upper axils, ± scabrid mainly along the angles with short strigose antrorse setae. Leaves opposite, rarely 3-whorled, shortly petiolate, pale green; lamina 2.5–6.5 × 0.4–2 cm, narrowly lanceolate to elliptic, smaller on the branches, acute at the apex and cuneate at the base, crenate-serrulate on the margins in the upper part and entire in the lower one third, with scattered short strigose hairs on the upper surface and sparsely strigose beneath with the hairs mainly on the nerves, glandular punctate on both surfaces; nerves impressed above, raised and paler beneath, the lower pair more pronounced and curving to half way up the blade. Spikes densely many-flowered, up to 1.2 cm long and ovoid in flower, increasing to 2.6 × 1.2 cm and cylindric-ovoid in fruit; peduncles solitary in the upper leaf axils, ascending, up to 5.3 cm long, subequalling the subtending leaf. Bracts appressed imbricate, 4–10 × 3–6 mm, longer than the flowers, broadly ovate, shortly acuminate, sparsely shortly pubescent with appressed white hairs outside, shortly ciliate at the margin, glandular outside, greenish-yellow, the lowermost not leaf-like. Calyx distinctly 2-lobed, $^1/_3$–$^1/_2$ as long as the corolla, uniformly hispidulous. Corolla cream to pale yellow; tube c. 4 mm long, pubescent and glandular outside, with pubescent throat; limb 2-lipped. Mericarps 2.25 × 1.75 mm, semiglobose, flat on the commissural face.

Botswana. SE: Kweneng Distr., c. 16 km NW of Lephephe (Lephepe) Village, Kalahari Sandveld Pasture Research Station, 23.v.1969, *Kelaole* 604 (SRGH); South East Distr., near St. Joseph's College, Kgale Hill (Khale) 24°45'S, 25°52'E, 1050 m, 31.x.1978, *O.J. Hansen* 3518 (K; S).
    Also in South Africa (Limpopo, KwaZulu-Natal, North West and Free State Provs.). Kalahari Sand wooded grassland with *Acacia fleckii* and *Combretum apiculatum*, pan margins and roadsides; c. 1050 m.

6. **Lippia woodii** Moldenke in Phytologia **2**: 318 (1947); op. cit. **13**: 176 (1966). —R. Fernandes in Bol. Soc. Brot., sér. 2, **59**: 261 (1987). —Verdcourt in F.T.E.A., Verbenaceae: 33 (1992).
Type: Malawi, Blantyre, *Buchanan* in *Medley-Wood* 6937 (F, 83373, holotype; NH, isotype).
    Lantana salviifolia sensu J.G. Baker in F.T.A. **5**: 276–277 (1900) quoad *Buchanan* 245 non Jacq. (1798).
    Lippia wilmsii sensu H. Pearson in F.C. **5**, 1: 196 (1901) pro parte. —sensu Moldenke in Phytologia **2**: 469 et 483 (1948); op. cit. **3**: 77 (1949); op. cit.: 457 et 458 (1951) quoad distr. Zambia, Zimbabwe et Mozambique; op. cit. **13**: 171 et 174 (1966) et op. cit. **40**: 202 et 204 (1978) pro parte, quoad saltem specim. in area Fl. Zamb. coll.; Fifth Summ. Verbenaceae pro parte, quoad saltem pags. 246, 248, 250 et 252 (1971). —sensu Wild in Kirkia **5**: 64 (1965); op. cit. **7**: 20 et 57 (1968–1969). —sensu Binns, First Check List Herb. Fl. Malawi: 103 (1968). —sensu Jacobsen in Kirkia **9**: 172 (1973), non H. Pearson (1901) sens. str.
    Lippia africana sensu Moldenke in Phytologia **2**: 469 (1948) quoad specim. in area Fl. Zamb. coll. —sensu Wild in Clark, Victoria Falls Handb.: 158 (1952). —sensu Binns, loc. cit, non Moldenke (1948).
    Lippia wilmsii var. *scaberrima** sensu Moldenke in Phytologia **3**: 458 (1951), quoad distr. in Zimbabwe; op. cit. **13**: 175 (1966) quoad saltem specim. zimbabwens. *Young* 882 et excl. specim. malawiens. *Whyte* s.n. (in Nyika Plateau coll.); Fifth Summ. Verbenaceae: 248 (1971), non Moldenke (1953) neque *L. africana* var. *scaberrima* Moldenke (1948).
    Lippia wilmsii var. *villosa* sensu Moldenke in Phytologia **13**: 176 (1966) quoad saltem *Maas-Geesteranus* 4786, non Moldenke (1953).

A perennial herb or subshrub up to 1 m high from a stout woody rootstock, leaves strongly aromatic when crushed. Stems annual or, in plants escaping annual fires, persisting for several years, erect, leafy, terete or subangular; indumentum hispid-strigose with patent or ± appressed tubercle based whitish hairs and small sessile glands, the setae markedly unequal acicular up to c. 2 mm long, very dense towards the stem and branch apices; annual stems slender, ± numerous, tufted, unbranched or shortly branched, rarely long branched from the base; older stems stouter, up to 5 mm or more in diameter at the base, ± branched, lower internodes up to 14.5 cm long. Leaves opposite-decussate, subsessile to shortly petiolate; lamina 1–9 (11.5) × 0.5–3(3.4) cm, narrowly elliptic to elliptic, acute to obtuse at the apex, cuneate at the base, crenulate to serrate on the margin, usually only entire in the basal ¼ to ⅓, scattered to ± densely short hispid-strigose on the upper surface and hispid beneath with the hairs mainly on the nerves, ± glandular beneath, mostly light green and often paler beneath; nerves 5–7 pairs ascending, impressed above and raised beneath; petiole 0–5 mm long; the lowermost pair of leaves much smaller, relatively broader and less acute. Spikes densely many-flowered, up to 1.2 × 0.7–1.2 cm, broadly and shortly ovoid in flower, increasing to 2.2 cm long in fruit becoming ovoid-oblong to cylindric; peduncles solitary, rarely 2 per leaf axil, in the upper c. 6 nodes, ascending, (0.6)1–5(6.5) cm long, usually shorter than the subtending leaves, slender, glandular and densely strigose or pubescent. Bracts appressed imbricate, 5–9 × 1.5–2.5 mm, longer than the flowers, lanceolate to ovate-lanceolate, acuminate to attenuate and subulate (the ensemble of the terminal part of the upper bracts forming a corona at the top of the flowering spikes), hispid and glandular outside, appressed setose within mainly towards the apex, persistent. Calyx 1–2 mm long, ± truncate at the apex or sometimes emarginate at the anterior face, thinly membranous, glabrous or sparsely setose along the anterior and posterior faces with a dense band of straight white setae on each side, ciliate at the upper margin. Corolla whitish, greenish-white to pale yellow, usually with a deep yellow centre, 2.5–4 mm long and narrowly funnel-shaped, the upper part of the tube densely spreading pubescent and glandular; limb not clearly 2-lipped, 1.5–2.6 mm wide, the largest lobe up to 1 mm long and wide. Fruits enveloped in the thin brown papery calyx, white pubescent outside; mericarps 2–2.25 × 1.5–1.75 mm, ± hemispherical, brownish, glabrous and slightly shining on the outer face, flat and white on the commissural face.

Zambia. B: Kaoma Distr., near Luena River, 20.xi.1959, *Drummond & Cookson* 6682 (E; K; LISC; SRGH). W: Solwezi Distr., c. 19.5 km from Solwezi, along road to Mwinilunga, 20.xi.1972, *Strid* 2511 (K). C: Lusaka Distr., c. 8 km east of Lusaka, 2.i.1956, *King* 256 (K). E: Chadiza Distr., Nsadzu Bridge, 900 m, 27.xi.1958, *Robson* 746 (BM; BR; K; LISC; SRGH). S: Kalomo Distr.,

* Comb. non rite publ.

Siantambo, Sichifulo Controlled Hunting Area, 3.i.1964, *B.L. Mitchell* 24/63 (B; LMU; SRGH).
Zimbabwe. N: Makonde Distr., Mangula (Mhangura), Molly South Hill, 3.ii.1965, *Wild & Simon* 6781 (BR; K; SRGH). W: Matobo Distr., Farm Besna Kobila, c. 1500 m, ii.1954, *O.B. Miller* 2163 (K; LISC; SRGH). C: cultivated land SE of Gweru (Gwelo), c. 1472 m, 21.xi.1966, *Biegel* 1555 (SRGH). E: Mutasa Distr., south of Penhalonga, 1934, *Gilliland* Q919 (BM; K). S: Masvingo Distr., Mutirikwi Recreational Park (Kyle Nat. Park), 10.xii.1970, *Basera* 236 (SRGH). **Malawi**. N: Mzimba Distr., Mtwalo Junction, near Ezondweni, c. 1280 m, 2.iii.1974, *Pawek* 8174 (K; SRGH). C: Dedza Mt., 23.x.1956, *Banda* 293 (BM; BR; K; SRGH). S: Mt. Zomba, c. 1472, c. 1920 m, xii.1896, *Whyte* s.n. (K). **Mozambique**. T: Moatize Distr., at 21 km from Zóbuè to Ntengo wa Mbalame (Metengobalame), c. 900 m, 11.i.1966, *Correia* 398 (LISC). MS: Manica Distr., Border Cantine östliche Mutare (Umtali), 10.xii.1913, *A. Peter* 51166 (B).

Also in Kenya and South Africa (Limpopo and Mpumalanga Provs.). Widespread, in miombo and open mixed deciduous woodlands, wooded grasslands and submontane grasslands, on dambo margins and in riverine vegetation, in sandy soils, Kalahari Sand and lateritic soil; also in disturbed ground beside roads; 500–2200 m.

Pearson referred the *L. woodii* specimens *Whyte* s.n. and *Buchanan* 1381, both from Malawi, Zomba Mt., and *Cecil* 219 from Zimbabwe, Nyanga to his species *L. wilmsii*. Perhaps for this reason nearly all material of *Lippia woodii* from the Flora Zambesiaca area has been confused with *L. wilmsii*, in herbaria and by subsequent authors. *L. wilmsii* is a South African species with which the following are considered to be synonymous: *L. africana* Moldenke, *L. rehmannii* H. Pearson, *L. pretoriensis* H. Pearson and *L. bazeiana* H. Pearson.

L. woodii differs from *L. wilmsii* by the larger leaves which are relatively longer and narrower and separated by longer internodes than in *L. wilmsii*; by the shorter peduncles, usually rather shorter than the subtending leaves (in *L. wilmsii* they equal or exceed the leaves); by the bracts relatively longer and narrower, attenuate into long subulate apices which give the top of the flowering spikes a corona-like appearance; and by the very important and distinct characters of the calyx: not so compressed as in *L. wilmsii*, somewhat plicate, usually undulate or truncate at the apex (in *L. wilmsii* the calyx is flat and usually ± deeply 2-lobed), with a ± glabrous median area on the anterior and usually also the posterior faces between the dense bands of long, stout, straight, white setae on the sides (in *L. wilmsii* the calyx is uniformly covered with not very dense, short, slender, dirty-white hairs).

The type of *L. wilmsii* var. *villosa* (Moldenke) Moldenke, [*Holm* 32 (S, holotype; K, photo) from Kenya (Mt. Elgon)], approaches *L. woodii* more closely than it does *L. wilmsii*, particularly in the indumentum, bract and calyx characters described above. However, more material needs to be seen before the affinity of *L. wilmsii* var. *villosa* can be determined with certainty.

Young 882, from Zimbabwe, Bindura, was cited as *L. wilmsii* var. *scaberrima* by Moldenke (op. cit.: 175, 1966). It is merely a form of *L. woodii* with larger leaves.

7. **Lippia praecox** Mildbr. ex Moldenke in Phytologia **4**: 292 (1953); op. cit. **12**: 356 (1965). — Verdcourt in F.T.E.A., Verbenaceae: 32 (1992). Type from Tanzania.

Pyrophyte with short annual stems from a large woody rootstock. Stems 1–numerous, 1–15 cm tall at flowering, increasing to 25 cm later in the season; indumentum densest on young growth, hispid, of unequal acicular setae. Leaves decussate, petiolate; lamina up to 4 × 2.5 cm, broadly elliptic to broadly obovate or rhomboid, rounded-obtuse to shortly acuminate at the apex, ± broadly cuneate at the base, crenulate to crenate-serrate in the upper half to two thirds and entire below, nervation impressed above and strongly raised beneath, ± densely strigose on upper surface, patent-hispid with hairs densest on the nerves beneath; petiole up to 7 mm long. Inflorescences produced before the leaves from nodes near ground level, later also from upper leaf axils. Spikes 5–12 mm long and wide, hemispherical, increasing to c. 15 mm long and becoming cylindrical; peduncles ascending, paired and up to 6.5 cm long from the lower nodes, shorter when produced later in the leaf axils, indumentum densely hispid-pubescent as in the stems; bracts appressed imbricate, the lowermost not leaf-like, 5–8.5 × 1.5–4 mm, ovate acuminate or lanceolate, becoming narrower and attenuate in the upper bracts, ± equalling the flowers they subtend, the indumentum densely hispid-pubescent as in the upper stems. Calyx c. 1.5–3 mm long, 2-lobed, ± densely white patent hairy outside. Corolla bright lemon-yellow or white with a yellow centre, or in var. *nyikensis* deep magenta; tube 3–5 mm long, pubescent in upper half and with pubescent throat; limb c. 2-lipped, 3–4 mm wide, the upper lip rounded entire, the lower one 3-lobed, the largest lobe 1.5–2 mm long, entire. Mericarps ovoid, very convex, 2–2.5(3) × 1.25–2 mm, brownish and slightly shining outside, flat and white inside.

128. VERBENACEAE 41

Var. **praecox**

Corollas bright lemon-yellow or white with a yellow centre.

Zambia. N: Mbala Distr., c. 3 km northwest of Mbala (Abercorn), 22.vii.1930, *Hutchinson & Gillett* 4012 (K); old aerodrome, Mbala (Abercorn), 13.xii.1954, *Richards* 3617 (K). W: Solwezi Distr., Solwezi Dambo, 20.ix.1930, *Milne-Redhead* 1155 (K).
Also in Tanzania. Appearing in recently burnt grassland, dry open woodlands and in open ground on dambo margins, sometimes in small colonies; 1350–1660 m.

Var. **nyikensis** (R. Fern.) G.V. Pope, comb. nov.*
  Lippia baumii var. *nyikensis* R. Fern. in Bol. Soc. Brot., sér. 2, **59**: 252 (1987), as "*niykensis*".
Type: Malawi, Nyika Plateau, road to Juniper Forest, 2496 m, *Pawek* 7872 (SRGH, holotype).
  Lippia baumii sensu Binns, First Check List Herb. Fl. Malawi: 103 (1968) pro parte quoad Distr. Rumphi.

Corollas magenta or pink-purple with a yellow centre.

Malawi. N: Rumphi Distr., Nyika Plateau, c. 24 km out on Kasaramba view road, 2432 m, 26.xii.1966, *Pawek* 752 (SRGH).
Known only from Malawi. In montane grassland.

4. PHYLA Lour.

Phyla Lour., Fl. Cochinch.: 66 (1790). —Atkins in Kadereit (ed.), Fam. Gen. Vasc. Pl. (Kubitzki, ed. in chief) **VII**: 462 (2004).

Prostrate perennial herbs with woody taproots; indumentum of sparse to closely spaced appressed medifixed hairs or plants glabrous. Stems trailing, rooting at the nodes; branches short, ascending. Leaves opposite, simple, shortly petiolate or subsessile, serrate. Inflorescences axillary, pedunculate, spicate; spikes short, dense, ovoid becoming cylindric after anthesis, with bracts closely imbricate. Flowers small, sessile, each solitary in the axil of a bract. Calyx membranous, dorsiventrally flattened, 2-lobed, persistent. Corolla usually white, becoming rose-coloured; tube straight or curved, cylindric or widening toward the apex; limb oblique, c. 2-lipped, with the (upper) adaxial lip entire, emarginate or 2-lobed, and the lower (abaxial) lip 3-lobed with the median lobe usually larger than the others. Stamens 4, included or somewhat exserted. Ovary 1-carpellate, carpel 2-locular, each locule 1-ovulate; style short, undivided; stigma thick, obliquely capitate. Fruit a schizocarp, compressed oblong, separating at maturity into two 1-locular mericarps; mericarps with a dry hard pericarp, semi ovoid, flat on the commissural face, rounded and smooth on the dorsal face, glabrous.

A genus of c. 10 species in tropical and subtropical America, with 1 species widespread throughout the warmer parts of the world.

Phyla nodiflora (L.) Greene in Pittonia **4**: 46 (1899). —Mogg in Macnae & Kalk, Nat. Hist. Inhaca Isl., Moçamb.: 152 (1958). —Brenan in Mem. New York Bot. Gard. **9**: 37 (1954). — F.W. Andrews, Fl. Pl. Anglo-Egypt. Sudan **3**: 197 (1956). —Moldenke in Fl. Madag., Verbenaceae: 13–14, t. 2 fig. 1–3 (1956). —Leon & Alain, Fl. Cuba **4**: 293 cum. fig. (1957). —Meikle in F.W.T.A., ed. 2, **2**: 437 (1963). —Backer & Bakhuizen van den Brink Jr., Fl. Java **2**: 597 (1965). —Friedrich-Holzhammer et al. in Merxmüller, Prodr. Fl. SW. Afrika, fam. 122: 9 (1967). —Binns, First Check List Herb. Fl. Malawi: 103 (1968). —Greenway in J. E. Afr. Nat. Hist. & Nat. Mus. **27**: 196 (1969). —van der Schijff, Check List Vasc. Pl. Kruger Nat. Park: 81 (1969). —J.H. Ross, Fl. Natal: 299 (1972). —Verdcourt in F.T.E.A., Verbenaceae: 25, fig. 4 (1992). TAB. **6**. Type from N America (Virginia).
  Verbena nodiflora L., Sp. Pl.: 20 (1753). —N.L. Burman, Fl. Indica: t. 6 fig. 1 (1768). — Willdenow, Sp. Pl., ed. 4, **1**: 117 (1797).

---

* Var. **nyikensis** (R. Fern.) G.V. Pope, comb. nov.
  Lippia baumii var. *nyikensis* R. Fern. in Bol. Soc. Brot., sér. 2, 59: 252 (1987), as "*niykensis*".
Type: Malawi, Nyika Plateau, *Pawek* 7872 (SRGH, holotype). G.V. Pope.

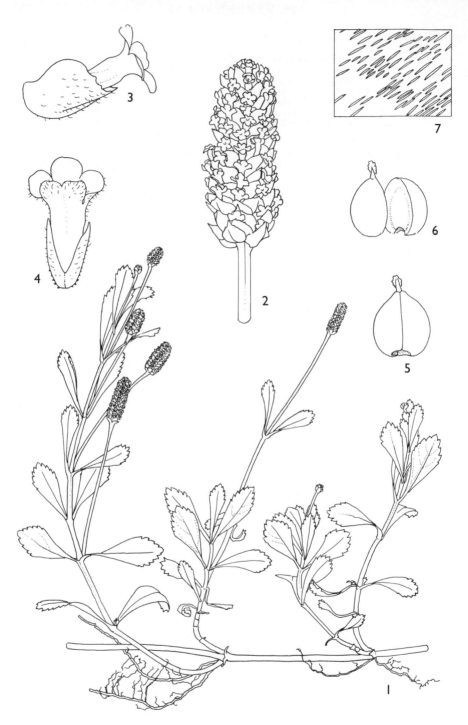

Tab. 6. PHYLA NODIFLORA. 1, habit (× ²⁄₃); 2, inflorescence (× 3), 1 & 2 from *Brummitt* 15400; 3, bract and flower (× 10); 4, flower, showing calyx (× 10); 5, fruit (× 10); 6, fruit, nutlets separated (× 10), 3–6 from *Fanshawe* 195; 7, detail of hairs on leaf (× 18), from *Brummitt* 15400. Drawn by Deborah Lambkin.

Verbena capitata Forssk., Fl. Aegypt.-Arab.: 10 (1775). Type from Egypt.
Blairia nodiflora (L.) Gaertn., Fruct. Sem. Pl. 1: 266, t. 56 (1788). Type as for *Phyla nodiflora*.
Zapania nodiflora (L.) Lam., Tabl. Encycl. [Illustr. Gen.] 1: t. 17 fig. 3 (1791). —Persoon, Syn. Pl. 2: 140 (1806). —Poiret, Encycl. Méth. Bot. 8: 839 (1808).
Lippia nodiflora (L.) Rich. in Michaux, Fl. Bor. Amer. 2: 15 (1803). —Walpers, Repert. Bot. Syst. 4: 49 (1845). —A. Richard, Tent. Fl. Abyss. 2: 168 (1847). —Schauer in A. de Candolle, Prodr. 11: 585 (1847). —Schweinf., Beitr. Fl. Aethiop.: 120 (1867). —J.G. Baker, Fl. Mauritius & Seychelles: 252 (1877); in F.T.A. 5: 279 (1900). —C.B. Clarke in J.D. Hooker, Fl. Brit. India 4: 563 (1885). —Kuntze, Revis. Gen. Pl., part 2: 508 (1891). —Gürke in Engler, Pflanzenw. Ost-Afrikas C: 338 (1895). —Briquet in Engler & Prantl, Nat. Pflanzenfam. 4, 3a: 152 (1897). —Hiern, Cat. Afr. Pl. Welw. 1, 4: 829 (1900). —H. Pearson in F.C. 5, 1: 193 (1901). —Gürke in Warburg, Kunene-Samb.-Exped. Baum: 349 (1903). — T. & H. Durand, Syll. Fl. Congol.: 435 (1909). —Muschler, Man. Fl. Egypt 2: 808 (1912). — R.E. Fries, Wiss. Ergebn. Schwed. Rhod.-Kongo-Exped.: 273 (1916). —A. Chevalier, Expl. Bot. Afr. Occ. Fr. 1: 503 (1920). —Broun & Massey, Fl. Sudan: 350 (1929). —Hutchinson & Dalziel, F.W.T.A. 2: 270 (1931). —P. de Sousa in Ann. Junta Invest. Ultramar 4, 1: 51 (1949). —Adams, Fl. Pl. Jamaica: 630 (1972). —Tutin in Fl. Eur. 3: 123 (1972).
Lippia nodiflora var. *repens* (Bertol.) Schauer in A. de Candolle, Prodr. 11: 586 (1847). — Briquet in Engler & Prantl, Nat. Pflanzenfam. 4, 3a: 152 (1897).
Lippia nodiflora var. *sarmentosa* (Willd.) Schauer in A. de Candolle, Prodr. 11: 585 (1847). —Klotzsch in Peters, Naturw. Reise Mossambique 6, part 1: 260 (1861).

Prostrate perennial herb with long trailing stems radiating from a woody taproot. Stems up to 2(3) m long, decumbent, terete and somewhat fleshy, drying ± angular and sulcate, rooting at the nodes; internodes 0.7–9 cm long; nodes somewhat thickened; branches 1–2 at each node, from leaf axils, up to 30 cm long, erect or the longer branches ± decumbent and sometimes rooting at the nodes. Indumentum of short appressed medifixed sharply pointed white hairs, sparse on old parts, ± dense on younger parts, mainly on leaves and floral bracts. Leaves somewhat fleshy, opposite, sessile or shortly petiolate, 0.8–7.2 × 0.4–2.5 cm, spathulate, oblanceolate or obovate, obtuse to rounded or subacute at the apex and sharply coarsely serrate at the upper part, cuneate and entire at the lower part; petiole up to 8 mm long. Spikes short, dense, subglobose or shortly-ovoid at flowering time, and up to 1–2.5(3) cm × 5–9 mm at maturity, cylindric; peduncles (0.5)1–11.5 cm long; bracts closely imbricate, up to c. 2.5 × 2 mm, broadly obovate to oblate-cuspidate, persistent and ± accrescent. Calyx compressed, membranous, deeply split almost to the base abaxially and to about the middle adaxially, persistent but hardly accrescent; lobes 1.5–2 mm long, hyaline or dark purple, shortly toothed on the keels. Corolla mauve-pink or white with a yellow throat, often white and purple in the same inflorescence, sometimes purple in bud, white when open, c. 2.5 mm long; tube 1.5–2 mm long and 1–1.5 mm broad at the mouth; limb 2–2.5 mm in diameter with the upper lip 2-lobed and erect, the lower lip larger, 3-lobed with the median lobe oblong. Mericarps 1.5–2 mm long, sub-hemispherical, flattened at the commissural face, obtuse or rounded at the apex, glabrous, enclosed within the persistent calyx.

Caprivi Strip: c. 53 km from Katima Mulilo on road to Linyanti, c. 960 m, 27.xii.1958, *Killick & Leistner* 3130 (K; SRGH). **Botswana**. N: Shorobe, 38 km NE of Maun, 19.iii.1965, *Wild & Drummond* 7179 (K; LISC; SRGH). SE: Mopipi Flats, c. 948 m, iv.1977, *Allen* X/417 (PRE). **Zambia**. N: Mbala Distr., Kasikalawi, west of Mpulungu, 10.iv.1961, *Phipps & Vesey-FitzGerald* 3033 (K; LISC; SRGH). C: Lusaka Distr., Kasoka Farm, c. 1152 m, ix.1929, *Sandwith* 206 (K; SRGH). S: Gwembe Distr., Zambezi Valley at lower end of Kariba Gorge, 16°22'S, 28°51'E, 400 m, 15.x.1972, *Kornaś* 2392 (K). **Zimbabwe**. N: Mazowe (Mazoe) Dam, 15.i.1971, *D.S. Mitchell* 1318 (S; SRGH). W: Lupane Distr., Gwayi (Gwaai) River Bridge, Bulawayo to Victoria Falls main road, at foot of centre bridge buttress, x.1973, *Best* 1063 (K; SRGH). C: Chegutu (Hartley), Cedrela, 12.xii.1969, *R.M. Hornby* 3459 (K; LISC; SRGH). S: Mwenezi Distr., Bubye R., near Bubye Ranch homestead, 9.v.1958, *Drummond* 5710 (SRGH). **Malawi**. N: Rumphi Distr., Mwenembwe (Mwanemba), 2.iii.1902, *McClounie* 78 (K). C: Dedza, Chinsen Village, Mtakataka, 13.xi.1967, *Salubeni* 888 (K; LISC; SRGH). S: Mangochi, by Shire R. at west end of bridge, 18.ii.1979, *Brummitt* 15400 (K; S). **Mozambique**. N: Angoche Distr., proximo da Missão Catolica de Malatane, 8.xi.1936, *Torre* 989 (COI; LISC). Z: Changara Distr., Boruma, vii.1891, *Menyharth* 744 (Z). MS: Beira Distr., c. 6.5 km north of Macúti Beach, 8.ix.1962, *Noel* 2476 (K; LISC; SRGH). GI: Aldeia da Barragem, bank of Limpopo R., 20.xi.1957, *Barbosa & Lemos* 8218 (COI; K; LISC; LMA). M: Maputo, Costa do Sol, near the crossing to Maotas, 12.vii.1974, *Marques* 2497 (K; LISC; LISU; M; SRGH).
Widespread in the tropics, subtropics and warmer temperate regions; throughout tropical Africa from Senegal, Central African Republic, Sudan, Ethiopia and Somalia to South Africa.

Lake and dam shores, river banks, floodplains, pan margins, coastal dunes and swamp margins, in damp or seasonally inundated soils, sands and heavy black clays and in mud; 0–1300 m.

Moldenke, in Fifth Summ. Verbenaceae: 252 (1971), has recorded *Phyla nodiflora* var. *reptans* (H.B. & K.) Moldenke for Moçambique. However, we have not been able to confirm this record.

5. CHASCANUM E. Mey.

Chascanum E. Mey., Comment. Pl. Afr. Austr.: 275 (1838) *nom. conserv.* —Moldenke in Repert. Spec. Nov. Regni Veg. **45**: 113–160, 300–319 (1938). —Gilbert et al. in Taxon **35**: 391 (1986). —Sebsebe & Verdcourt in Kew Bull. **43**: 319 (1988). —Atkins in Kadereit (ed.), Fam. Gen. Vasc. Pl. (Kubitzki, ed. in chief) **VII**: 463 (2004). *Plexipus* Raf., Fl. Tellur. **2**: 104 (1837). —R. Fernandes in Bol. Soc. Brot., sér. 2, **57**: 265–273 (1984).

Chascanum E. Mey. sect. *Pterygocarpum* Walp., Repert. Bot. Syst. **4**: 39 (1845).
Svensonia Moldenke in Repert. Spec. Nov. Regni Veg. **41**: 129 (1936).

Erect or spreading perennial herbs or undershrubs, ± woody at the base, ± densely pubescent, the indumentum often with gland-tipped hairs. Leaves opposite-decussate, simple, entire, serrate or dentate, lobed or pinnatipartite, sessile or petiolate. Flowers sessile or shortly pedicellate in terminal spikes or racemes, slightly zygomorphic. Calyx narrowly tubular, 5-angled, with a green longitudinal rib on each angle and hyaline whitish between the ribs, slightly oblique, 5-dentate or 5-apiculate, the teeth ± unequal; fruiting calyx dilating and splitting longitudinally at the front, from base to apex. Corolla white or yellowish, funnel-shaped or hypocrateriform; tube slender straight or curved, somewhat widened at the apex, glabrous outside; limb ± oblique, 5-lobed or 5-partite with lobes subequal or unequal, glabrous outside, papillose-hairy at the base as well as in the throat. Stamens 4, didynamous, included, inserted in the upper part of the corolla tube; filaments short, shortly glandular hairy; anthers ovate, basifixed. Ovary cylindric, glabrous, 1-carpellate, carpel 2-locular, each locule with one basal anatropous ovule; style terminal, filiform, glabrous; stigma ± oblique, divided at the apex into an anterior stigmatiferous lobe and a posterior small tooth. Fruit a schizocarp enclosed within the persistent calyx, shorter than the calyx, oblong, hard, glabrous, formed by 2 equal 1-seeded mericarps, ribbed on the outer surface, sometimes winged in the upper ¹/₃, minutely verruculose-asperulous on the seed part inside, separating from one another. Seeds linear, without albumen.

A genus of c. 25 species, mainly African, extending to Madagascar, Arabia, Pakistan and India.

1. Mericarps separating from one another when ripe, each with an oblique pit at the base; bracteoles absent · 2
– Mericarps not separating from one another when ripe, the fruit with a pit in lower half of the anterior face; bracteoles present · 5
2. Indumentum without gland-tipped hairs · 3
– Indumentum with gland-tipped hairs, at least on the inflorescence · · · · · · · · · · · · · · · · 4
3. Leaves sessile or indistinctly petiolate; lamina linear-lanceolate or lanceolate, up to 10(12) mm wide, usually entire or sometimes with a few teeth near the apex; vegetative indumentum of short retrorse hairs, not interspersed with long shaggy hairs; calyx 6–10 mm long · 1. *schlechteri*
– Leaves distinctly petiolate; lamina ovate-oblong or elliptic, up to 22 mm wide, regularly serrate in the upper ²/₃; vegetative indumentum of uniformly short bristles interspersed with long white shaggy hairs; calyx 13.5–14 mm long · · · · 2. *angolense* subsp. *zambesiacum*
4. Leaf lamina width ± equalling length, up to c. 2.5(3) cm wide and 2.5(4) cm long, broadly obovate to flabelliform, rounded or obtuse at the apex and cuneate at the base · 3. *hederaceum*
– Leaf lamina width much less than length, up to 1.4 cm wide and 2.5 cm long, narrowly obovate to oblanceolate, acute at the apex and tapering to the base · · · · 4. *adenostachyum*
5. Leaves pinnatipartite to pinnatilobed, with the segments or lobes usually as wide as the rhachis; schizocarp with anterior and posterior surfaces prominently raised reticulate and the pit on the anterior face ± half as long as the fruit · · · · · · · · · · · · · · · · 5. *pinnatifidum*
– Leaves deeply dentate or dentate-serrate; schizocarp with the posterior surface obscurely reticulate-striate and the pit on the anterior face almost as long as the fruit · · 6. *pumilum*

1. **Chascanum schlechteri** (Gürke) Moldenke in Phytologia **1**: 18 (1933); op. cit. **4**: 450 (1953); op. cit. **7**: 375 (1961); op. cit. **31**: 237 (1975); in Repert. Spec. Nov. Regni Veg. **45**: 130 (1938); Résumé Verbenaceae: 446 (1959). Type from South Africa (Mpumalanga).
   Bouchea schlechteri Gürke in Notizbl. Königl. Bot. Gart. Berlin **3**: 75 (1900). —H. Pearson in F.C. **5**, 1: 201 (1901).
   Plexipus schlechteri (Gürke) R. Fern. in Bol. Soc. Brot., sér. 2, **57**: 273 (1984).

A suffrutex up to c. 40 cm high; indumentum (except on corolla) of minute or very short subappressed, retrorse, somewhat stiff white hairs. Stems annual from a woody rootstock, 1–several, branched from the base, erect, slender; branchlets 0 or short. Leaves sessile or somewhat petiolate; lamina linear-lanceolate or lanceolate, up to 10(12) mm wide, attenuate to apex and base, entire, membranous, dark brownish on drying; lateral nerves 1–4 on each side, ascending, slightly prominent beneath. Spikes terminal, 7–25 cm long, many-flowered, with the calyces appressed to the inflorescence axis; bracts 3.5–5 mm long, lanceolate, attenuate, acute; bracteoles absent. Calyx 8–10 mm long and 1–1.2 mm in diameter; teeth unequal, 0.3–1 mm long, subulate. Corolla white with a slender tube abruptly expanded into a flat limb; tube 24–27 mm long, widening towards the apex, curved; limb with 5 broadly oblong or obovate lobes, 3.6–6.5 × 4–7 mm, rounded or emarginate at the apex, the upper lobes smallest, the lower lobe broadest. Filaments of the upper stamen pair 2–2.6 mm long, filaments of the lower stamen pair 0.5–0.8 mm long. Ovary 1.4 × 0.4 mm, oblong. Schizocarp c. 5 × 1.75 mm; mericarps subcylindric, flattened on the commisural face, the back reticulate in the upper $^{1}/_{3}$ and striate in the lower $^{2}/_{3}$.

Leaves sessile, linear-lanceolate, 1.5–5 × 0.25–0.75(1) cm, entire, very acute at the apex · · · · ·
· forma *schlechteri*
Leaves subsessile, attenuate into a short subpetiolar base, lanceolate, up to 1.2 cm wide, subacute, with some remote, short, acute teeth towards the apex · · · · · · · · · · forma *torrei*

Forma schlechteri

Mozambique. M: Namaacha Distr., Goba, prox. rio Maiuána, 4.xi.1960, *Balsinhas* 192 (K; LMA). Also in South Africa (Mpumalanga). In wooded grassland and open woodland with *Combretum* sp., *Ziziphus* sp., *Strychnos* sp., *Acacia* sp. etc., in red stony soils; c. 140 m.

Forma torrei (Moldenke) R. Fern. in Bol. Soc. Brot., sér. 2, **63**: 296 (1990). Type: Mozambique, Magude Distr., near Chobela, *Torre* 7039 (LISC, holotype).
   Chascanum schlechteri var. *torrei* Moldenke in Bol. Soc. Brot., sér. 2, **40**: 121 (1966); in Phytologia **31**: 237 (1975). Type as above.
   Plexipus schlechteri (Gürke) R. Fern. var. *torrei* (Moldenke) R. Fern. in Bol. Soc. Brot., sér. 2, **57**: 273 (1984).

Mozambique. M: Magude Distr., near Chobela, 3.i.1948, *Torre* 7039 (LISC).
Known only from Mozambique. In *Acacia* woodland; c. 30 m.

2. **Chascanum angolense** Moldenke in Repert. Spec. Nov. Regni Veg. **45**: 142 (1938). Type from Angola.
   Plexipus angolensis (Moldenke) R. Fern. in Bol. Soc. Brot., sér. 2, **57**: 266 (1984).

Subsp. **zambesiacum** (R. Fern.) R. Fern. in Bol. Soc. Brot., sér. 2, **63**: 295 (1990). Type: Mozambique, Vilankulo Distr., Muabsa, 24.iv.1973, *Balsinhas* 2482 (LISC, holotype; K; LMA).
   Plexipus angolensis subsp. *zambesiacus* R. Fern. in Bol. Soc. Brot., sér. 2, **57**: 267 (1984). Type as above.

A shrub or perennial herb, up to 1.2 m high; indumentum (except on corolla) of minute or uniformly short, erect or somewhat retrorse, stiff white hairs interspersed with scattered to numerous long white shaggy hairs up to c. 2 mm long. Stem woody with ascending, straight, densely leafy branches. Leaves petiolate; lamina up to 6 × 2.2 cm, narrowly elliptic, acute or obtuse at the apex, cuneate at the base, serrate in the upper $^{1}/_{2}$ to $^{2}/_{3}$, entire toward the base, ± sparsely pubescent on both surfaces with short bristles interspersed with scattered longer hairs; midrib and lateral nerves somewhat impressed above, prominent beneath; petiole up to 1.5 cm long, slender. Spikes terminal, straight, slender, 14–25 cm long, many-flowered, with the calyces appressed to the inflorescence axis, up to 45 cm long in fruit; bracts 5–6 mm long, narrowly lanceolate, attenuate-acute, becoming somewhat recurved at the apex, the upper bracts subulate forming a bristle-like tuft at the apex of flowering spikes. Calyx 13.5–14 mm long at anthesis; teeth 1–1.75 mm long, unequal. Corolla white with a slender tube abruptly expanded into a flat limb; tube c. 30 mm long, slender, straight or slightly curved, glabrous outside, sparsely pubescent within; limb densely papillose-hairy in the throat. Anthers of the upper stamens c. 1.25 mm long, those of the lower ones slightly smaller. Mericarps 5.5 mm long and c. 1.3 mm thick at the base, rounded at the apex, reticulate in the upper $^1/_4$, striate below, separating at maturity.

Malawi. S: Chikwawa Distr., Murukanyama foothills, near Mikombo R., 10 km SSW of Ngabu, 180 m, 22.iv.1980, *Brummitt & Osborne* 15532 (K). **Mozambique**. GI: Vilankulo Distr., between the cross of roads Vilankulo–Mambone–Mabote (Vilanculos–Mamboni–Maboti), c. 10 km from the junction, fl. & fr. 28.iii.1952, *Barbosa & Balsinhas* 5044 (BM; LMA).
Known only from Malawi and Mozambique. Grassland and thickets near rivers; 10–200 m.
The typical subspecies, from Angola, is a more robust plant with relatively shorter, acute, sharply serrate leaves and shorter petioles (up to 5 mm long), and with smaller flowers (corolla tube 20 mm long, not c. 30 mm long as in subsp. *zambesiacum*).
C. hildebrandtii Vatke, a related taxon, is an annual herb from East Africa, with leaves relatively shorter and broader and more deeply and bluntly toothed, the apex obtuse or rounded, the bracts not apically recurved and not forming a bristle-like tuft at the apex of the flowering spikes, the calyx teeth rather shorter (0.25–0.5 mm long) and the mericarps narrowing progressively from the base to the acute apex.

3. **Chascanum hederaceum** (Sond.) Moldenke in Torreya **34**: 9 (1934); in Repert. Spec. Nov. Regni Veg. **45**: 144 (1938); in Phytologia **4**: 444 (1953); op. cit. **7**: 359 (1961); op. cit. **30**: 205 (1975); Résumé Verbenaceae: 445 (1959). —Hutchinson, Botanist South. Africa: 400 (1946). —Letty, Wild Fl. Transvaal: 281, t. 140 fig. 3 (1962). —van der Schijff, Check List Vasc. Pl. Kruger Nat. Park: 81 (1969). Type from South Africa (Transvaal).
   Bouchea hederacea Sond. in Linnaea, **23**: 36 (1850). —H. Pearson in F.C. **5**, 1: 200 (1901).
   Plexipus hederaceus (Sond.) R. Fern. in Bol. Soc. Brot., sér. 2, **57**: 270 (1984).

Perennial herb 15–45 cm high, with a slender woody taproot; indumentum (except on corolla) puberulous, densest on the inflorescence, of patent gland-tipped short hyaline bristles, sometimes with few to numerous scattered eglandular soft setae interspersed. Stems 1–several, decumbent to erect, simple or branched, leafy. Leaves petiolate; lamina 0.8–3(4) × 0.7–2.5(3) cm, obovate to very broadly obovate, sometimes rhomboid-flabelliform, rounded to obtuse at the apex, tapering gradually to ± abruptly into the petiole, dentate to serrate-dentate in the upper $^1/_2$ to $^2/_3$, entire toward the base, ± sparsely pubescent on both surfaces with gland-tipped bristles; midrib and lateral nerves somewhat impressed above, prominent beneath; petiole 1–7 mm long. Spikes terminal, straight, slender, 4–30 cm long, many-flowered, with the calyces appressed to the inflorescence axis; peduncle up to 2 cm long; bracts 4–6 × 1–2 mm, acute, ciliate; bracteoles absent. Calyx erect, 8.8–12.4 × 1.4–1.6 mm, minutely appressed puberulous within towards the apex; teeth 0.2–1.3 mm long, subequal or unequal, subulate; fruiting calyx somewhat wider (2–3 mm in diameter), gibbous at base, more conspicuously costate and plicate at the apex, stramineous. Corolla creamy-white, with a slender tube abruptly expanded into a flat limb; tube 2–2.5(3) cm long, usually curved; lobes 4–4.5 mm long and wide, oblong or obovate, obtuse or emarginate at the apex, undulate or crisped at the margins. Filaments 1.3–1.5 mm long in the upper pair of stamens, 0.7–1 mm long in the lower pair; anthers c. 0.8 × 0.5 mm, ovate. Ovary 2.2 × 0.5 mm, narrow-oblong or cylindric. Style up to 2.5 cm long after anthesis, often persisting for a short time as an exerted slender thread from the calyx tube. Mericarps separating when ripe, 4–6.5 × 0.9 mm, subcylindric, somewhat flattened at the commissural face, the back longitudinally ridged or striate below and reticulate above, more distinctly so on the upper $^1/_4$, at first stramineous, finally black, with basal pit.

Indumentum of short hairs only, mostly gland-tipped · · · · · · · · · · · · · · · · · var. *hederaceum*
Indumentum of short, usually gland-tipped hairs intermixed with few to many long (up to 3
   mm) spreading, white eglandular setae, densest on stems, branches, petioles and venation
   of the leaf lamina · var. *natalense*

Var. **hederaceum**

Botswana. N: North East Distr., near Tshesebe (Tsessebe), 8.iii.1965, *Wild & Drummond* 6815 (K; SRGH). SE: South East Distr., at railway-road crossing c. 5 km north of Gaborone, 18.ix.1977, *O.J. Hansen* 3188 (C; GAB; K; PRE, n.v.). **Zambia**. C: Kabwe (Broken Hill), iv.l909, *Rogers* 8159 (K). **Zimbabwe**. W: Bulilima Mangwe (Bulalima-Mangwe), Embakwe, 28.iii.1942, *Feiertag* in *GHS* 45393.

Also in Namibia and South Africa (Mpumalanga, KwaZulu-Natal and Cape Prov.). Mopane and mixed deciduous woodlands, and in short grassland with scattered shrubs, on sandy soil; 960–1200 m.

Var. **natalense** (H. Pearson) Moldenke in Torreya **34**: 9 (1934); in Repert. Spec. Nov. Regni Veg. **45**: 147 (1938); in Phytologia **4**: 444 (1953); op. cit.: **7**: 359 (1961); op. cit.: **30**: 206 (1975); Résumé Verbenaceae: 446 (1959). —J.H. Ross, Fl. Natal: 299 (1972). Syntypes from South Africa (KwaZulu-Natal).

Bouchea wilmsii Gürke in Notizbl. Königl. Bot. Gart. Berlin **3**: 74 (1900). —H. Pearson in F.C. **5**, 1: 200 (1901). Syntypes from South Africa (Mpumalanga).
Bouchea hederacea var. *natalensis* H. Pearson in F.C. **5**, 1: 200 (1901).
Denisia wilmsii (Gürke) Kuntze in Deutsch. Bot. Monatschr. **21**: 173 (1903).
Chascanum wilmsii (Gürke) Moldenke in Phytologia **1**: 18 (1933).
Plexipus hederaceus var. *natalensis* (H. Pearson) R. Fern. in Bol. Soc. Brot., sér. 2, **57**: 270 (1984).

Zimbabwe. W: Matobo Distr., Hope Fountain Mission, c. 1400 m, 21.xi.1973, *Norrgrann* 419 (K; SRGH). **Mozambique**. MS: lower Umswiriswe R., 1906, *Swynnerton* 1938 (BM; K). M: Matutuíne Distr., Catuane near Mazeminhama R., 5.x.1968, *Balsinhas* 1351 (LISC; LMA).

Also in South Africa (Mpumalanga, KwaZulu-Natal and Cape Prov.). In sand-veld and with short grasses in *Acacia* woodland, and in disturbed ground, in sandy soils; 40–1400 m.

4. **Chascanum adenostachyum** (Schauer) Moldenke in Torreya **34**: 8 (1934); in Repert. Spec. Nov. Regni Veg. **45**: 140 (1938); in Phytologia **4**: 440 (1953); op. cit. **7**: 354 (1961); op. cit. **30**: 201 (1975); Résumé Verbenaceae: 446 (1959). Type from South Africa (Cape Prov.).

Bouchea adenostachya Schauer in A. de Candolle, Prodr. **11**: 560 (1847). —H. Pearson in F.C. **5**, 1: 199 (1901).
Bouchea longipetala H. Pearson in F.C. **5**, 1: 199 (1901). Type from South Africa (Mpumalanga).
Chascanum longipetalum (H. Pearson) Moldenke in Phytologia **1**: 18 (1933).
Plexipus adenostachyus (Schauer) R. Fern. in Bol. Soc. Brot., sér. 2, **57**: 266 (1984).

Perennial herb up to 65 cm high, with a slender woody taproot; plant (except on corolla) pubescent, indumentum of minute or uniformly short, erect or somewhat retrorse, stiff gland-tipped hairs interspersed with scattered to numerous long white eglandular hairs up to c. 2 mm long, the gland-tipped hairs more frequent on the upper surface of the leaves and on the inflorescence. Stems 1–several, erect, simple or branched, leafy. Leaves subsessile to shortly petiolate; lamina 1–2.5 × 0.4–1.4 cm, lanceolate or oblanceolate, acute at the apex and irregularly and sharply 2–9 dentate in the upper half or $^1/_3$, attenuate-cuneate and entire in the lower half or $^2/_3$, the base ± decurrent into the petiole, subchartaceous, rigid on drying, ± sparsely pubescent on both surfaces; midrib and lateral nerves somewhat impressed above, prominent beneath; petiole 0–4 mm long. Spikes terminal, straight, 4.5–20 cm long, many-flowered, with the calyces closely appressed to the axis; bracts 4–8 mm long, lanceolate, usually attenuate and acute, hyaline-whitish, ciliate; bracteoles absent. Calyx erect, 7.5–10 × 1.3 mm, puberulous with gland-tipped hairs; teeth 0.3–1 mm, unequal. Corolla pale yellow, with a slender tube abruptly expanded into a flat limb; tube c. 2.7 cm long, usually curved; lobes 4–5 × 3.5 mm, broadly elliptic or obovate. Filaments c. 1.4 mm long in the upper pair of stamens, c. 0.9 mm long in the lower pair. Mericarps separating when ripe, 5–6 × 0.7–0.9 mm, subcylindric, flat on the commissural face, the back reticulate in the upper half and striate towards the base, black when mature.

Botswana. SE: c. 3 km south of Lobatse, east of railway, Farm Springfield, 17.i.1960, *Leach & Noel* 151 (K; SRGH). **Zimbabwe**. W: Bulawayo, 1908, *Chubb* in *Rogers* 102 (BM; Z).

Also in South Africa (Limpopo, North West and Northern Cape Provinces). Stony hillsides and in deep sand; 1190–1350 m.

5. **Chascanum pinnatifidum** (L.f.) E. Mey., Comment. Pl. Afr. Austr.: 277 (1838). —Walpers, Repert. Bot. Syst. **4**: 39 (1845). —Moldenke in Repert. Spec. Nov. Regni Veg. **45**: 303 (1938), excl. specim. *Eyles* 1142; in Phytologia **4**: 448 (1953); op. cit. **7**, 7: 373 (1961); op. cit. **31**: 234 (1975); Résumé Verbenaceae: 446 (1959); Fifth. Summ. Verbenaceae: 858 (Taxon. Index, 1971). —Friedrich-Holzhammer et al. in Merxmüller, Prodr. Fl. SW. Afrika, fam. 122: 3 (1967). —van der Schijff, Check List Vasc. Pl. Kruger Nat. Park: 81 (1969). TAB. **7**. Type from South Africa (Cape Prov.).

    Buchnera pinnatifida L.f., Suppl. Sp. Pl.: 288 (1782). —Murray, Syst. Veg., ed. 14: 572 (1784). —Thunberg, Prodr. Fl. Cap. part 2: 100 (1800). —Persoon, Syn. Pl. **2**: 149 (1807). Type as above.

    Bouchea pinnatifida (L.f.) Schauer in A. de Candolle, Prodr. **11**: 560 (1847). —Bocquillon in Adansonia **3**: 237 (1862–63). —Bentham & Hooker, Gen. Pl. **2**: 1145 (1876). —Briquet in Engler & Prantl, Nat. Pflanzenfam. **4**, 3a: 154 (1897). —Pearson in F.C. **5**, 1: 205 (1901), excl. *Rehmann* 5311. Type as above.

    Bouchea pumila sensu R. Good in J. Bot., Suppl. 2 (Gamopet.): 140 (1930) non (E. Mey.) Schauer (1847).

    Chascanum pumilum sensu Moldenke in Repert. Spec. Nov. Regni Veg. **45**: 319 (1938) quoad saltem spec. *Gossweiler* 73; Résumé Verbenaceae: 146 (1958); in Phytologia **31**: 236 (1975) quoad *Barbosa & Correia* 9102, non E. Mey. (1838).

    Plexipus pinnatifidus (L.f.) R. Fern. in Bol. Soc. Brot., sér. 2, **57**: 272 (1984).

Spreading, perennial herb 20–60 cm high, with a woody taproot up to 20 cm long; plant (except on corolla) puberulous, the indumentum of minute or uniformly short, erect or somewhat retrorse, stiff eglandular hairs. Stems up to 8 mm thick at the base, with few to many main branches from near the base; branches erect or spreading, terete or obtusely 4-angled, ± densely leafy and again branched giving the plant a bushy habit, the branchlets opposite. Leaves subsessile to shortly petiolate; lamina 1–5 × 1–3 cm, deeply pinnatipartite to pinnatilobed with segments or lobes usually as wide as the rhachis, sometimes broader; leaf segments 5–7, ± spaced on the rhachis, 3.5–20 × 1–2 mm, narrow-oblong to linear or sometimes oblong-triangular, obtuse or subacute, entire, hispidulous on both surfaces, chartaceous and bright green, usually turning olive-green or brownish on drying. Inflorescences of 1–many-flowered spikes 1–6.5 cm long or of many-flowered racemes up to 15 cm long, terminal on branches and branchlets, the short inflorescences being overtopped by lateral branchlets also bearing terminal inflorescences; bracts 4–5 mm long, linear-lanceolate, acute or obtuse, membranous-margined; bracteoles c. 1 mm long, subulate, puberulent. Calyx 10–14 mm long and c. 1.4 mm in diameter, straight, puberulous outside; teeth 0.5–0.9 mm long, slightly unequal, subulate; fruiting calyx up to 3 mm in diameter, oblong-ovoid, somewhat gibbous at the base behind, yellowish, splitting nearly to the apex, plicate above, the teeth conduplicate and connivent. Corolla cream or yellowish-green, with a slender tube abruptly expanded into a flat limb; tube 1.8–2.5 cm long and c. 0.75 mm in diameter, straight. Ovary c. 1.4 × 0.7 mm; style c. 16 mm long. Schizocarp (5)6.5–7.5 long, somewhat broader and thicker at the base, ± rounded at the apex, prominently reticulate-alveolate, shining; pit (2)3–4.5 mm long, ± half as long as the schizocarp length, with an irregularly dentate margin; the 2 mericarps not separating when ripe.

Inflorescences 1–6.5 cm long, usually few-flowered, with the flowers ± closely spaced; pedicels 0 or very short; schizocarps up to c. 3 mm wide at the base · · · · · · · · · · · var. *pinnatifidum*
Inflorescences up to 15 cm long, many-flowered, with the flowers ± widely spaced; pedicels distinct; schizocarps narrower · var. *racemosum*

Var. **pinnatifidum**

Botswana. N: Ngamiland Distr., near Bushman's Pits, 930 m, 27.iii.1961, *Richards* 14883 (K). SW: Ghanzi, Farm 48, 9.iv.1969, *de Hoogh* 235 (K; SRGH). SE: road to Gaborone, ix.1967, *Lambrecht* 299 (K; LISC; SRGH). **Zimbabwe**. W: Bulawayo, Lochview road, 26.x.1975, *L.C. Cross* 253 (BR; K; SRGH). E: Chimanimani Distr., Changadzi (Changazi) R., c. 5 km east of Birchenough Bridge, i.1938, *Obermeyer* in Herb. Tvl. Mus. 37545 (PRE). S: Beitbridge, between Customs Post and Limpopo R., 25.iii.1959, *Drummond* 6007 (BR; K; LISC; SRGH); Gwanda Distr., Doddieburn Ranch, near dam on Mtshibizini (Tsibizini) R., 10.v.1972, *G.V. Pope* 725 (K; SRGH).

    Also in Angola, Namibia and South Africa (Limpopo, North West, Free State and Northern Cape Provinces). Dry deciduous tree savanna with *Combretum, Grewia, Acacia, Commiphora spp.* and *Colophospermum mopane*, in open short grassland, on pan margins and on stony hillsides, in sandy soil; 500–1350 m.

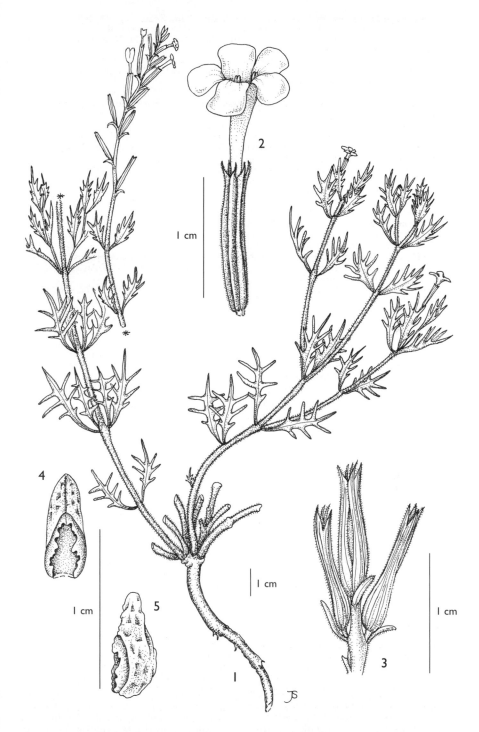

Tab. 7. CHASCANUM PINNATIFIDUM. 1, habit, from *Cross* 253; 2, flower; 3, fruiting calyces; 4, fruit, front view; 5, fruit, side view, 2–5 from *Wild* 5015. Drawn by Judi Stone.

Var. **racemosum** Schinz ex Moldenke in Repert. Spec. Nov. Regni Veg. **45**: 306 (1938); in Phytologia **4**: 449 (1953); op. cit. **7**: 374 (1961); op. cit. **31**: 235 (1975); Résumé Verbenaceae: 446 (1959). —van der Schijff, Check List Vasc. Pl. Kruger Nat. Park: 81 (1969). Type from South Africa.

Bouchea pinnatifida sensu H. Pearson in F.C. **5**, 1: 206 (1901), quoad *Rehmann* 5311.

Plexipus pinnatifidus var. *racemosus* (Schinz ex Moldenke) R. Fern. in Bol. Soc. Brot., sér. 2, **57**: 272 (1984). Type as above.

Besides the longer racemes and widely spaced flowers, this variety differs from the typical variety in the longer pedicels and narrower fruit, not so prominently reticulate-alveolate.

Zimbabwe. W: Bulawayo, c. 1472 m, xii.1902, *Eyles* 1142 (BM; SRGH; Z). **Mozambique**. GI: Massingir Distr., 32 km west from Albufeira de Massingir (Lagoa Nova), right bank of rio dos Elefantes, fl. & fr. 20.iv.1972, *Louša & Rosa* 250 (LMA).

Also in Namibia and South Africa (Limpopo, North West and Northern Cape Provinces). Dry deciduous tree savanna; 145–1450 m.

6. **Chascanum pumilum** E. Mey., Comment. Pl. Afr. Austr.: 277 (1838). —Walpers, Repert. Bot. Syst. **4**: 38 (1845). —Moldenke in Repert. Spec. Nov. Regni Veg. **45**: 317 (excl. saltem *Gossweiler* 73 (1938); in Phytologia **4**: 449 (1953); op. cit. **7**: 374 (1961); op. cit. **31**: 236 (1975) excl. *Barbosa & Correia* 9102; Résumé Verbenaceae: 446 (Taxon. Index, 1959) sed excl. p. 146. —Friedrich-Holzhammer et al. in Merxmüller, Prodr. Fl. SW. Afrika, fam. 122: 3 (1967). Type from South Africa (Cape Prov.).

Bouchea pumila (E. Mey.) Schauer in A. de Candolle, Prodr. **11**: 560 (1847). —Bentham & Hooker, Gen. Pl. **2**: 1145 (1876). —Briquet in Engler & Prantl, Nat. Pflanzenfam. **4**, 3a: 154 (1897). —H. Pearson in F.C. **5**, 1: 205 (1901).

Bouchea pubescens Schauer in A. de Candolle, Prodr. **11**: 560 (1847). Type from South Africa (Cape Prov.).

Bouchea pumila var. *subcanescens* Sond. in Linnaea, **23**: 86 (1850). Type from South Africa (Cape Prov.).

Bouchea namaquana var. *latifolia* H. Pearson in Ann. S. African Mus. **9**: 184 (1913). Syntypes from South Africa (Cape Prov.).

Bouchea lignosa Dinter in Repert. Spec. Nov. Regni Veg. Beih. **53**: 53 (1928) *nom. nud.*

Chascanum incisum var. *canescens* Moldenke in Repert. Spec. Nov. Regni Veg. **45**: 309 (1938); Résumé Verbenaceae: 446 (1959); in Phytologia **7**: 370 (1961). Type from Namibia.

Chascanum lignosum Dinter ex Moldenke in Repert. Spec. Nov. Regni Veg. **46**: 2 (1939); in Phytologia **4**, 7: 447 (1953); Résumé Verbenaceae: 446 (1959). Type from Namibia.

Plexipus pumilus (E. Mey.) R. Fern. in Bol. Soc. Brot., sér. 2, **57**: 272 (1984).

Perennial herb up to 30 cm high, with a long slender woody taproot; the indumentum of minute or uniformly short, erect or somewhat retrorse, stiff eglandular white hairs. Stems up to 7 mm thick at the base, branched; branches spreading, terete, leafy and again branched giving the plant a bushy habit, the branchlets opposite. Leaves petiolate, somewhat fleshy when fresh, rather thick and rigid when dried, grey-green to dark green on both surfaces; lamina 1–3.5 × 0.6–2.2 cm, elliptic to ovate-oblong or subspathulate, rounded to obtuse at the apex, tapering ± abruptly or gradually into the petiole, dentate to serrate-dentate in the upper $^1/_2$ to $^2/_3$, with 2–4 teeth on each side, entire toward the base, puberulous on both surfaces; midrib somewhat impressed above, prominent beneath; petiole 0.5–2.5 cm long, flatish. Inflorescences of 1–few-flowered spikes 1–3.5 cm long, terminal on branches and branchlets, and overtopped by lateral branchlets also bearing terminal inflorescences (dichasial-like); peduncle 0 or very short; bracts c. 3 × 1 mm, ovate or lanceolate; bracteoles 1–1.5 mm long, subulate. Calyx 11–14 mm long and c. 1.4 mm in diameter, puberulous outside and minutely hairy inside; teeth c. 0.2 mm long, subulate, subequal; fruiting calyx up to 3 mm in diameter, oblong-ovoid, somewhat gibbous at the base, splitting nearly to the apex, chartaceous when dry. Corolla white, with a slender tube abruptly expanded into a flat limb; tube 1.4–2.7 cm long and c. 0.7 mm in diameter at the base, up to c. 2.2 mm in diameter above, straight; lobes 2.9–3.5 × 1.7–2 mm, emarginate. Filaments c. 1.4 mm long in the upper pair of stamens, c. 0.7 mm long in the lower pair. Ovary c. 9 mm long and 0.5 mm in diameter; style 1–2 cm long. Schizocarp 6–7 mm long and 2–2.5 mm wide, oblong-ovoid, reticulate-alveolate at the apex and around the pit, faintly reticulate-striate at the posterior face; pit 4.5–5.5 × 2.7 mm, deeply excavate, bordered by a prominent-toothed margin; the 2 mericarps not separating when ripe.

Botswana. S: Kgalagadi Distr., c. 72 km north of Kang, fl. & fr. 18.ii.1960, *Wild* 5044 (BM; K; LD; M; SRGH). SE: Southern Distr., c. 26 km east of Khakhea (Khakea) Village, iii.1950, *O.B. Miller* B/1015 (PRE).

Also in Namibia and South Africa (North West and Northern Cape Provinces). Dry deciduous tree savanna with *Combretum, Grewia* and *Acacia spp.* in open short grassland, in Kalahari Sand and on pan margins; 900–1200 m.

6. STACHYTARPHETA Vahl

Stachytarpheta Vahl, Enum. Pl. **1**: 205 (1804) *nom. conserv.* —R. Fernandes in Bol. Soc. Brot., sér. 2, **57**: 87–111 (1984). —Atkins in Kadereit (ed.), Fam. Gen. Vasc. Pl. (Kubitzki, ed. in chief) **VII**: 463 (2004).

Valerianoides Medic., Philos. Bot. **1**: 177 (1789), as "*Valerianodes*".

Vermicularia Moench, Meth. Suppl.: 150: (1802).

Annual or perennial herbs or shrubs, glabrous to densely hairy. Leaves opposite or alternate, simple, serrate, dentate or crenate, usually petiolate. Inflorescence an elongate terminal spike; bracts present. Flowers sessile, crowded, each partially sunk in a hollow in the rhachis and subtended by an acuminate bract. Calyx persistent, closely appressed, not enlarging in fruit; tube narrow, dorsally compressed, 5-ribbed, often plicate, membranous or herbaceous, very shortly 4–5-dentate with subequal teeth, or bifid with 2-dentate lobes or entire. Corolla blue, purple, red or white, funnel-shaped or hypocrateriform; tube cylindrical, narrow, straight or curved, pubescent inside at the throat; limb ± spreading, oblique, 5-lobed, the lobes broad, subequal or unequal. Fertile stamens 2, anterior, inserted at the upper half of corolla tube, included; staminodes 2 or 0; filaments short; anthers with linear-oblong divaricate thecae. Ovary 2-locular, dorsally compressed, on a short gynophore; loculi lateral, each with 1 erect anatropous; style filiform with a capitate or subcapitate stigma. Fruit a schizocarp; schizocarp oblong, included within the persistent calyx, splitting at maturity into 2 mericarps; mericarps hard, shortly beaked, striate on the back and usually pitted near the apex. Seeds without endosperm.

A genus of about 112 species (cf. Moldenke, Fifth Summ. Verbenaceae: 907–912, 1971) of tropical and subtropical America, a few naturalized or native in the tropics of the Old World.

Cultivated taxa

Stachytarpheta mutabilis (Jacq.) Vahl —Biegel, Check-list Ornam. Pl. Rhod. Parks & Gard. [Rhodesia Agric. J., Research Report No. 3]: 100 (1977).

Shrub 2–2.5 m high, with densely pubescent crenate-dentate leaves to 14 × 10 cm and spikes to 30(50) cm long and c. 6 mm wide (excluding corollas); corolla rose-coloured with tube tinged blue and throat often violet; stigma blue.

Zimbabwe. Cultivated ornamental in Harare gardens.

Stachytarpheta urticifolia Sims (= *Cymburus urticifolius* Salisb. *nom. illegit.*)

Herb or slender subshrub 0.5–2 m high, with ± erect branched glabrous or very sparsely pubescent stems from a long taproot, woody at the base. Leaves 4–12.5 × 2–7 cm, broadly elliptic to ovate, distinctly acute at the apex and regularly and sharply serrate on the margins with spreading teeth, glabrous. Spikes slender, elongate, many-flowered; bracts lanceolate; corollas dark purple-blue, mauve or royal-blue with a light or white throat.

Mozambique. M: Maputo, Jardim Tunduru (Jardim Vasco da Gama), 22.v.1974 *Balsinhas* 2710 (LMA).

It is much used as a hedge plant in Nairobi, Kenya.

1. Corolla limb 1–2 cm wide; inflorescence rhachis stout, over 5 mm wide when dry (up to 10 mm wide after flowering, fide *Moldenke*) (cultivated) · · · · · · · · · · · · · · · · · · S. *mutabilis*
- Corolla limb less than 1 cm wide; inflorescence rhachis more slender, c. 3 mm wide when dry (up to 7 mm wide after flowering, fide *Moldenke*) · 2
2. Bracts narrowly triangular, mostly less than half as wide as the fruiting calyx; inflorescence ± pubescent; corolla tube exserted 0–2 mm · 2. *cayennensis*

– Bracts lanceolate, long acuminate, mostly more than half as wide as the fruiting calyx;
  inflorescence glabrous; corolla tube exserted up to c. 7 mm · 3
3. Leaves usually obtuse at the apex, crenate-dentate, the teeth short and blunt with apices
  excentric; calyx without peltate glands · 1. *jamaicensis*
– Leaves sharply acute at the apex, spreading dentate, the teeth sharply acute with apices
  central; calyx with 1–2 peltate glands near the apex (cultivated) · · · · · · · · · · S. *urticifolia*

1. **Stachytarpheta jamaicensis** (L.) Vahl, Enum. Pl. **1**: 206 (1804). —Sims in Bot. Mag. **44**: t.
   1860 (1817). —Schauer in A. de Candolle, Prodr. **11**: 564 (1847) pro parte, as
   "*Stachytarpha*". —Briquet in Engler & Prantl, Nat. Pflanzenfam. **4**, 3a: 154 (1895) pro parte.
   —Hutchinson & Dalziel, F.W.T.A. **2**: 277 (1931) pro parte quoad *Vogel* 30, *Mann* 89, *Scott
   & Elliot* 3833 et *Johnston* s.n. —Moldenke in Lilloa **4**: 300 (1939); in Fl. Madag.,
   Verbenaceae: 22, fig. III, 1–2 (1956). —Roberty, Petite Fl. Ouest-Afr.: 180 (1954) pro parte.
   —Adams, Fl. Pl. Jamaica: 632 (1972). —H. Moldenke & A. Moldenke in Rev. Fl. Ceylon **4**:
   253 (1983). —R. Fernandes in Bol. Soc. Brot., sér. 2, **57**: 100 (1984). —Verdcourt in
   F.T.E.A., Verbenaceae: 19 (1992). Type a specimen in the Linnean herb. (S, lectotype,
   microfiche 7.13).
       Verbena jamaicensis L., Sp. Pl.: 19 (1753); ed. 2, **1**: 27 (1762). —Jacquin, Observ. Bot. **4**: 6,
   t. 85 (1771). —Willdenow in L., Sp. Pl., ed. 4, **1**: 115 (1797).
       Valerianoides jamaicensis (L.) Medic., Philos. Bot. **1**: 177 (1789), as "*Valerianodes*". —
   Kuntze, Revis. Gen. Pl., part 2: 509 (1891). —Britton, Fl. Bermuda: 313, fig. 334 (1918,
   facsim. 1965).
       Zapania jamaicensis (L.) Lam., Tabl. Encycl. [Illustr. Gen.] **1**: 59, No. 255 (1792).
       Stachytarpheta indica sensu Schauer in A. de Candolle, Prodr. **11**: 564 (1847) pro parte,
   as "*Stachytarpha*". —sensu J.G. Baker in F.T.A. **5**: 284 (1900) pro max. parte, excl. *Scott Elliot*
   4162. —sensu Danser in Ann. Jard. Bot. Buitenz. **40**: 5 (1929). —sensu Meeuse in Blumea
   5: 70 (1942). —sensu Brenan in Kew Bull. **5**: 225, fig. B (1950). —sensu Hepper in
   F.W.T.A., ed. 2, **2**: 434, fig. 305 M-R (1963). —sensu Backer & Bakhuizen van den Brink Jr.,
   Fl. Java **2**: 598 (1965), non *Verbena indica* L. (1759) neque *Stachytarpheta indica* (L.) Vahl
   (1804).
       Abena jamaicensis (L.) Hitchc. in Rep. Missouri Bot. Gard. **4**: 117 (1893). Type as above*.

A perennial herb or subshrub, 0.3–1.2 m high, woody at the base, ± sparsely
pubescent mainly on the young parts, glabrescent on older parts. Stem and branches
pilose at the nodes, often purplish. Leaves petiolate, lamina 2–9(11) × 1.2–5.5 cm,
elliptic, oblong, obovate or spathulate, obtuse to rounded or somewhat acute at the
apex, attenuate and decurrent into the petiole or ± rounded and abruptly narrowed
and narrowly tapering-cuneate into the petiole, crenate-dentate on the margin with
7–15 teeth on each side, entire below, glabrous on both surfaces or with sparse hairs
on the nerves beneath; petiole up to c. 1 cm long. Spikes slender, 14–45(50) cm
long, up to 4 mm thick in fruit, glabrous; bracts 5–8 × 2–2.5 mm, ± as broad as the
fruiting calyx, lanceolate or ovate-lanceolate, apically attenuate, not longer than the
fruiting calyx, striate, scarious on the margins, glabrous, not thickened or a very little
so at the base. Calyx 5–6 × 2.5 mm, tubular, ellipsoid, bifid and without peltate
glands at the apex, 4-toothed, the 2 central teeth rather shorter. Corolla blue, violet
or nearly white; tube 8–11 mm long, slightly curved, exserted up to c. 5 mm above
the calyx; limb c. 9 mm in diameter, the lobes c. 3 mm long. Style exserted 3–4 mm
above the calyx after the fall of the corolla. Fruiting calyx ± completely embedded
in the furrows of the rhachis. Mericarps c. 7 mm long.

Mozambique. Z: Namacurra Distr., c. 17 km to Macuze (Macusi) from the junction of the
road to Maganja da Costa, 29.i.1966, *Torre & Correia* 14240 (LISC).
  Native to, and widely distributed throughout, tropical America. Introduced to and naturalized
in Africa, Madagascar, Indian Ocean Islands, tropical Asia, Australasia and Oceania. Seasonally
wet floodplain grasslands; c. 10–40 m.

----

* *Stachytarpheta marginata* Vahl (Enum. Pl. **1**: 207, (1804)) and *S. pilosiuscula* H.B. & K. (Nov.
Gen. & Sp. **2**: 279, (1818)) are both treated as synonyms of *S. jamaicensis* by Schauer (loc. cit.)
and Moldenke (loc. cit., 1939, 1983). As I have not seen the respective types, and considering
the confusion by many authors between some species of this genus, I prefer not to consider
them here. Specimen 455 of Willdenow's Herb., determined on the sheet as *S. pilosiuscula* and
cited by Schauer under *S. jamaicensis*, seems to me, on examination of microfiche, to be *S.
urticifolia* Sims.

Tab. 8. STACHYTARPHETA CAYENNENSIS. 1, apical portion of flowering branch (× 1/5), from *Lye* 1186; 2, portion of spike, showing flowers lying in excavations of the rhachis (× 4), from *Scott Elliot* 4162 and *Goldsmith* 2/77; 3, bract (× 6), from *Lye* 1186; 4, flower (× 6); 5, flower opened, showing 2 stamens and 2 staminodes (× 6); 6, fruit, 2 nutlets showing the minutely tuberculate commissural inner faces (× 6), 4–6 from *Scott Elliot* 4162. Drawn by Deborah Lambkin.

2. **Stachytarpheta cayennensis** (Rich.) Vahl, Enum. Pl. 1: 208 (1804), as "*cajanensis*". —Persoon, Syn. Pl. 2: 140 (1806), as "*cajanensis*". —Walpers, Repert. Bot. Syst. 4: 5 (1845). —Schauer in A. de Candolle, Prodr. 11: 562 (1847); in Martius, Fl. Bras. 9: 199 (1851), as "*Stachytarpha*". —Briquet in Engler & Prantl, Nat. Pflanzenfam. 4, 3a: 154 (1895). —Urban, Symb. Antill. 4: 533 (1911). —Danser in Ann. Jard. Bot. Buitenz. 40: 2 (1929). —Moldenke in Lilloa 4: 299 (1939). —A. Meeuse in Blumea 5: 69 (1942). —Brenan in Kew Bull. 5: 223, fig. C (1950). —Leon & Alain, Fl. Cuba 4: 296 (1957). —Hepper in F.W.T.A., ed. 2, 2: 434, fig. 305 G–L (1963). —Backer & Bakhuizen van den Brink Jr., Fl. Java 2: 598 (1965). — Adams, Fl. Pl. Jamaica: 632 (1972). —Verdcourt in F.T.E.A., Verbenaceae: 18 (1992). TAB. 8. Type from French Guiana.

   Verbena cayennensis Rich. in Acta Soc. Hist. Nat. Paris, 1: 105 (1792).
   Verbena dichotoma Ruiz & Pavon, Fl. Peru, 1: 23, t. 34 fig. b (1798). Type from Peru.
   Stachytarpheta dichotoma (Ruiz & Pavon) Vahl, Enum. Pl. 1: 207 (1804). —Persoon, op. cit.: 139 (1806). —Schauer op. cit.: 561 (1847), as "*Stachytarpha*" pro parte? —H. Moldenke & A. Moldenke in Rev. Fl. Ceylon 4: 261 (1983).
   Zapania cayennensis (Rich.) Poir., Encycl. Méth. Bot. 8: 842 (1808), as "*cajanensis*".
   Zapania dichotoma (Ruiz & Pavon) Poir., loc. cit.
   Valerianoides cayennensis (Rich.) Kuntze, Revis. Gen. Pl., part 2: 510 (1891).
   Abena cayennensis (Rich.) Hitchc. in Rep. Missouri Bot. Gard. 4: 117 (1893).
   Stachytarpheta indica sensu J.G. Baker in F.T.A. 5: 284 (1900) pro parte quoad *Scott Elliot* 4162, non (L.) Vahl (1804).
   Stachytarpheta jamaicensis sensu Hutchinson & Dalziel, F.W.T.A. 2: 277 (1931) pro parte quoad specim. *Scott Elliot* 4162, *Dawodu* 364 et *Dennett* 502, non (L.) Vahl (1804).
   ?*Stachytarpheta australis* Moldenke in Phytologia 1: 470 (1940). Type from Argentina.
   Stachytarpheta australis var. *neocaledonica* sensu Moldenke, in Fl. Madag., Verbenaceae 24–25 (1956) quoad saltem *Decary* 15338 Moldenke (1949)*.

A perennial herb or much branched subshrub, 0.4–1.5(2.5) m high, woody at the base, glabrous or shortly pubescent and with a few longer hairs. Stem and branches subterete, hispid and scattered pilose mainly on the young parts, glabrescent on older parts. Leaves petiolate, usually blackish on drying; lamina 2–8(9) × 1.6–3.5(4.5) cm, ovate to elliptic, obtuse or rounded to subacute at the apex, attenuate and decurrent into the petiole or ± rounded and abruptly narrowed and narrowly cuneate at the base, crenate-serrate on the margin with teeth usually longer than deep, ± scabridulous on upper surface, glabrous or hairy on the nerves beneath, sometimes with some peltate glands near the base; petiole 1–1.5 cm long. Spikes slender, 20–45 cm long with rhachis up to 3.5 mm thick in fruit, sparsely but distinctly pubescent to almost glabrous; bracts 3.5–5 × 0.5–1 mm, less than half as wide as the fruiting calyx, narrowly triangular and tapering to a subulate apex, ± equalling or slightly shorter than the fruiting calyx, very narrowly scarious and ± ciliate on the margin, thickened at the base. Calyx 4–5 mm long, tubular, ellipsoid with 1–3 peltate glands near the apex, 4-toothed with the teeth subequal. Corolla pale blue, pale lilac or white; tube 4–6 mm long, about equalling the calyx or exserted up to 2 mm; limb 4–7 mm in diameter, the lobes 1.5–2 mm long. Style exserted 1.5–2.5 mm from the calyx after the fall of the corolla. Fruiting calyx somewhat embedded in the furrows of the rhachis. Mericarps 3–4 mm long, nearly black at maturity.

Zimbabwe. E: Chipinge Distr., Chirinda Forest, near Chako Township, v.1977, *Goldsmith* 2/77 (BR; K; M; S). **Mozambique**. MS: Sofala, xii.1907, *Coomans* (LD).
Widespread and native in tropical America, now widely naturalized throughout the tropics. Grassland, road sides and disturbed areas; 10–1200 m.
Moldenke in Fifth Summ. Verbenaceae: 629, 1971, considered *S. dichotoma* (Ruiz & Pavon) Vahl to be synonymous with *S. cayennensis*, but in 1983 Moldenke & A. Moldenke (op. cit.: 261–263) treated these taxa as separate species, commenting (op. cit. 264, 1983) that *S. dichotoma* may be a variety of *S. cayennensis*. Examination of the types of *S. dichotoma* (Peru, Ruiz & Pavon, MA) and of *S. cayennensis* (French Guiana, *Leblond* 356, G) confirms that the plants are indeed synonymous, the first being a plant with larger leaves.

---

* Moldenke & A. Moldenke (op. cit.: 262, 1983) referred some other synonyms to this species, whose types were not seen by me.

7. PRIVA Adans.

Priva Adans., Fam. Pl. **2**: 505 (1763). —Moldenke in Repert. Spec. Nov. Regni Veg. **41**: 1–76 (1936). —R. Fernandes in Bot. Helvetica **95**, 1: 33–45 (1985). —Atkins in Kadereit (ed.), Fam. Gen. Vasc. Pl. (Kubitzki, ed. in chief) **VII**: 464 (2004).

Perennial herbs with woody rootstocks, rarely with tuberous roots. Stem and branches 4-angled. Leaves opposite or subopposite, usually decussate, simple, petiolate or sessile, crenate-serrate or serrate, sometimes laciniate or lobed at the base. Inflorescences usually of terminal or axillary spike-like racemes; bracts small. Flowers small, shortly pedicellate, spirally alternate or pseudo-secund on the mostly elongate rhachis, each solitary in the axil of a narrow bract. Calyx tubular, 5-ribbed and slightly 5-plicate at anthesis, shortly and subequally 5-dentate or subtruncate and 5-denticulate at the apex, membranous, usually covered with hooked hairs intermixed with setae, becoming inflated about the fruit and ± narrowed and beaked at the mouth. Corolla hypocrateriform or funnel-shaped, usually white or sometimes blue, rose, red or purple; tube longer than the calyx, cylindric, straight or slightly curved, widened at the throat; limb oblique, c. 2-lipped, with the abaxial lip 3-lobed and the adaxial lip 2-lobed. Stamens 4, included; filaments inserted about or above the middle of the corolla tube; anthers ovate or oblong, dorsifixed. Ovary 2-carpellate, each carpel 2-locular, each locule 1-ovulate; ovules anatropous, erect, attached laterally at or near the base; style terminal, filiform, subequalling the lower stamens, 2-lobed at the apex with a short, stigmatic anterior lobe and an acute horny sterile posterior tooth. Fruit a schizocarp included within a persistent inflated calyx, separating at maturity into 2 bilocular mericarps (or mericarps 1-locular by abortion); mericarps with a dry and hard pericarp, linear or oblong, flat concave or deeply excavate on the commissural face, raised-reticulate or ribbed on the lateral faces, echinate or ridged on the dorsal face rarely smooth, glabrous, puberulous or shortly pubescent. Seeds without endosperm; cotyledons oblong, somewhat thick.

A genus of about 20 species in tropical and subtropical America (Florida and Texas to Paraguay and West Indies), Asia (Arabia, India, Sri Lanka, etc.) and Africa. The c. 10 species occurring in Africa all belong to sect. *Priva*.

Leaves with petiole up to 4.5 cm long; lamina 2–10 × 1.3–8 cm, each margin 10–17-toothed, without basal lobes; mericarps with only 1 row of spines on each side of the dorsal face, the spines ± connected by a hyaline membrane · 1. *flabelliformis*
Leaves sessile, or shortly petiolate with the petiole usually less than 0.3 cm long; lamina 1–4.5 × 0.4–2 cm, each margin 2–5-toothed, sometimes also 2-lobed at the base; mericarps with 3 or more rows of spines or tubercles in a band on each side of the narrow dorsal face, the spines not connected by a hyaline membrane · 2. *africana*

1. **Priva flabelliformis** (Moldenke) R. Fern. in Bot. Helvetica **95**, 1: 37 (1985). —Verdcourt in F.T.E.A., Verbenaceae: 21 (1992). TAB. **9**. Type from Tanzania.
   Priva leptostachya sensu Gürke in Engler, Pflanzenw. Ost-Afrikas **C**: 338 (1895) pro parte quoad distrib. saltem in Malawi et Mossambique. —sensu Kobuski in Ann. Missouri Bot. Gard. **13**: 10 (1926) quoad specim. malawiens. et mossambicens. —sensu J.G. Baker in F.T.A. **5**: 285 (1900) quoad specim. malawiens. et mossambicens., non Jussieu (1806).
   Priva meyeri sensu Moldenke in Repert. Spec. Nov. Regni Veg. **41**: 19 (1936) quoad saltem *Snowden* 34, *Swynnerton* 269 et 517; in Phytologia **5**: 110 (1954) quoad specim. tanzaniens. et mossambicens; op. cit. **44**: 108–109 (1979) quoad saltem specim. sudanens., zimbabwens. et mossambicens.; Résumé Verbenaceae: 143, 145, 149 et 151 (1958). —sensu Mogg in Macnae & Kalk, Nat. Hist. Inhaca Isl., Moçamb.: 152 (1958), non Jaubert & Spach (1855).
   Priva cordifolia var. *abyssinica* sensu Moldenke in Repert. Spec. Nov. Regni Veg. **41**: 45–47 (1936) quoad specim. ugandens. pro parte, zimbabwens., malawiens. et mossambicens. (etiam natalens.?); in Phytologia **5**: 70 (1954) pro parte; in Fl. Madag., Verbenaceae: 31 (1956). —sensu van der Schijff, Check List Vasc. Pl. Kruger Nat. Park: 81 (1969), non *P. cordifolia* (L.f.) Druce var. *abyssinica* (Jaubert & Spach) Moldenke sens. str.
   Priva cordifolia var. *australis* Moldenke in Repert. Spec. Nov. Regni Veg. **41**: 47 (1936); in Phytologia **5**: 70 (1954); op. cit. **49**: 62 (1981); Résumé Verbenaceae: 154 (1958). Type from South Africa (Natal).

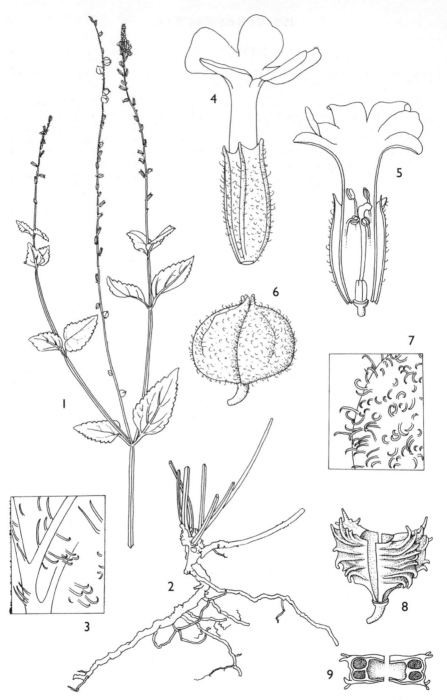

Tab. 9. PRIVA FLABELLIFORMIS. 1, apical portion of branch with flowers and fruit (× ⅓), from *Pope* 463; 2, rootstock, from *Best* 203; 3, detail of leaf lower surface showing hairs (× 20), from *Pope* 463; 4, flower (× 6); 5, flower, opened to show stamens (× 6), 4 & 5 from *P.A. Smith* 2608; 6, inflated calyx at fruiting stage (× 8), from *Wild* 1911; 7, detail of calyx surface, showing hooked hairs of varying size (× 20); 8, fruit, 2 nutlets showing excavated commissural faces and spines on the dorsal faces (× 8), 7 & 8 from *Brass* 17847; 9, diagram of transverse section through the nutlets. Drawn by Deborah Lambkin.

Priva cordifolia var. *flabelliformis* Moldenke in Repert. Spec. Nov. Regni Veg. **41**: 47 (1936); in Phytologia **5**: 71 (1954); op. cit. **44**: 92 (1979); op. cit. **49**: 62 (1981); Résumé Verbenaceae: 468 (Taxon. Index, 1959). —Robyns, Fl. Sperm. Parc Nat. Alb. **2**: 139 (1947). —Brenan in Mem. New York Bot. Gard. **9**: 37 (1954).

Perennial, rather weak erect or straggling herb 0.30–2 m tall, with 1–several herbaceous stems from a woody rootstock, branching from the base. Stems and branches square, pubescent with fine short spreading hooked hairs or glabrescent, internodes up to c. 10 cm long; nodes with a transverse band of acicular setae on the faces between pairs of opposite petioles. Leaves 2–10 × 1.3–8 cm, ovate or ovate-lanceolate, acute or subobtuse at the apex, subtruncate or rounded then cuneate at the base into a 0.9–4.5 cm long petiole, crenate-serrate or serrate on the lateral margins with 10–17 teeth on each side, entire on the basal margin, puberulous with short hooked hairs interspersed with longer tubercle based bristles on the upper surface, the indumentum similar but denser on the lower surface particularly on the midrib and nerves. Racemes slender, 9–17 cm long with young flowers densely spaced towards the apex, and the progressively older flowers becoming more laxly spaced towards the base on the elongating axis; rhachis eventually up to 55 cm long; pedicels erect in flower and up to 0.75–1 mm long and reflexed in fruit; bracts 2–3 mm long, linear-lanceolate. Calyx persistent; tube 5–6 mm long, truncate with 5 small teeth at the apex, densely puberulous with short hooked hairs interspersed with a few longer tubercle based bristles, becoming inflated and subspherical with a narrowed beak-like apex in fruit. Corolla usually white or sometimes pale pink or pale mauve with purple or red lines on the lower lip, not persistent; tube ± twice as long as the calyx, glabrous outside, hairy within; lobes c. 5 × 2 mm, rounded at the apex. Filaments hairy, c. 1 mm long in the lower stamens, c. 1.5 mm long in the upper pair; anthers c. 0.5 mm long. Style 4–4.5 mm long at anthesis, persisting and elongating in fruit, included within the inflated calyx. Schizocarp ± round in outline, 3–4 mm long and 4.5 mm wide. Mericarps glabrous or shortly pubescent, oblong, deeply excavate on the commissural face, and with 2 longitudinal combs of spines on the flat dorsal face, one on either side; spines (6)8–11(14), conical, straight ± curved or hooked, united for ¼ to ¾ of their length by a hyaline membrane, or sometimes all nearly free, when the membrane is well developed the mericarps seem to be winged on each side of the dorsal face; dorsal face transversely ridged; lateral face strongly ribbed, the ribs sometimes somewhat reticulate; the commissure bordered by a hyaline membrane.

Botswana. N: Okavango R., east bank, 1 km south of Caprivi Strip border, 27.iv.1975, *Biegel, Müller & Gibbs Russell* 5017 (K; SRGH). **Zimbabwe**. N: Mazowe (Mazoe) Citrus Estate, fl. & fr. 28.x.1970, *B.F. Searle* 85 (K; SRGH). W: Matobo Distr., Farm Besna Kobila, c. 1500 m, fl. & fr. i.1954, *O.B. Miller* 2052 (SRGH). C: Shurugwi (Selukwe), Wanderer Valley road, c. 1440 m, fl. & fr. 28.iv.1968, *Biegel* 2618 (K; SRGH). E: Chipinge Distr., Chirinda Forest along road verges, 1120 m, fl. & fr. iv.1967, *Goldsmith* 43/67 (B; BR; LD; S; SRGH). S: Chivi Distr., c. 16 km north of Runde (Lundi) Bridge on the Masvingo (Fort Victoria) to Beitbridge road, 704 m, fl. & fr. 5.xii.1961, *Leach* 11303 (K; LISC; M; SRGH). **Malawi**. S: Thyolo Distr., Nswadzi R., 840 m, fl. & fr. 27.ix.1946, *Brass* 17847 (K; SRGH). **Mozambique**. Z: Morrumbala Mt., west side of mountain at the base, fl. & fr. 12.xii.1971, *Pope & Müller* 591 (K; LISC; LMA; SRGH). MS: Sussundenga Distr., between Moribane and Dombe, fl. & fr. 5.vi.1971, *G.V. Pope* 463 (K; SRGH). GI: Xai-Xai Distr., between Xai-Xai (João Belo) and Praia Sepúlveda, 19.viii.1966, *Lavranos* 4694 (PRE). M: Matola, near Matola R., right side of the road, 5.iii.1974, *A. Marques* 2429 (K; LISC; LMU; M).

Also in Dem. Rep. Congo, southern Ethiopia, Uganda, Tanzania, South Africa (KwaZulu-Natal and Transvaal) and in Madagascar. Mixed high rainfall woodland, evergreen forest margins, riverine vegetation and in long grasses on stream banks, usually in moist localities, often in damp soil; sea level to 1500 m.

The hooked hairs of the indumentum enable the plant, and especially the fruit, to adhere to clothing.

Some Flora Zambesiaca plants of this species have been incorrectly referred to *P. meyeri* Jaub. & Spach by Moldenke, or have been determined as such by various herbaria. Specimens so named include: Zimbabwe [*Swynnerton* 269 and 517 (BM)] and Mozambique [*Balsinhas & Marques* 769 (LMA); *Barbosa* 1151 (LISC); *Barbosa & Lemos* 7869 (K; LMA); *Junod* 277 (LISC; PRE); *Lavranos* 4694 (PRE); *Marques* 2429 (M); *Mogg* 28374 (K) and 29286 (LISC); *Simão* 154b (LISC; LMA); *Torre* 2425 (LISC)]. However, this Flora Zambesiaca material has mericarps with spines and ornamentation characteristic of those of *P. flabelliformis*. The mericarps typical of *P. meyeri* lack spines and the dorsal and lateral faces are almost smooth.

Other specimens of *P. flabelliformis* have been referred to *P. cordifolia* var. *abyssinica* (Jaub. & Spach) Moldenke; namely Zimbabwe [*Swynnerton* 6635 (BM)], Malawi [*Buchanan* 887 (BM); *Whyte* s.n. (K)] and Mozambique [*Kirk* 225 and s.n. (K); *Marques* 2429 (K); *Rogers* 45210 (K); *E. Sousa* 5 (LISC); *Swynnerton* 2123 (BM)]. However, the indumentum, shape of leaves and calyces, and ornamentation of the mericarps, exclude them from *P. abyssinica*, where the mericarps always have completely free spines disposed in several rows.

2. **Priva africana** Moldenke in Repert. Spec. Nov. Regni Veg. **41**: 36 (1936); in Phytologia **5**: 63 et 105 (1954); op. cit. **43**: 332 (1979); op. cit. **49**: 60 (1981); Résumé Verbenaceae: 154 (1958). —van der Schijff, Check List Vasc. Pl. Kruger Nat. Park: 81 (1969). —R. Fernandes in Bot. Helvetica **95**, 1: 36 (1985). Type from South Africa (Transvaal).

Perennial herb up to 40 cm tall with 1–several stems from a woody rootstock, somewhat bushy and branching from the base with the main branches sometimes equalling the stem in length. Stem and branches erect or ascending, acutely 4-angular, hispid-pubescent with a mixture of small to large patent bristle-like hairs mostly hooked at the apex, some remaining straight and tapering to a fine apex, or ± glabrescent; nodes with a transverse band of acicular setae on the faces between pairs of opposite petioles. Leaves subsessile or with a petiole up to 3 mm long, scabridulous and rather stiff; lamina 1–4.5 × 0.4–2 cm, ovate-lanceolate, triangular-ovate to ± hastate, acute at the apex, broadly cuneate to rounded and ± strongly lobed at the base, coarsely serrate on the lateral margins with 2–5 teeth on each side, entire on the basal margin, sparsely hispid-pubescent on upper surface, more densely hispid-pubescent beneath particularly on the midrib and nerves, the indumentum similar to that on the stem and branches; basal lobes up to 5 mm long and with 1–2 teeth on each margin. Racemes slender, 2.5–10 cm long with young flowers densely spaced towards the apex, and the progressively older flowers becoming more laxly spaced towards the base on the elongating axis; rhachis eventually up to 28 cm long; pedicels up to 1.5 mm long and erect in flower, up to 3 mm long and reflexed in fruit, rather densely puberulous-hirsute; bracts 3–4 mm long, lanceolate, acute. Calyx persistent; tube 7–10 mm long, obsoletely 5-dentate at the apex, hispid-pubescent with a mixture of small to large patent bristle-like hairs mostly hooked at the apex, some remaining straight and tapering to a fine apex, becoming inflated and subspherical with a narrowed beak-like apex in fruit. Corolla white with mauve tint, to purple, not persistent; tube 8–9 mm long, glabrous outside, hairy within; lobes obovate, rounded at the apex, the lower median one up to 4 × 3.5 mm, the others smaller. Filaments hairy, c. 1.3 mm long; anthers c. 0.7 × 1 mm. Style 4–4.5 mm long at anthesis, persisting and elongating in fruit, included within the inflated calyx. Fruit pendent. Schizocarp c. 3–4 mm long and 4.5 mm wide, round in outline. Mericarps glabrous or minutely pubescent, c. 3.5–4 × 2 mm, somewhat laterally compressed hemispherical, deeply excavate on the commissural face, and with 2 broad longitudinal bands of spines on the curved dorsal face; spines numerous, almost covering the dorsal surface, 0.2–0.9 mm long, conical, sometimes slightly hooked at the apex; dorsal face transversally ridged; lateral face strongly ribbed; the commissure bordered by a hyaline membrane.

Botswana. SE: 93 km NW of Serowe, 24.iii.1965, *Wild & Drummond* 7271 (K; LISC; SRGH); north of Mabeleapudi (Mabele-a-podi), 27.i.1969, *A.D. Buerger* 1029 (PRE). **Zimbabwe**. S: Beitbridge Distr., Shashe-Limpopo R. confluence, 22.iii.1959, *Drummond* 5940 (SRGH). **Mozambique**. M: between Changalane and Catuane, Lubemba Sul Estate, 21.xii.1952, *Myre & Carvalho* 1423 (K; LISC; LMA; SRGH).
Also in South Africa (Northern Province). In open grassland and *Acacia/Combretum/Colophospermum mopane* scrub, and in thickets; 37–1060 m.

8. DURANTA L.

Duranta L., Sp. Pl.: 637 (1753); Gen. Pl., ed. 5: 284 (1754). —Verdcourt in F.T.E.A., Verbenaceae: 48 (1992). —Munir in J. Adelaide Bot. Gard. **16**: 1–16 (1995). — Atkins in Kadereit (ed.), Fam. Gen. Vasc. Pl. (Kubitzki, ed. in chief) **VII**: 467 (2004).

Shrubs or trees, erect or subscandent, unarmed or spinose; stem branched, woody. Leaves simple, opposite or verticillate, deciduous, petiolate. Flowers few to many, pedicellate, in terminal and axillary racemes or panicles; bracts small. Calyx cylindric-tubular or subcampanulate, 5-ribbed, 5-dentate, persistent, accrescent in fruit. Corolla tubular, hypocrateriform; tube straight or curved; limb 2-lipped with 5 unequal lobes. Stamens 4, usually also a staminode, slightly didynamous, inserted at or above the middle of corolla tube, included; filaments subulate; anthers sagittate, dorsifixed, erect. Ovary globose, 4-carpellate with carpels 2-locular, each locule with a single erect anatropous ovule; style equalling or shorter than the corolla tube; stigma subcapitate, unequally 4-lobed. Fruit fleshy, drupaceous with four 2-seeded pyrenes, enclosed within the accrescent calyx. Seeds erect, without albumen.

A subtropical American genus of about 35 species. One species is widely cultivated throughout the tropics and subtropics and now often naturalized.

Duranta erecta L., Sp. Pl.: 637 (1753). —Hiern, Cat. Afr. Pl. Welw. **1**, 4: 831 (1900). —Urban, Symb. Antill. **4**: 536 (1911); op. cit. (Fl. Doming.) **8**: 599 (1921). —Caro in Rev. Argent. Agrom. **23**: 6, fig. 1 (1956). —Backer & Bakhuizen van den Brink Jr., Fl. Java **2**: 599 (1965). —Verdcourt in F.T.E.A., Verbenaceae: 48, fig. 7 (1992). —Munir in J. Adelaide Bot. Gard. **16**: 5, fig. 1 (1995). —M. Coates Palgrave, ed. 3 of K. Coates Palgrave, Trees South. Africa: 976 (2002). TAB. **10**. Type: Plumier plate subsequently published as t. 79 in Plumier, Plantarum Americanum, ed. J. Burman **4**: 70 (1756), Groningen Univ. Library, (lectotype, chosen by Verdcourt, loc. cit.)*.

Duranta repens L., Sp. Pl.: 637 (1753). —Kuntze, Revis. Gen. Pl., part 2: 507 (1891). — Hutchinson & Dalziel, F.W.T.A. **2**: 269 (1931). —Moldenke in Lilloa **4**: 313 (1939); in Fl. Madag., Verbenaceae: 40, t. 5 fig. 1–4 (1956); Résumé Verbenaceae: 455 (Taxon. Index, 1959). —Brenan, Check-list For. Trees Shrubs Tang. Terr.: 639 (1949). —F.W. Andrews, Fl. Pl. Anglo-Egypt. Sudan **3**: 196 (1956). —White, For. Fl. N. Rhod.: 368 (1962). —van der Schijff, Check List Vasc. Pl. Kruger Nat. Park: 81 (1969). —van der Schijff & Schoonraad in Bothalia **10**, 3: 496 (1971). —Sebald in Stuttg. Beitr. Naturk., n° 244: 12 (1972). — Adams, Fl. Pl. Jamaica: 632 (1972). —H. Moldenke & A. Moldenke in Rev. Fl. Ceylon **4**: 278 (1983). Type: "Plumier, Nov. Pl. Amer. Gen.: 30, t. 17 (1703)" (fide *Verdcourt*, loc. cit.).

Duranta plumieri Jacq., Enum. Syst. Pl. Ins. Carib.: 26 (1760); Select. Stirp. Amer. Hist.: 186, t. 176 fig. 76 (1763). —Linnaeus, Sp. Pl., ed. 2: 888 (1762). —Willdenow, Sp. Pl., ed. 4, 3: 380 (1800). —Walpers, Repert. Bot. Syst. **4**: 79 (1845). —Schauer in A. de Candolle, Prodr. **11**: 615 (1847). —Gürke in Engler, Pflanzenw. Ost-Afrikas **C**: 338 (1895). —J.G. Baker in F.T.A. **5**: 287 (1900). —H. Pearson in F.C. **5**, 1: 210 (1901). —Gibbs in J. Linn. Soc., Bot. **37**: 463 (1906). —Mildbraed, Wiss. Ergebn. Zweit. Deutsch. Zentr.-Afrika-Exped., Bot. **2**: 68 t (1922). —Broun & Massey, Fl. Sudan: 351 (1929). *Nom. illegit.*, the types of both Linnaean species are cited in synonymy.**

Spreading, drooping, trailing or subscandent often thorny shrub or small tree up to 6 m high; branchlets slender, 4-angular or terete, young shoots appressed pubescent; spines paired, 0.8–2.7 cm long, axillary or inserted just above the axils, spreading, straight or slightly curved, very sharp. Leaves opposite or verticillate, petiolate; lamina 1.5–7(8) × 1–3(4) cm, ovate, obovate, ovate-lanceolate or elliptic, acute or obtuse at the apex, cuneate or rounded at the base, entire or serrate on the margins above the middle, glabrous or finely appressed pubescent on both surfaces more densely so on the nerves; petiole 0.5–1.5 cm long, slender, appressed-pubescent. Racemes many-flowered, 3–16 cm long, single or in lax panicles, the terminal raceme longer than the lateral ones; pedicels 1–5 mm long, alternate, opposite or subopposite; bracts and bracteoles 1.5–2 mm and 1 mm long respectively, linear, those of the lower flowers usually larger and leaf-like. Calyx tube 3–4 mm long, ± sparsely appressed-puberulous; teeth up to 1 mm long, triangular with midrib produced as a mucro. Corolla lilac to violet-blue or sometimes white; tube longer than the calyx, curved, puberulous outside more densely so in the upper part, glandular-pubescent within; limb spreading, 7–9 mm in diameter, glandular-pubescent on the upper surface, densely puberulous beneath, lobes 3–5 × 2.5–3.5 mm, oblong to orbicular, densely puberulous. Filaments glandular-pubescent, c. 1.25 mm long. Fruit yellow to orange-coloured, shiny, 6–11 mm in diameter, globose or obpyriform, completely enveloped by the accrescent calyx the

---

* Hiern united *D. repens* and *D. erecta*, both of equal date, and used the latter name.
** For a more complete synonymy see A.A. Munir (op. cit.: 5–7). This author refers also to another type (neotype) for *D. repens* L.: *Patrick Browne* s.n., s.d. (Herb. Linn. 806.2 (LINN).

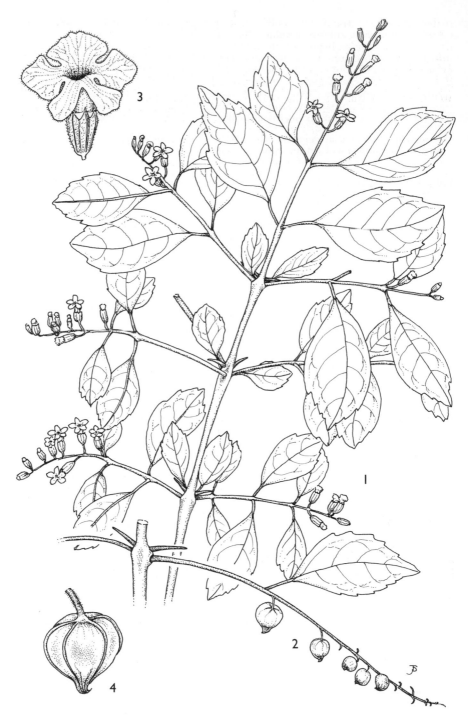

Tab. 10. DURANTA ERECTA. 1, apical portion of flowering branch (× ²/₃), from *Rutherford-Smith* 458, 2, fruiting branchlet (× ²/₃), from *Leach* 11652; 3, flower (× 2); 4, fruit (× 2), 3 & 4 from *Rutherford-Smith* 458. Drawn by Judi Stone.

apical part of which forms a curved beak; pyrenes 3.5–5.5 × 2.5–3.5 mm. Seeds 2.5–3 × 1–1.5 mm, cuneiform, white, smooth.

Botswana. N: Ngamiland Distr., Ng-gokha (Nqogha) Island, Nqogha R., 1.vi.1976, *P.A. Smith* 1744 (K; SRGH). SE: Southern Distr., Kanye, Pharing Gorge, fl. & fr. iii.1944, *O.B. Miller* B/299 (PRE; SRGH). **Zambia**. N: Mbala (Abercorn), garden of old International Red Locust Control Services office, fl. & fr. 22.v.1952, *Siame* 220 (K, cult.). C: Lusaka Forest Nursery, fr. 4.iii.1952, *White* 2182 (FHO, cult.). S: Choma Distr., Kanchomba Agricultural Station, 16.ii.1960, *White* 7105 (FHO; K). **Zimbabwe**. N: Shamva Distr., Harare to Shamva road, c. 69 km from Harare, 9.xii.1960, *Rutherford-Smith* 458 (BR; K; LISC; M; SRGH). W: Matobo Distr., Matopos near Whitewaters, 6.vi.1954, *Plowes* 1745 (K; LISC; SRGH). C: Goromonzi Distr., cultivated at Umwindsidale, c. 25 km east of Harare, 15.vi.1965, *Müller* 148 (K; SRGH). E: Mutare Distr., commonage east of Mutare (Umtali), fl. 1.xi.1958, *Chase* 7009 (BR; K; SRGH). S: Masvingo Distr., west of Mutirikwi (Kyle) Dam, 5.xii.1960, *Leach & Chase* 10551 (BM; COI; K; M; SRGH). **Malawi**. C: Dedza, Chongoni Forest Reserve, fl. & fr. 6.ii.1968, *Salubeni* 955 (K; SRGH). S: Blantyre, fl. & fr. 26.v.1941, *E. Lawrence* s.n. (PRE). **Mozambique**. N: Meluco Distr., Muaguide, fl. 6.iv.1948, *Pedro & Pedrógão* 5070 (LMA). MS: Gondola Distr., Chimoio, Vandúzi R., 28.iv.1948, *Andrada* 1205 (LISC). M: Marracuene, between Matola and Umbelúzi, fl. 29.iv.1947, *Pedro & Pedrógão* 912 (LMA; LMU).

Native of tropical America but cultivated and naturalized throughout the tropics and subtropics. In moist places, in riverine vegetation, gully forest and on granite outcrops, occasionally in miombo woodlands, often on termite mounds; 50–2000 m.

Introduced as a garden ornamental now widespread as a scattered naturalized shrub or tree. Known as the 'golden dew drop'.

129. LAMIACEAE*

By R. Fernandes**

Annual or perennial herbs, shrubs, or trees, often aromatic. Stems frequently square in cross section. Leaves opposite and decussate or sometimes whorled, very rarely alternate, usually simple rarely pinnately or palmately dissected or compound, usually crenate or serrate, sometimes entire or more deeply toothed, petiolate or sessile, stipules absent. Inflorescences terminal or axillary, branched or simple, thyrsoid with determinate cymes arranged along an indeterminate axis, often with paired cymes congested into verticils but sometimes appearing racemose by the reduction of cymes to a single flower, or spicate or capitate by reduction of both the internodes of the inflorescence axis and cyme axes; bracts persistent or not, occasionally coloured; bracteoles present or absent. Flowers zygomorphic occasionally actinomorphic, usually hermaphrodite, hypogynous, usually showy, often with a fleshy nectariferous disk, sometimes resupinate, usually many subtended by a bract, occasionally 1 in bract axils (in reduced cymes), usually pedicellate, rarely sessile. Calyx gamosepalous, actinomorphic to bilabiate, persistent, tubular to broadly campanulate or spreading, sometimes bearded at throat; lobes usually 5, sometimes 2 or 3, and often enlarged in fruit, sometimes closing throat. Corolla gamopetalous, tubular, 5-lobed, typically bilabiate with posterior lip 2–4-lobed and anterior lip 1–3-lobed, variously coloured; tube cylindrical, parallel sided, amplified or constricted distally, or saccate below, straight or curved, sometimes annulate within. Stamens 4, didynamous (anterior pair longer) or of equal length, or sometimes reduced to 2, epipetalous, free or rarely fused (monodelphic), usually exserted; filaments hairy or glabrous, rarely appendiculate; anthers basi- or dorsifixed, sometimes with a prominent connective, usually with 2-thecae, rarely with one aborted, introrse; thecae dehiscing longitudinally, rarely apically, parallel to divaricate, sometimes confluent at apex or synthecous. Gynoecium 2-carpellate,

---

* For family delimitation see notes preceding the Verbenaceae treatment above.

Subfamilies Viticoideae and Ajugoideae are dealt with in this part of Volume 8. Subfamilies Scutellarioideae, Lamioideae and Nepetoideae will be treated in Volume 8 part 8. The remaining subfamilies do not occur in the Flora Zambesiaca area. They are the Symphorematoideae from Asia and the Prostantheroideae from Australia.

** *Vitex* by F. Sales, *Teucreum* by A. Paton.

fused to form a pistil; ovary superior, 2-locular but appearing 4-locular due to ovary
wall intrusions (false septa), slightly to deeply 4-lobed; ovules 1 in each apparent
locule, anatropous, erect, placentation axile (often appearing basal) with ovules
laterally attached on the face of the false septum, subterminal (just short of inrolled
carpel margins); style 1, gynobasic (arising from central depression of ovary lobes)
to terminal, usually bifid at apex; stigmas minute at stylar branch tips. Fruit a
schizocarp splitting into 4 dry nutlets or drupaceous, undivided or 4-lobed with
usually 4 pyrenes, sometimes fewer by abortion, subtended by or enclosed within
persistent calyx; endosperm present, scant or absent; embryo generally straight,
rarely bent.

A cosmopolitan family of c. 258 genera, with around 7000 species mainly in temperate zones,
and particularly diverse in the Mediterranean region. Around 35 genera and 300 species are
found in the Flora Zambesiaca region.

This family contains many herbs and shrubs used as ornamentals and trees for timber.
Species belonging to genera of the subfamilies Viticoideae and Ajugoideae introduced into the
Flora Zambesiaca area, which are grown in cultivation but which have not become naturalized
are mentioned below, and are included in the key to the genera of these subfamilies.

Cultivated taxa of subfamilies Viticoideae and Ajugoideae

Callicarpa macrophylla Vahl, Symb. Bot. **3**: 13, pl. 53 (1794). —Biegel, Check-list Ornam. Pl.
Rhod. Parks & Gard. [Rhodesia Agric. J., Research Report No. 3]: 32 (1977). —H.
Moldenke & A. Moldenke in Rev. Fl. Ceylon **4**: 297 (1983). Type from India[*].

Shrub up to 3.5 m high. Branchlets, petioles, lower surface of leaves, peduncles, branches
of cymes and calyx, very densely greyish woolly tomentose; hairs branched. Leaves decussate,
petiolate; lamina 6–23 × 2.5–9.7 cm, oblong or oblong-ovate, acuminate or sometimes
attenuate towards an acute apex, cuneate at the base, sharply serrate on the margin (teeth
small), discolorous. Cymes solitary at each axil, spreading-dichotomous (often branching up
to 8 times), 4–8 × 7–11 cm, dense, many-flowered forming floriferous masses on the
branchlets. Calyx 1.3–1.6 mm long, oblong-campanulate, distinctly 4-dentate. Corolla lilac or
purple, hypocrateriform; tube 1–2 mm long; limb 4-partite. Fruit c. 2 mm in diameter,
subglobose, white, puberulent or glabrous, surrounded by the cupuliform or patelliform,
fructiferous calyx.

Zimbabwe. C: Harare, Golden Stairs Nursery, 18.i.1974, *Biegel* 4476 (SRGH, cult.).
Native to India, China, Hong Kong, Hainan, Myanmar and Thailand. Cultivated in some
countries, mainly in botanic gardens.

Caryopteris incana (Thunb.) Miq. in Ann. Mus. Bot. Lugd.-Bat. **2**: 97 (1865). Type from Japan.
Nepeta incana Thunb., Fl. Jap.: 244 (1784).
Barbula sinensis Lour., Fl. Cochinch.: 367 (1790). Type from China.
Mastacanthus sinensis (Lour.) Endl. ex Walpers, Repert. Bot. Syst. **4**: 3 (1845).
Caryopteris mastacanthus Schauer in A. de Candolle, Prodr. **11**: 625 (1847). —J.D. Hooker
in Bot. Mag. **111**: t. 6799 (1885). —Henriques in Bol. Soc. Brot. **3**: 144 (1885). —Hemsley
in J. Linn. Soc., Bot. **26**: 263 (1890). —P. Dop in Fl. Gén. Indo-Chine **4**, 7: 885, fig. 90, 3–8
(1935). Type from China.
Caryopteris sinensis (Lour.) Dippel, Handb. Laubholzk. **1**: 59 (1889).

Perennial aromatic herb or shrub up to c. 1.5 m high, appressed whitish-tomentose on
branches, peduncles and inflorescence. Leaf lamina very variable, 2.5–7.5 × 1.5–2 cm, ovate,
oblong-ovate or oblong-lanceolate, acute or obtuse at the apex, rounded or cuneate at the base,
coarsely serrate except at the base, chartaceous and somewhat greyish-green, very shortly
pubescent and glandular on the upper surface, white to silvery appressed tomentose beneath;
petiole 0.6–1.8 cm long, slender, tomentose. Cymes axillary in upper leaves, dense, 1.5–3 cm
in diameter, subglobose; peduncles 1–3.7 cm long, slender. Calyx c. 2.5–3 mm long, 5-lobed;
lobes lanceolate, acute, subequalling the tube. Corolla bright blue, lavender-blue, mauve or
purple, pubescent; tube longer than the calyx, subcylindric; limb c. 8 mm in diameter, with the
4 upper lobes spreading, rounded-ovate, obtuse, the lower lobe ± twice as long as the others,
deflexed, fimbriate. Stamens 2–3-times longer than the corolla. Ovary pubescent; style
subequalling the stamens, glabrous. Capsule c. 2 mm long, hirsute.

Zimbabwe. C: Harare, Marlborough Nursery, 2.ii.1974, *Biegel* 4708 (SRGH, cult.).

---

[*] Further synonymy and bibliography can be consulted in Moldenke & A. Moldenke, op. cit.:
297–298.

Native to E Asia (China, Macau, Hong Kong, Korea, Japan, Taiwan), it is cultivated as an ornamental in some countries.

Caryopteris odorata (D. Don) B.L. Rob. in Proc. Amer. Acad. Arts **51**: 531 (1916). —C.B. Clarke in J.D. Hooker, Fl. Brit. India **4**: 597 (1885). —H. Moldenke & A. Moldenke in Rev. Fl. Ceylon **4**: 484 (1983). —Verdcourt in F.T.E.A., Verbenaceae: 3 (1992). Type from Bengal. *Caryopteris wallichiana* Schauer in A. de Candolle, Prodr. **11**: 625 (1847).

A shrub 1–5 m tall, ± canescent-pubescent when young, nearly glabrous when older. Leaves opposite-decussate, petiolate, odorous when crushed; lamina 3–10 × 1–3.5 cm, ovate or oblong-lanceolate to lanceolate, long-acuminate at the apex, serrate or entire at the margin, very densely glandular-punctate and pubescent mainly on the venation beneath. Inflorescences axillary, shorter than the leaves but running together terminally and spike-like, incanous-pubescent. Calyx 2–4 mm long, deeply divided into 5(6) triangular, narrow teeth. Corolla blue to blue-violet or bluish-purple to light violet, light mauve, purplish, lavender, lilac, or with some lobes white or all white; tube 8–12 mm long; limb 1.2–2 cm wide, spreading, with the 4 upper lobes subequal, oblong, rounded at the top, and the lower one longer (up to 8 mm long) and broader. Stamens exserted but shorter than the style. Capsule 4–6 mm in diameter, globular, pubescent, separating into 4 valves, each winged along one margin, and 1-seeded.

Malawi. S: Derek Arnall's garden, Chowe Estate at the edge of Namizimu Forest Reserve on the escarpment above Mangochi, c. 1088 m, 28.vii.1981, *Chapman* 5817 (K), "small tree, 4 m tall, shapely form with spreading branches, free flowering, the flowers pinkish-purple. A most attractive ornamental plant".

Native to northern Pakistan, Nepal, Bhutan, Sikkim and northern India to Thailand, and China. Cultivated as an ornamental in southern Asia, Egypt, Kenya and elsewhere.

Gmelina

Three species of *Gmelina*, introduced from Asia and cultivated in the Flora Zambesiaca area are separable as follows:

1. Bracts conspicuous, ovate or orbicular, 1.5–4 × 1–3.5 cm, persistent, green or coloured, membranous, imbricate; a straggling or scandent, usually spinose shrub · · · 3. *philippensis*
 – Bracts lanceolate or linear-lanceolate, smaller, caducous, never coloured, not membranous; a shrub or tree, spinose or not · 2
2. Calyx dentate; leaves large, 10–25 × 7.5–18 cm, broadly ovate, long-acuminate or caudate at the apex, subcordate or truncate at the base; inflorescences large, 7.5–39 cm long; a tree up to 20 m high, not spinose · 1. *arborea*
 – Calyx truncate; leaves smaller, 3–9 × 3–5 cm, ovate or elliptic, acute at the apex, obtuse or rounded at the base; inflorescences small, 2–5 cm long; a shrub or a small tree with spinose branches · 2. *villosa*

1. **Gmelina arborea** Roxb., Pl. Corom. **3**: 41, Pl. 246 (1815); Fl. Ind., ed. 2, **3**: 84 (1832). — Brenan, Check-list For. Trees Shrubs Tang. Terr.: 639 (1949). —White, For. Fl. N. Rhod.: 368 (1962). —Huber, Hepper & Meikle in F.W.T.A., ed. 2, **2**: 432 (1963). —van der Schijff & Schoonraad in Bothalia **10**, 3: 496 (1971). —J.H. Ross, Fl. Natal: 300 (1972). —Biegel, Check-list Ornam. Pl. Rhod. Parks & Gard. [Rhodesia Agric. J., Research Report No. 3]: 58 (1977). —H. Moldenke & A. Moldenke in Rev. Fl. Ceylon **4**: 390 (1983). —Verdcourt in F.T.E.A., Verbenaceae: 2 (1992). Type a cultivated plant in Calcutta Botanical Garden.
 Premna arborea (Roxb.) Roth, Nov. Pl. Sp.: 287 (1821). Type as above.
 Gmelina rheedii Hook. in Bot. Mag. **74**: t. 4395 (1848). Type a cultivated plant at Royal Botanic Gardens of Kew.
 Gmelina arborea var. *glaucescens* C.B. Clarke in J.D. Hooker, op. cit.: 582 (1885). Type from India.

An unarmed deciduous tree up to 20 m tall; trunk straight, bole clear to 10 m. Branchlets tawny pubescent soon glabrous and smooth greyish to brownish with conspicuous circular lenticels, striate. Leaves ± discolorous, 10–25 × 7.5–18 cm, broadly-ovate or ovate-cordate, acuminate or caudate at the apex, subcordate to rounded or truncate at the base with 2 glands, entire, toothed or lobed on turions or young plants, sparsely to densely lepidote and glaucous-green and glabrous to velvety stellate-pubescent or -tomentose beneath; with 4–5 lateral nerves on each side of the midrib, the lowest pair basal; petiole 5–15 cm long. Panicles many-flowered, 7.5–39 cm long, erect, with short-cymes, the flowers appearing before or with the young leaves; bracts 1–1.5 cm long, linear or linear-lanceolate, caducous. Calyx c. 5 mm long, campanulate, shortly 5-dentate, fulvous-tomentose, somewhat accrescent in fruit. Corolla yellow or brilliant orange to reddish or brownish-yellow or pinkish-brown to salmon or apricot-coloured, 2.5–4 cm long, densely pubescent outside; limb 2-lipped, with the upper lip deeply divided into 2 oblong, obtuse lobes, the lower lip 3-lobed with the median lobe longer and broader than the lateral ones. Stamens ± exserted, one pair sometimes sterile. Drupe 1.5–2.5 cm long, ovoid on obovoid-pyriform, orange-yellow at maturity. Seeds 1–2 (by abortion).

Zambia. C: Kabwe (Broken Hill) Nursery, seedling, 10.viii.1954, *Cooling* 54 (FHO). **Zimbabwe**. C: Harare, cultivated as ornamental in Avondale Garden, 27.ii.1972, *Biegel* 4113 (K). **Malawi**. N: Karonga town, planted as a street tree, 480 m, 15.vii.1970, *Brummitt* 12133 (K). **Mozambique**. T: Estima, c. 340 m, 19.x.1973, *Macêdo* 5301 (LMA). MS: Beira, 23.iv.1962, *Balsinhas & Macuácua* 578 (LISC).

Native to Pakistan, Bhutan, India, Myanmar, Thailand, Indo-China, Malaysia, Polynesia and S China. Introduced in tropical South America and elsewhere. In Africa it is cultivated as a garden ornamental and as a street tree. In Malawi it has been planted in plantations as fuel wood mostly for use in tobacco barns (fide *Brummitt* 12133).

2. **Gmelina villosa** Roxb., Hort. Beng.: 46 (1840). —Verdcourt in F.T.E.A., Verbenaceae: 3 (1992).

A spinose shrub or small tree; leaves 3–9 × 3–5 cm, ovate or elliptic, acute at the apex, obtuse or rounded at the base with ovate or elliptic leaves, glabrous above when mature and tomentose-villose beneath; inflorescences short, fulvous-tomentose, bracts linear-lanceolate or lanceolate, caducous, never coloured, not membranous; calyx truncate and obsoletely dentate, provided with several large flattish glands; corolla yellow, 3–4 cm long.

Mozambique. MS: Beira, 23.iv.1962, *Balsinhas & Macuácua* 578 (COI; LISC; LMA; LMU). Sometimes cultivated as a garden ornamental.

3. **Gmelina philippensis** Cham. in Linnaea **7**: 109 (1832). —Verdcourt in F.T.E.A., Verbenaceae: 3 (1992). (*Gmelina hystrix* Schult. ex Kurz.).

A straggling or scandent, usually spinose shrub, up to c. 3 m high. Leaves 1.5–10 × 1.5–6 cm, ovate or elliptic to rhomboidal-elliptic, shortly acuminate at the apex, cuneate at the base, glabrous and shiny above, glaucous and puberulous only on the nerves beneath; inflorescence 10–20 cm long with the flowers subtended by large, conspicuous, coloured bracts (in the sheet seen, lightly brown-mauve to green, with light brown-mauve markings), bracts 1.5–4 × 1–3.5 cm, ovate or orbicular, persistent; corolla bright yellow or lemon-yellow, 4.5–5.5 cm long.

Zimbabwe. C: Harare, c. 1480 m, 5.iv.1977, *Biegel* 5461 (K), grown from cuttings of plants cultivated at Mombassa.

A native to the Philippine Islands, it is sometimes cultivated as a garden ornamental in the Flora Zambesiaca area.

Holmskioldia sanguinea Retz., Observ. **6**: 31 (1791). —Schauer in De Candolle, Prodr. **11**: 696 (1847). —Walpers, Repert. **3**: 894 (1844). —White, For. Fl. N. Rhod.: 568 (1962). —H. Moldenke & A. Moldenke in Fl. Ceylon **4**: 477 (1983). —Verdcourt in F.T.E.A., Verbenaceae: 4 (1992).

A straggling shrub or small tree up to 10 m high, laxly branched, branches 4–angled, glabrous. Leaves 3–12 × 1.5–8 cm, ovate or ovate-oblong, acuminate, nearly entire or shallowly dentate, rounded at the base, membranous, subglabrous above, pubescent beneath on the nervation; petiole 0.5–2.5 cm long. Cymes in short branchlets. Fruiting calyx 1.5–2.5 cm in diameter, ± flat and saucer-shaped, reticulate, yellow, orange, bright to brick-red. Corolla coloured as the calyx. Corolla tube 17.5–20 mm long, funnel-shaped, slightly curved; corolla limb short, oblique, 2-lipped, with the median lobe c. twice as long as the other. Fruit globose, 4-lobed, brown, verrucose-glandular.

Zambia. Cultivated in Lusaka Forest Nursery, fl. 3.iii.1953, *White* 2158 (FHO). **Zimbabwe**. Cultivated in Kia Ora Nursery, Highlands, Harare, c. 1460 m, fl. x.1969, *L.H. Smith* 23 (K). **Malawi**. S: Mulanje Distr., Fort Lister Gap, foot of Mchese Mt., c. 1070–1300 m, 16.iv.1970, *Brummitt & Banda* 9881 (S; K). **Mozambique**. N: Ribáuè, Posto Agronómico, fl. & fr. juv. 8.v.1948, *Pedro & Pedrógão* 3253 (LMA; LMU). MS: Manica Village, Quinta e Horta, fl. 26.vi.1949, *Pedro & Pedrógão* 6971 (LMA, cult.). M: Maputo, José Cabral's Park, fl. 24.v.1973, *Balsinhas* 2514 (K; LMA).

Native from India and Pakistan. Cultivated in many countries and sometimes naturalized (southern Asia, Mauritius, Indonesia and West Indies). One specimen seen from the Flora Zambesiaca area, *Brummitt & Banda* 9881 from Malawi, could be naturalized; the label says "Presumably an escape from cultivation, but well established in woodland near avenue of *Widdringtonia* and *Eucalyptus*, at least 1 km from the nearest house or hut. In *Brachystegia-Uapaca* woodland".

Tectona grandis L.f., Suppl. Sp. Pl.: 151 (1781). —Moldenke in Phytologia **1**: 154-165 (1935). —Brenan, Check-list For. Trees Shrubs Tang. Terr.: 641 (1949). —Gossweiler, Fl. Exot. Angola: 182 (1950). —Moldenke in Phytologia **5**: 112–120 (1954). —H. Moldenke & A. Moldenke in Rev. Fl. Ceylon **4**: 304 (1983). —Verdcourt in F.T.E.A., Verbenaceae: 2 (1992). Type from India.

Tectona theka Lour., Fl. Cochinch.: 137 (1790). Type from Indo-China.
Theka grandis (L.f.) Lam., Tabl. Encycl. [Illustr. Gen.] 1: 111, t. 136 (1792).
Jatus grandis (L.f.) Kuntze, Revis. Gen. Pl., part 2: 508 (1891).

A large tree up to 50 m tall, densely cineraceous- or ochraceous-furfuraceous-tomentose. Leaves 11–95 × 6–50 cm, broadly elliptic, acute or shortly acuminate at the apex, acute or attenuate and prolonged into the winged petiole at the base or sessile and clasping, entire or repand denticulate at the margin, chartaceous, dark green, very rough or minutely bullate above, light yellow-green and densely stellate-tomentose beneath (often with a red coloration on rubbing), glabrescent, drooping; petiole short, stout, winged, densely ochraceous-furfuraceous, or absent. Inflorescences terminal and in the axils of the uppermost leaves, massive, usually c. 40 × 40 cm, the terminal ones often larger, densely cineraceous- or ochraceous-furfuraceous; cymes opposite, divaricate, distant, branched, many-flowered; peduncles often elongate; pedicels 1–4 mm long; bracts up to 15 × 4 mm, linear-lanceolate, attenuate. Calyx 4–4.5 × 3–3.5 mm, 5–7-dentate or-lobed, with teeth 1.5–2.5 mm long and ovate or ovate-oblong, obtuse, often reflexed, yellow-green, densely furfuraceous-tomentellous. Corolla white or with the lobes rose-coloured; tube 1.5–3 mm long and c. 1.5 mm wide; limb 5–7-partite, the segments 2.5–3 mm, obovate-elliptic, rounded at the top, erect or reflexed. Filaments 2.5–4 mm long, ampliate and flattened below, glabrous; anthers ovate or oblong. Ovary 1.2–2 mm long, ovate or conical, pubescent; style 3.6–5.2 mm long, pubescent. Fruit 1.2–2 cm in diameter, subglobose or tetragonal, umbilicate and 4-lobed at the top, densely tomentose with hairs irregularly branched, light brown or ochraceous, enclosed in the inflated calyx; fruiting calyx up to 2.5 cm in diameter, bladder-like, light brown and brittle on drying, irregularly plaited and crumpled.

Mozambique. MS: Chimoio, in the estate of Mr. Rodrigues, 10.x.1954, *Pedro* 4943 (LMA). M: arboretum of Matola (Forestry Services, cult.), 9.vi.1949, *Cardoso & Myre* 727 (LMA).

Native to India, Myanmar, Thailand, Malaysia and Java, it is cultivated in many tropical countries of Asia and Africa for its valuable wood. In Mozambique it is cultivated on sandy, reddish soil.

In FHO there is a sheet with three seedlings, obtained from seeds sown 29.iii.1958, at a nursery in Samfya (Zambia, N:). However, F. White does not refer to this species as being cultivated in Zambia in the Forest Flora of Northern Rhodesia.

Information on teak has been summarized by Krishna Murthy (Bibliography on teak, Dehra Dun, 1981).

Key to genera of Lamiaceae subfamilies Viticoideae and Ajugoideae

1. Fruits not dividing into 4 distinct nutlets (mericarps) or rarely splitting late in development to produce 4 nutlets which have a conspicuous scar or areole (the area where adjacent nutlets were united) which covers at least two-thirds of the nutlet length · · · · · · · · · · · ·2
   – Fruits splitting into 4 nutlets with minute or small scars or areoles which cover less than half of the nutlet length (subfamilies *Scutellarioideae*, *Lamioideae* and *Nepetoideae*) · · · · · · · · · · ·
   · see Volume 8 part 8
2. Leaves compound, with 3–5 leaflets · 2. **Vitex**
   – Leaves simple, sometimes deeply lobed or 3–5-partite but not compound · · · · · · · · · · ·3
3. Corolla 1-lipped with tube deeply cleft between the upper lobes (opposite the lip) and the stamens exserted through the cleft; fruit dry ·4
   – Corolla ± regular or irregular and 2-lipped; fruit drupaceous or if dry (capsule) then the lowest and largest corolla lobe fimbriate ·6
4. Calyx limb very accrescent, patelliform or rotate, papyraceous, rigid, reticulate-nerved and sometimes coloured at fruiting stage · 6. **Karomia**
   – Calyx not as above, tubular or funnel-shaped, thick-textured · · · · · · · · · · · · · · · · · ·5
5. Anther thecae widely diverging apically, fused and confluent at the base; fruit composed of 4 dry nutlets with areole more than half as long as the nutlet · · · · · · · · · · 7. **Teucrium**
   – Anther thecae parallel, not confluent; fruit drupaceous, 3–4-lobed or a drupaceous schizocarp · 4. **Rotheca**
6. Calyx limb very accrescent, patelliform or rotate, papyraceous, rigid, reticulate-nerved and coloured at fruiting stage · **Holmskioldia sanguinea**
   – Calyx not as above, tubular or funnel-shaped, thick-textured · · · · · · · · · · · · · · · · · · 7
7. Corolla ± regular ·8
   – Corolla ± irregular (2-lipped) · 11
8. Calyx and corolla 5–7-merous; stamens 5–6; fruit completely enclosed in the inflated accrescent calyx; a tall tree with very large leaves (cultivated) · · · · · · · · · · · · · · **Tectona**
   – Calyx and corolla 4(5)-merous; stamens 4; fruit not enveloped by the calyx · · · · · · · · · · 9
9. Stigma depressed-capitate or peltate; plants cultivated · · · · · · · · · · · · · · · **Callicarpa**
   – Stigma bifid; plants native or cultivated · 10

10. Corolla tube short, 2–4 mm long; fruit unlobed · 1. **Premna**
– Corolla tube usually long; fruit (2)4-lobed · · · · · · · · · · · · · · · · · · · 3. **Clerodendrum**
11. Fruit dry, 4-lobed; lowest and largest corolla lobe fimbriate; plant a cultivated garden ornamental · **Caryopteris**
– Fruit drupaceous; corolla lobes not fimbriate · 12
12. Flowers solitary on axillary peduncles; peduncles becoming spinescent after the fall of the fruit; corolla reddish · 5. **Kalaharia**
– Flowers grouped in inflorescences; peduncles never spinescent · · · · · · · · · · · · · · · · · 13
13. Endocarp divided into four 1-locular pyrenes (or two 2-locular pyrenes) · · 3. **Clerodendrum**
– Endocarp undivided giving a 4-locular pyrene · 14
14. Corolla tube up to 4 mm long, cylindric; stamens inserted at the throat of corolla tube; exocarp red, purple or black · 1. **Premna**
– Corolla tube 20 mm or more long, funnel-shaped; stamens inserted at the base of the dilated part of corolla tube; exocarp yellow; plants cultivated · · · · · · · · · · · · · · **Gmelina**

Subfam. VITICOIDEAE

1. PREMNA L.

Premna L., Mant. Pl. Alt.: 154, 252 (1771) *nom. conserv.* —R. Fernandes in Bol. Soc. Brot., sér. 2, **63**: 297 (1990). —Verdcourt in F.T.E.A., Verbenaceae: 68 (1992). —Harley et al. in Kadereit (ed.), Fam. Gen. Vasc. Pl. (Kubitzki, ed. in chief) **VII**: 194 (2004).
Appella Adans., Fam. Pl. **2**: 84 et 519 (1763).

Trees shrubs climbers or subshrubs, glabrous to densely hairy. Leaves petiolate, opposite or 3–4-whorled, simple, entire or dentate, often with minute spherical glands and aromatic when crushed. Inflorescences terminal, usually many-flowered, with flowers in cymes corymbosely, racemosely or paniculately arranged, rarely the inflorescences axillary. Flowers hermaphrodite or polygamous, small and dull-coloured. Calyx cupuliform or campanulate, usually actinomorphic, truncate to 3–5-dentate or 2–5-lobed, or sometimes c. 2-lipped, somewhat accrescent in fruit. Corolla hypocrateriform; tube short, infudibular or cylindric, villous or with a ring of hairs; limb 2-lipped, with the posterior lip 2-fid or emarginate and the anterior 3-fid or 3-partite, with the lobes subequal or the median one the largest, or ± regularly 4-fid. Stamens 4, didynamous or subequal, inserted near or at the throat of corolla tube, included or ± exserted; anthers dorsifixed, ovoid or subglobose, with divergent thecae. Ovary 2-locular with 2 ovules per loculus, or 4-locular by means of false septa and each locule 1-ovulate; style filiform subequalling the stamens; stigma 2-fid. Fruit drupaceous, small, with a thin fleshy mesocarp and bony endocarp; pyrene 4-locular or 2–3-locular by abortion; calycine cup persistent, often venose Seeds oblong, without albumen.

A genus of c. 226 species and infraspecific taxa (fide *Moldenke & A. Moldenke*, 1983)*, native mainly in Old World tropics and subtropics, Africa, Asia, Australia, Oceania and extending into China and Japan.

1. Branches and leaves with stellate or dendritically branched hairs; inflorescences terminal and axillary · 2
– Branches and/or leaves glabrous or with simple or multicellular hairs, not stellate or branched; inflorescences terminal · 3
2. Leaves elliptic to oblanceolate, 3–7.5(10) × 1–3.5(4) cm, sharply acuminate at the apex; inflorescences of dense ovoid clusters, 1–2 × 1.5 cm, mostly on short lateral shoots; calyx tubular with distinct triangular teeth c. 1 mm long · · · · · · · · · · · · · · · · 1. *tanganyikensis*
– Leaves relatively broader, 2.8–14(25) × 2.3–9(14) cm, rounded or shortly acuminate at the apex; inflorescences paniculate, up to 3.5(7) × 5(16) cm; calyx cup-shaped, with short or obsolete teeth · 2. *schliebenii*

---

* 50 species according to Mabberley (1997) and 200 according to Verdcourt (1992).

3. Inflorescence, calyx and drupe with gland-tipped hairs; inflorescences few-flowered, racemiform; calyx lobes (1.5)2–2.5 mm long, oblong; leaves up to 4 × 2.5 cm (usually smaller) · 3. *mooiensis* forma *glandulifera*
– Inflorescence, calyx and drupe without gland-tipped hairs; inflorescences many-flowered, corymbiform; calyx lobes shorter; leaves usually larger · 4
4. Petiole up to 6 cm long, usually more than ⅓ as long as the leaf blade, slender; leaves usually ± long acuminate, thin and softly membranous, drying light to pale green; corolla (3.5)5–6 mm long (total length) · 4. *senensis*
– Petiole up to 3.5 cm long, usually less than ⅓ as long as the leaf blade; leaves not acuminate or shortly acuminate, thicker and more rigid, drying dark green at least above; corolla usually smaller · 5
5. Leaf blades up to 8.5 × 5.7 cm, discolorous, dark green above, golden-yellow-brown velvety beneath; petiole 0.2–1.2(2) cm long, usually less than ¼ as long as the leaf blade; corymbs small, up to 3.5 × 4.3 cm · 5. *velutina*
– Leaf blades usually larger, concolorous, dark green on both surfaces; petiole up to 3.5 cm long; corymbs usually larger · 6
6. Leaf blades up to 15(17) × 9(14) cm, usually glabrous above with some hairs on the main nerves and in lower nerve axils beneath and at base of the midrib; petiole up to 3 cm long, less than ⅓ as long as the leaf blade · 6. *serratifolia*
– Leaf blades smaller, evenly and ± sparsely hairy on both surfaces; petiole up to 3.5 cm long, ⅓ as long as the leaf blade or slightly longer · 7. *sp*

1. **Premna tanganyikensis** Moldenke in Phytologia **7**: 83 (1959). —Verdcourt in F.T.E.A., Verbenaceae: 80 (1992). Type from Tanzania.

Scandent shrub or small tree, 2–4 m tall; indumentum on the branches, petioles, inflorescence branches and calyx of stellate or dendritically branched hairs. Older branches with a pale green, thinly corky fissured epidermis, glabrous; young branches and branchlets slender, densely fulvous-tomentose to glabrescent; lateral shoots abbreviated, 0.5–3 cm long. Leaves opposite-decussate, somewhat closely spaced, longer than the internodes, 3–7.5(10) × 1.7–3.2(4) cm, elliptic to oblanceolate, ± sharply acuminate at the apex, cuneate at the base, subentire to shallowly and ± distantly dentate or only so in the upper half, light green, slightly discolorous (darker above), stellate-pubescent on both surfaces when young, later glabrescent, the hairs densest on the nervation beneath, with minute spherical yellow glands beneath; lower pair of nerves basal; petiole 3–7(10) mm long, slender, densely ferruginous pubescent with branched hairs. Inflorescences short dense terminal and axillary ovoid clusters mostly on the short lateral branches, 1–2 cm long and c. 1.5 cm wide, shorter than the subtending leaf; bracts 2–6 mm long, linear; peduncles up to 13 mm long; pedicels 0.5–1 mm long. Calyx densely ferruginous pubescent; tube c. 3 mm long, tubular, with teeth 1 mm long, triangular. Corolla yellow or yellowish-brown, 2-lipped, 6–6.5 mm long, stellate-pubescent outside; upper lip ± cucullate, suborbicular in outline, very shortly emarginate; lower lip 3-lobed, the lobes oblong, rounded at the top. Stamens exserted, shorter than the corolla lips; anthers c. 0.5 mm long, pale. Drupes c. 3.5 mm in diameter, black, glabrous.

Mozambique. N: Palma, monte a ca. 2 km sul do rio Rovuma, a 16 km de Nangade, c. 220 m, 18.iv.1964, *Torre & Paiva* 12141 (COI; LISC; LMU; M; WAG).
Also in Tanzania. In granitic rupideserta and in thickets, on sandy soils; c. 220 m.

2. **Premna schliebenii** Werderm. in Notizbl. Bot. Gart. Berlin-Dahlem **12**: 89 (1934). —Brenan, Check-list For. Trees Shrubs Tang. Terr.: 641 (1949). —Verdcourt in F.T.E.A., Verbenaceae: 81 (1992). TAB. **11**, fig. B. Type from Tanzania.

A scandent or many-stemmed shrub to c. 4 m high (or a 1–3-stemmed, laxly branched tree to 16 m tall); bark smooth, finely longitudinally fissured; indumentum of dendritically branched stellate hairs often intermixed with longer hairs in which the branches are reduced to barbs. Older branches whitish or yellow-brown, glabrescent, leaf scars prominent; young shoots densely yellowish-rusty stellate-tomentose. Leaves opposite, clustered toward the end of the branchlets, petiolate; blades (2.8)4.5–14(25) × (2.3)3.7–9(14) cm, suborbicular to oblong or broadly elliptic, rounded or shortly acuminate at the apex, ± cuneate or slightly auriculate at the base, entire repand or occasionally denticulate, light green above, pale yellowish-green beneath, with small yellowish stellate hairs above and on the prominent venation beneath, to completely velvety tomentose beneath; nervation impressed above, prominent beneath; petiole 0.5–1(1.5) cm long, densely tomentose. Inflorescences terminal (or axillary), corymbose (or paniculate), many-flowered, up to 3.5(7) × 8(12) cm, the axes yellowish ferruginous tomentose with stellate and branched hairs intermixed; bracts 3–12 mm, lanceolate; peduncles to 15 mm long, pedicels c. 1 mm long. Calyx 2.5–3 mm long and 2.5 mm in diameter, cup-shaped with 5 short or obsolete lobes, stellate tomentose (with glandular hairs intermixed). Corolla white or yellowish; tube cylindric, ± equalling the calyx or slightly longer, glabrous; limb c. 2-lipped, the lower lip 3-lobed with lobes orbicular, the median lobe slightly larger, the upper lip longer, obtuse and shortly emarginate at the apex, concave, all puberulous and with minute glandular hairs on the outside and ciliolate at the margins. Stamens exserted from the corolla tube but shorter than the limb; filaments c. 2–4 mm long, subequal; anthers c. 0.5 mm long. Drupes c. 4–6 mm in diameter, subglobose, gland-dotted and, when young, with gland-tipped hairs.

Mozambique. N: Pemba (Porto Amélia), prox. S. Paulo de Mahate Mission, fl. 27.x.1942, *Mendonça* 1082 (BR; COI); Ancuabe Distr., localidade de Metoro, Aldeia Comunal de Namatuco, caminho para Mecaruma a 10 km, fr. 31.i.1984, *Groenendijk, de Koning, Maite & Dungo* 902 (K; LMU).
Also in Tanzania. Open riverine forest and dry rocky areas in granite sand; c. 10–345 m.
This species is closely related to *P. chrysoclada* (Boj.) Gürke.
Specimens from Mozambique differ from typical *P. schliebenii*, as described by Werdermann, in habit (recorded as a shrub from Mozambique, not a tree to 16 m); and also in leaf size (up to c. 14 × 9 cm, not up to 25 × 14 cm as in Tanzanian material) and shape (rounded not subacute or subacuminate at the apex, and cuneate not slightly auriculate at the base), and in the inflorescence (terminal not axillary, up to 3.5 × 5 cm not 6 × 8 cm, and corymbose not paniculate), and in the calyx (stellate-tomentose not also glandular-hairy). These specimens may belong to a single species occurring in evergreen forest and also dry rocky areas. However, more material from Tanzania and Mozambique needs to be examined in order to establish whether or not two taxa are involved: one a large-leaved tree from Tanzania, the other a shrub from Mozambique.

3. **Premna mooiensis** (H. Pearson) G. Piep. in Bot. Jahrb. Syst. **62**, Beibl. Nr. 142: 80 (1929). —
     M. Coates Palgrave, ed. 3 of K. Coates Palgrave, Trees South. Africa: 977 (2002). Type from South Africa.
     Vitex mooiensis H. Pearson in F.C. **5**: 212 (1901); in Hooker's Icon. Pl. **28**: t. 2705 (1901).

Forma **glandulifera** (Moldenke) R. Fern. in Bol. Soc. Brot., sér. 2, **63**: 299 (1990). Type: Mozambique, Maputo, Goba, 6.xii.1942, *Mendonça* 1654 (LISC, holotype).
     Vitex mooiensis var. *rudolphii* H. Pearson in Hooker's Icon. Pl. **28**: t. 2705 (1901), with respect to text and type only; in F.C. **5**, 1: 719 (1912). Type: Mozambique, Moamba Distr., Ressano Garcia, *Schlechter* 11935 (UBI?, holotype).
     Premna holstii Gürke var. *glandulifera* Moldenke in Bol. Soc. Brot., sér. 2, **40**: 121 (1966). Type as for *P. mooiensis* forma *glandulifera*.
     Premna mooiensis var. *rudolphii* (H. Pearson) Moldenke in Fifth Summ. Verbenaceae: 257 (1971).
     Premna mooiensis sensu Moldenke, Fifth Summ. Verbenaceae: 252 (1971) non (H. Pearson) G. Piep. (1929).

A much branched shrub 3–4 m high or a small tree 4–5 m high with trunk up to 20 cm in diameter; indumentum of short, curved-antrorse or spreading, simple hairs. Older branches greyish, striate, longitudinally fissured, glabrous; young branches and branchlets puberulous; nodes prominent. Leaves up to 4 × 2.5 cm, ovate or subrhombic, obtuse to acute or shortly acuminate at the apex, cuneate at the base, usually entire or sometimes subdentate, sparsely puberulous on both surfaces or puberulous only on the nerves; petiole up to 12 mm long, slender, puberulous. Inflorescences terminal, with cymes racemosely arranged on a slender axis, small, few-flowered; branches and pedicels somewhat filiform, puberulous with gland-tipped hairs interspersed, the latter more numerous towards the apex of the pedicels. Calyx 3–4.5 mm long, ± densely covered with gland-tipped hairs mainly at the base; lobes (1.5)2–2.5 × 1 mm, oblong, obtuse; fruiting calyx up to 5 mm long. Flowers aromatic; corolla whitish with ciliate-fimbriate lobes. Drupe obovoid, exserted from the fruiting calyx, ± densely glandular-hairy, mainly toward the apex.

Mozambique. M: Goba, frontier, 12.xii.1947, *Barbosa* 707 (LISC).
Known only from Mozambique. In deciduous bush and thickets on dry rocky hillsides and in seasonally dry drainage lines, in stony soil; c. 300 m.
Torre 2037, also from Goba, with a lax indumentum of gland-tipped hairs, is intermediate with forma *mooiensis* which lacks glandular hairs and occurs in South Africa (KwaZulu-Natal and Mpumalanga).

4. **Premna senensis** Klotzsch in Peters, Naturw. Reise Mossambique **6**, part 1: 263 (1861). —
Gürke in Engler, Pflanzenw. Ost-Afrikas **C**: 338 (1895). —J.G. Baker in F.T.A. **5**: 292 (1900).
—Sim, For. Fl. Port. E. Africa: 93 (1909). —Brenan, Check-list For. Trees Shrubs Tang. Terr.:
641 (1949). —White, For. Fl. N. Rhod.: 370 (1962). —Richards & Morony, Check List Fl.
Mbala: 239 (1969). —R. Fernandes in Bol. Soc. Brot., sér. 2, **63**: 300 (1990). —Verdcourt in
F.T.E.A., Verbenaceae: 78 (1992). —M. Coates Palgrave, ed. 3 of K. Coates Palgrave, Trees
South. Africa: 978 (2002). TAB. **11**, fig. A. Type: Rios de Sena, *Peters* s.n. (B†).
Ehretia tetrandra Gürke in Bot. Jahrb. Syst. **28**: 311, 461 (1900). —J.G. Baker in F.T.A. 4,
2: 22 (1905). Type from Tanzania.
Premna viburnoides sensu J.G. Baker in F.T.A. **5**: 292 (1900) quoad specim. *Kirk* s.n.
(Lower Shire Valley). —sensu Moldenke, Fifth Summ. Verbenaceae: 252 (1971), non A.
Rich.(1850) nec Wall. ex Schauer (1847).
Premna hildebrandtii sensu Moldenke, Fifth Summ. Verbenaceae: 248 (1971), non Gürke
(1895).
Clerodendrum kibwesensis sensu Moldenke Fifth Summ. Verbenaceae: 251 (1971); in
Phytologia **61**: 405 (1986) quoad *Hornby* 2499, non Moldenke (1952).

A lax, spreading or scandent, much branched shrub 2–5 m high or a weakly erect
1–3-stemmed tree to 6 m tall; bark pale to dark brown or grey, smooth, finely
longitudinally fissured, peeling in filiform strips; indumentum spreading pilose of
simple or multicellular hairs. Older branches glabrescent, scattered lenticellate;
lateral shoots abbreviated, divaricate on the long branches of previous season,
slender, densely pubescent; nodes prominent. Leaves opposite, sometimes 3-
whorled, very variable in size and shape, thin, aromatic, petiolate; blade 1.8–12.5 ×
1.4–11.5 cm, ovate, oblong or obovate to suborbicular, rounded or distinctly
acuminate at the apex, rounded-truncate to somewhat cuneate at the base, entire,
undulate or crenate or irregularly dentate at the upper half, membranous (soft on
drying), light or pale green on both surfaces or cinereous beneath, sparsely to
densely velvety pubescent on both surfaces, very densely so when young, the
indumentum denser on the lower surface mainly on the nerves; petiole 1–6 cm
long, often up to half as long as the blade, slender (sometimes nearly filiform),
densely pubescent. Inflorescences of small open corymbs, mostly terminal on short
lateral shoots, 2–6 × 2.5–8.5 cm, with slender often nearly filiform branches,
pubescent; peduncles 1–3 cm long, pubescent; pedicels up to 2.5 mm long, usually
shorter than the calyx; bracts linear or subulate. Calyx 1.5–2 × 2 mm, cup-shaped,
distinctly to obsoletely 4-lobed, the lobes obtuse, white pubescent, glandular.
Corolla usually white or whitish, smelling when crushed, (3.5)5–6 mm long, sub-2-
lipped; tube c. 3.5 mm long (c. twice as long as the calyx), glabrous outside, lanate-
villous inside near the throat; upper lip entire or emarginate, the lower one 3-lobed,
the lobes oblong, obtuse, the median lobe the largest, 2–2.5 × 1.25–2 mm. Stamens
subequal, exserted, slightly exceeding the corolla lobes; filaments c. 3.5 mm long;
anthers c. 0.4 mm long, purple-brown. Style c. 5 mm long. Fruit dark blue or
blackish-purple when mature, 5–8 × 4–8 mm, subglobose; fruiting calyx c. 5 mm in
diameter, cup-shaped, 4-lobed.

Caprivi Strip. Mpalela Island (Mpilila), 960 m, fr. 13.i.1959, *Killick & Leistner* 3374 (K; M;
SRGH). **Botswana.** N: Chobe Distr., near Serondela Forest Office, fringing forest of Chobe R., 960
m, fl. xi.1951, *O.B. Miller* B/1201 (SRGH). **Zambia.** B: Kaoma (Mankoya), fl. 23.xi.1952, *White* 2124
(BR; K; FHO). N: Mbala Distr., Mpulungu, fl. 20.xi.1950, *Bullock* 3489 (K). W: Ndola Distr., 80 km
south of Ndola, fl. 1.xii.1952, *White* 3817 (FHO; K). C: Lusaka Distr., c. 13 km from Chilanga along
road to Kafue town, c. 1200 m, 4.xi.1972, *Strid* 2449 (LD). E: Petauke Distr., Sasare, road to gold
mine, 700 m, fl. 9.xii.1958, *Robson* 877 (BM; K; LISC). S: Livingstone Distr., Victoria Falls Trust, fl.

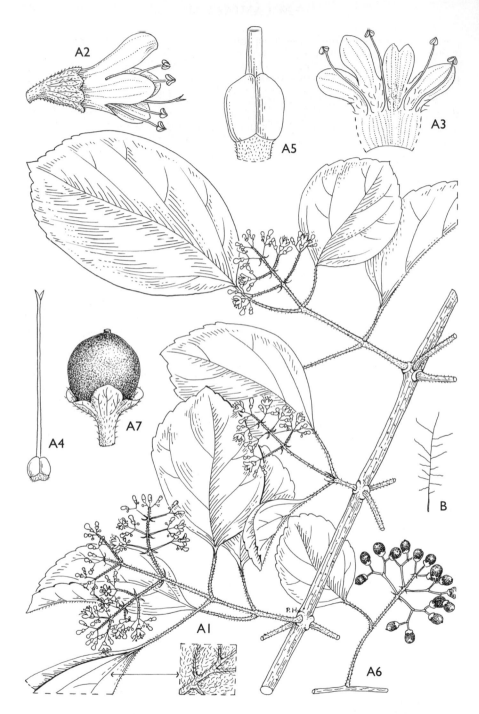

Tab. 11. A. —PREMNA SENENSIS. A1, portion of flowering branch (× ²⁄₃); A2, flower (× 4); A3, corolla, opened out (× 4); A4, ovary and style (× 6); A5, ovary (× 18), A1–A5 from *Bullock* 3489; A6, portion of fruiting branch (× ²⁄₃); A7, fruit (× 4), A6 & A7 from *Fanshawe* 2779. B. —PREMNA SCHLIEBENII, dendritically branched hair, much enlarged, from *Groenendijk, de Koning, Maite & Dungo* 902. Drawn by Pat Halliday.

28.ii.1956, *Gilges* 684A (K; SRGH). **Zimbabwe.** N: Mount Darwin Distr., riverine fringe of Senga R., c. 640 m, fr. 28.i.1960, *Phipps* 2455 (BR; SRGH). W: Hwange Distr., Victoria Falls National Park, Princess Christian Island, fl. 26.ii.1976, *Langman* 18 (E; SRGH). **Malawi.** N: Nkhata Bay Distr., Chikale Beach, c. 510 m, fl. 26.xi.1967, *Pawek* 1570 (SRGH). C: Dedza Distr., Njola's Village, Mua, fl. 15.xii.1969, *Salubeni* 1440 (K; SRGH). S: Machinga Distr., Chikala Hills slopes on the Lake Chilwa Plain, fl. juv. 28.xi.1973, *Banda* 1199 (SRGH). **Mozambique.** N: Muite, Milhana, on way to Catamaze, near Natitima Mt., fl. 13.xii.1967, *Macêdo* 2949 (LMA). T: Cahora Bassa Distr., Chitima (Estima), left bank of Cafuco R., fl. 12.ii.1973, *Macêdo* 5375 (LISC; LMA). MS: Gorongosa Distr., between Zangorga and Kanga-N'Thole, near the road of Gorongosa (Paiva de Andrade Village), fl. 21.xi.1956, *Gomes e Sousa* 4330 (COI; FHO; K; LISC; SRGH).

Also in Kenya, Tanzania, Dem. Rep. Congo and Burundi. Riverine woodland and thickets, lake shore vegetation, and on termite mounds in miombo woodland; 200–1270 m.

5. **Premna velutina** Gürke in Engler, Pflanzenw. Ost-Afrikas C: 338 (1895). —J.G. Baker in F.T.A. **5**: 291 (1900). —Brenan, Check-list For. Trees Shrubs Tang. Terr.: 641 (1949). — Verdcourt in F.T.E.A., Verbenaceae: 74 (1992). Syntypes from Tanzania.

A spreading straggling or scandent shrub up to c. 4 m or a small tree to c. 4 m; indumentum of spreading simple or multicellular hairs. Older branches usually leafless glabrescent with a pale corky epidermis; young branches ± densely yellowish tomentose, usually with the leaves clustered toward the apex and bearing a terminal inflorescence. Leaves opposite, shortly petiolate, very caducous; blades 1.5–8.5 × 1.3–5.7 cm, broadly elliptic or broadly ovate, ovate-oblong to suborbicular, obtuse, rounded or very shortly acuminate at the apex, rounded to subcordate or distinctly cordate at the base, entire or serrate towards the apex or in the upper half with teeth narrowly dentate and apiculate, papery, discolorous the upper surface dark green and slightly shiny, the lower surface yellowish-green or yellowish-cinereous, sparsely shortly pilose above, ± densely tomentose beneath, the main nerves beneath with tawny, antrorse, appressed hairs; petiole 0.2–1.2(2) cm long, usually less than ¼ of the length of the blade, densely yellowish-tomentose. Flowers usually in small terminal paniculate corymbs up to 3.5 × 4.3 cm, up to 5 cm in diameter in infructescences, the axis, branches and pedicels densely yellowish-tomentose; peduncles 0.6–1.5 cm long; pedicels 0.5–0.75 mm long. Calyx green, 1.5–2 mm long, cup-shaped, unequally 5-toothed, somewhat pilose, the teeth obtuse with the largest c. 0.75 mm long and the smaller ones c. 0.5 mm long. Corolla white or yellowish, glabrous outside, c. 3.5 mm long; tube c. 2 mm long, densely hairy at the throat; limb unequally 4-lobed. Stamens with filaments 1.5–2 mm long and transversely oblong, c. 0.5 mm broad; anthers somewhat exserted. Drupes orange to black, c. 3.75 × 3.75 mm, subglobose, narrowing slightly towards the base, sitting in shallowly-lobed venose sparsely pubescent cups 4–6 mm in diameter formed by the fruiting calyx.

Mozambique. N: Macomia Distr., mouth of R. Messalo (M'salu), fl. x.1911, *C.E.F. Allen* 43 (K). Also in Kenya and Tanzania. Coastal hills and riverine vegetation.

6. **Premna serratifolia** L., Mant. Pl. Alt.: 253 (1771). —Munir in J. Adelaide Bot. Gard. **7**: 13 (1984). —Nicholson et al., Interpr. van Rheede's Hort. Malab.: 262 (1988). —Verdcourt in F.T.E.A., Verbenaceae: 73 (1992)*. Type Herb. Linnaeus 7824 (LINN, lectotype).
   Cornutia corymbosa Burm.f., Fl. Ind.: 132, t. 141 fig. 1 (1768), non *Premna corymbosa* Rottler & Willd. (1803). Type from Sri Lanka.
   Premna integrifolia L., Mant. Pl. Alt.: 252 (1771). —C.B. Clarke in J.D. Hooker, Fl. Brit. India **4**: 574 (1885). —Briquet in Engler & Prantl, Nat. Pflanzenfam. **4**, 3a: 170 (1897). — Dop in Fl. Gén. Indo-Chine **4**: 818 (1935), *nom illegit. (superfl.).* Type as for *Cornutia corymbosa* Burm.f.
   Premna corymbosa Rottl. & Willd. in Ges. Naturf. Fr. Berlin, Neue Schrift **4**: 187 (1803). —Meeuse in Blumea **5**: 72 (1942). Type from India.
   Premna obtusifolia R. Br., Prodr. Fl. Nov. Holl. **1**: 512 (1810). —Bentham, Fl. Austr. **5**: 58 (1870). —Fosberg in Taxon **2**: 88–89 (1953). —Fosberg & Renvoize, Fl. Aldabra: 224, fig. 35, 6 (1980). —H. Moldenke & A. Moldenke in Rev. Fl. Ceylon **4**: 334 (1983). —R. Fernandes in Bol. Soc. Brot., sér. 2, **63**: 299 (1990). Types from Australia.

---

* For further bibliography and synonymy see: Meeuse (loc. cit.), Moldenke (op. cit.: 605–610, 1971), Moldenke & A. Moldenke (op. cit.: 335–337, 1983) and Nicholson et al. (loc. cit.).

Premna corymbosa sensu Brenan, Check-list For. Trees Shrubs Tang. Terr.: 640 (1949) non *P. corymbosa* Rottl. & Willd. (1803).
Premna obtusifolia var. *serratifolia* (L.) Moldenke in Phytologia **28**: 403 (1974), *comb. illegit.*
Premna obtusifolia forma *serratifolia* (L.) Moldenke in Phytologia **36**: 438 (1977). —H. Moldenke & A. Moldenke in Rev. Fl. Ceylon **4**: 342 (1983), *comb. illegit.*
Premna integrifolia var. *obtusifolia* (R. Br.) C. P'ei in Mém., Sc. Soc. China **3**: 75 (?), *comb. illegit.*

A much branched bushy shrub or tree up to 8 m tall, sometimes with sarmentose branches; indumentum of spreading simple or multicellular hairs. Branches striate, glabrescent; branchlets densely short-pubescent or appressed pilose; leaf-scars prominent. Leaves opposite; blades 2.5–13.5(15) × 2–9 cm or more, broadly elliptic, elliptic-oblong or obovate-elliptic to suborbicular, rounded obtuse to subacute or shortly acuminate at the apex, cuneate or rounded to subcordate at the base, entire or crenate or subserrate to coarsely serrate above the middle, chartaceous to firmly membranaceous, dark green on both surfaces, glabrous or appressed pilose on the midrib and the main lateral nerves, or some hairs at the base of the midrib and in the lower nerve axils beneath; lateral nerves 3–5 per side, slender, ± prominent above, prominent beneath; petiole 0.5–3 cm long, usually slender, canaliculate, pubescent or puberulent, rarely hispid. Inflorescences terminal, corymbose, 3–8 × 4–10(14) cm or larger; peduncles 2–2.5 cm long, appressed-pilose; bracts 1–3 mm long, lanceolate or linear-lanceolate. Flowers unpleasantly aromatic. Calyx 2.6 × 2.8 mm, campanulate, 2-lipped, with the upper lip entire or 2-lobed and the lower lip entire or shallowly 2–3-dentate, slightly puberulous and glandular on the outside. Corolla pale yellowish, whitish-green or greenish, 2-lipped; tube c. 3.5 mm long, cylindrical or ± funnel-shaped, pubescent outside and at the throat; limb upper lip 1.5 × 1.4 mm, entire or 2-lobed, lower lip 3-lobed, with the median lobe c. 1.5 × 2 mm and slightly concave, the lateral lobes a little smaller, glabrous or appressed puberulous outside. Stamens didynamous, slightly exserted; filaments c. 1.8 mm long, filiform, wider and pilose at the base; anthers c. 0.6 mm long, ovoid. Style c. 4.8 mm long. Ovary incompletely 4-locular. Fruit black, c. 4 × 4 mm, subglobose, sitting in a shallowly cupuliform accrescent fruiting calyx 3–3.5 mm in diameter.

Mozambique. N: Mogincual, 9.xi.1936, *Torre* 976 (LISC). Z: Namacurra Distr., Macuze (Macuse) R., 23.viii.1962, *Wild & Pedro* 5884 (K; LISC; SRGH). MS: Búzi Distr., Nova Sofala, 21.vi.1961, *Leach & Wild* 11120 (K; LISC; SRGH).
Also in Tanzania, Aldabra Islands, Madagascar, Mauritius and Seychelles, Sri Lanka, India, Indo-China, S China, Philippines, Australia and Islands of Oceania. On sand dunes and in coastal bush, also in bush at the margins of mangrove swamps; sea level–30 m.
The specimen *Kirk* s.n. (K), collected in March 1861 from Rovuma River, [cited by Baker (loc. cit.) as collected in German East Africa], is a more robust plant differing from other material from Mozambique in the bristle-like indumentum on the leaf blades as well as the nerves, in the absence of the characteristic tufts of soft hairs in the nerve axils, and in the stouter petioles.

7. **Premna sp.**
Premna hildebrandtii sensu Moldenke, Fifth Summ. Verbenaceae: 252 (1971), non Gürke (1895).
Premna sp. A —R. Fernandes in Bol. Soc. Brot., sér. 2, **63**: 301 (1990).

A shrub 3–4 m tall; branches with pronounced nodes; flowers apparently borne on leafless branches. Leaves (of which only 3 individual fallen leaves seen) 6.8–10 × 4–6.9 cm, elliptic-oblong, shortly acuminate at the apex, rounded-truncate at the base, entire or only slightly undulate-crenate toward the apex, uniformly hairy on both surfaces, with hairs straight and ± spreading, those on the nerves appressed antrorse, the blades sub-chartaceous and darkly concolorous on drying; petiole 2.6–3.5 cm long, slender, puberulous. Corymbs terminal, somewhat dense, up to 4.5 × 6 cm, the inflorescence branches, pedicels and peduncle densely shortly sub-appressed hairy; peduncle 1–2 cm long; bracts linear-subulate, the lower ones 5–7.5 mm long, the upper shorter. Flowers not opened; calyx puberulous.

Mozambique. N: Palma Distr., between Nangade and Palma, at 17 km from Palma, 20.x.1942, *Mendonça* 1004 (LISC).
Known only from Mozambique. In open deciduous forest; c. 60 m.
Moldenke, in 1966, determined *Mendonça* 1004 as *P. hildebrandtii* Gürke, in an annotation on

the specimen label. However, this material differs from *P. hildebrandtii* in its sparser indumentum (not velvety tomentose on the leaves as in *P. hildebrandtii*), with further differences seen in the leaf shape, the relatively shorter petioles and the smaller inflorescences, the latter corymbose and not paniculate as in *P. hildebrandtii*. Moldenke (loc. cit.), perhaps on the basis of *Mendonça* 1004, cited *P. hildebrandtii* as occurring in Mozambique (Cabo Delgado), but as yet no material of this species is known from Mozambique. He also recorded *P. hildebrandtii* for Zimbabwe (op. cit.: 248), but all Zimbabwean material seen of the genus *Premna* belongs to *P. senensis* Klotzsch. *Mendonça* 1004 approaches *P. serratifolia* and *P. velutina* in the shape and colour of the leaves and in the similarity of the corymbs, but differs from both in the leaf indumentum and the relatively longer petioles.

2. VITEX L.

By F. Sales

Vitex L., Sp. Pl.: 638 ['938'] (1753); Gen. Pl. ed. 5: 285 (1754). —Pieper in Bot. Jahrb. Syst. **62**, Beibl. 141 ['142']: 1–91 (1928). —Moldenke in Phytologia **5** & **6** (1955–1958)*; Rev. Fl. Ceylon, ed. Dassanayake **4**: 348 (1983). —Verdcourt in F.T.E.A., Verbenaceae: 50–68 (1992). —Bredenkamp & Botha in Bothalia **26**: 141–151 (1996) [with full citations of the Moldenke series of papers on *Vitex*]. — Sales in Kew Bull. **56**: 189–207 (2001). —Harley et al. in Kadereit (ed.), Fam. Gen. Vasc. Pl. (Kubitzki, ed. in chief) **VII**: 195 (2004).

Shrubs or trees, more rarely scrambling lianes, deciduous or evergreen, bark smooth or vertically fissured. Leaves opposite, rarely in whorls of 3, digitately compound with (1)3–7 leaflets, usually petiolate; leaflets entire or variously dentate or crenate, often with yellow gland-dots and aromatic; petiolules present or absent. Inflorescences terminal or axillary, mostly of dichasial cymes, occasionally of cymes aggregated into lax panicles, slender or sturdy; bracteoles inconspicuous, linear, narrowly-lanceolate or narrowly-spathulate. Calyx hemispherical, campanulate or obconical, 5-lobed (4-lobed in *V. schliebenii*), toothed or truncate, persistent and ± accrescent in fruit. Corolla ± zygomorphic (regular in *V. schliebenii*); tube usually creamy-white, straight or curved, shortly cylindrical to infundibuliform; limb blue lilac mauve or pale yellow (*V. chirindensis*), upper lip 2-lobed, lower lip 3-lobed, (limb 4-lobed in *V. schliebenii*), spreading, the lower-lip median lobe usually larger than the laterals and sometimes more strongly coloured. Stamens 4, didynamous, epipetalous, included or exserted from the corolla tube; filaments broadened at the base, glabrous or with glandular or eglandular hairs. Ovary initially c. 2-locular, later 4-locular, with a single ovule in each locule, gland-dotted or not, glabrous or apically hairy; style terminal, filiform, bifid, shortly or distinctly exserted (*V. schliebenii*). Fruit a drupe or a nut with a hard 4-locular endocarp; fruiting calyx firm and ± cupular to saucer-shaped below the drupe, or enlarged chartaceous campanulate about the nut. Seeds 1–4, oblong to obovoid, without endosperm.

A genus of more than 200 species, mainly in the Old and New World tropics and a few in temperate areas. Earlier authors, such as Pieper (Africa) and Moldenke (global), recognized many more species than are generally accepted today.

Pieper treated the African species as belonging to two subgenera, namely the subgenera *Vitex* and *Holmskioldiopsis*, and such infrageneric classification is retained here. Both subgenera are represented in the Flora Zambesiaca area with *species 1–13* belonging to subgenus *Vitex* and *species 14* and *15* to subgenus *Holmskioldiopsis* W. Piep. *Species 1* (and *V. zanzibarensis* and *V. trifolia*) belong to section *Vitex* [*Terminales* Briq.], and *species 2–13* to section *Axillares* Briq. of the subgenus *Vitex*.

Herbarium collections of the widespread, and usually polymorphic, African species (e.g. *V. doniana*, *V. ferruginea*, *V. fischeri*, *V. madiensis*, *V. mombassae*, *V. payos*) are abundant. In contrast the southern African taxa (*V. zeyheri*, *V. obovata*, *V. harveyana*) are less well represented and need

---

* Moldenke published numerous papers towards a revision of *Vitex* presenting much information, mostly compiled but also taken from a worldwide survey. Details of the papers are: Phytologia **5**: 142–174, 186–224, 257–280 & 293–336 (1955); **5**: 465–507, **6**: 13–64 & 70–128 (1957); **6**: 129–192 & 197–231 (1958). These papers are selectively cited in the present account, omitting repetition of most of the compilations and the unhelpful confused misidentifications.

further research to fully understand their variation. Often missing from field notes are details of leaf, flower and fruit colour — which changes at maturity and on drying. It is also very useful to have a description of the bark and general observations on the habit of the plant.

Cultivated taxa

Shrubs which have been introduced into the Flora Zambesiaca region as garden ornamentals include:

Vitex agnus-castus L., Sp. Pl.: 638 (1753). —Biegel, Check-list Ornam. Pl. Rhod. Parks & Gard. [Rhodesia Agric. J., Research Report No. 3]: 107 (1977). —Verdcourt in F.T.E.A., Verbenaceae: 50 (1992). —Sales in Kew Bull. **56**: 204 (2001). Type from Italy.

Deciduous shrub. Leaflets 5–7, aromatic, narrowly-lanceolate, petiolulate, glabrous above and white-tomentose beneath with tomentum obscuring the whole lower surface. Inflorescences terminal, spiciform, comprised of many-flowered cymes in narrow panicles, usually with additional spiciform inflorescences from the axils of the upper leaves. Flowers small, pale violet; corolla limb lower lip glabrous.

Although reported by Biegel (loc. cit) to be cultivated as a garden ornamental no specimens have been seen from the Flora Zambesiaca area.

Vitex trifolia L. var. **bicolor** (Willd.) Moldenke, Known Geogr. Distrib. Verbenaceae, ed. 2: 79 (1942). —Moldenke in Rev. Fl. Ceylon, ed. Dassanayake **4**: 386 (1983). —Verdcourt in F.T.E.A., Verbenaceae: 52 (1992).

 Vitex trifolia L., Sp. Pl.: 638 (1753). —Ross, Fl. Natal: 300 (1972). Type from India.
 Vitex negundo L., Sp. Pl.: 638 (1753) auct. —Sim, For. Fl. Port. E. Africa: 94 (1909).
 Vitex bicolor Willd., Enum. Pl.: 660 (1809). Type from eastern India. [=*V. negundo*]

Small deciduous shrub. Leaflets 1–5, lanceolate, discolorous, glabrous above and white-tomentose beneath with the tomentum obscuring the whole lower surface. Inflorescences terminal and axillary in upper leaf axils, comprised of much branched panicles of many-flowered cymes. Flowers small, blue, lavender or purple, 3–5 mm in diameter; corolla limb lower lip with a semicircular line of dense hairs at the base near the throat; drupe c. 5 mm long and wide.

Mozambique. N: Cabo Delgado, seaside vegetation, *Groenendijk & Dungo* 578. M: Maputo (Lourenço Marques), Jardim Tunduru (Jardim Vasco da Gama), *Balsinhas* 2260.

The taxonomy and nomenclature of *V. trifolia* s.l. is complex. A discussion of its synonymy and distribution has been provided by Moldenke (loc. cit., 1983). Here, I have followed Verdcourt (loc. cit., 1992) in calling our taxon *V. trifolia* var. *bicolor*, although in much African literature the name *V. negundo* has been used. Verdcourt considered *V. trifolia* var. *bicolor* to be native in East Africa, but in the Flora Zambesiaca area I have only seen two apparently cultivated/naturalized specimens from Mozambique (*Groenendijk & Dungo* 578; *Balsinhas* 2260). The main area of distribution of *V. trifolia*/*V. negundo* is the Indian subcontinent and tropical and subtropical SE Asia.

1. Inflorescences terminal, compound panicles, usually together with additional axillary ones from upper axils; young stems clearly 4-ridged or flattened* · 2
– Inflorescences all axillary, simple or compound dichasia, if simple panicles then calyces and corollas 4-lobed (*V. schliebenii*), if apparently terminal then very short; young stems ± terete · 3
2. Fruit completely covered in a bright yellow dense glandular coat; young stems clearly 4-ridged, square in cross-section; leaves ± scabridulous above; petioles not winged · 1. *buchananii*
– Fruit not covered in a yellow layer; young stems and twigs laterally compressed; leaves smooth; petioles often winged · V. *zanzibarensis***
3. Calyx distinctly broadly lobed; fruiting calyx campanulate with spreading lobes; fruit a nut, up to c. 3 × 5 mm, ± contained within the enlarged persistent calyx tube · · · · · · · · · · · 4
– Calyx shallowly lobed; fruiting calyx cupular or saucer-shaped; fruit a drupe, up to 4 × 3.5 cm but usually smaller, sitting on the enlarged calyx · 5
4. Leaflets greyish-green with a white velvety indumentum, elliptic to narrowly obovate (Botswana) · 14. *zeyheri*

---

* *V. trifolia* and *V. agnus-castus*, cultivated as ornamentals in our area, key out here because of their terminal inflorescences and are distinguished by their markedly discolorous leaves.
* *Although Verdcourt (F.T.E.A. Verbenaceae: 54 (1992)) lists "probably in Mozambique" in the external range of this Kenyan–Tanzanian species, I have seen no specimens from Mozambique.

– Leaflets dark green, glabrous or white-pubescent, lanceolate to obovate (Mozambique) · ·
· 15. *obovata*

5. Inflorescences lax with slender peduncles, few-flowered; median petiolules ± distinct · · 6
– Inflorescences ± compact with short or long sturdy peduncles, many-flowered; median
petiolules distinct or absent · 7

6. Leaflets 3; calyx pubescent, gland-dotted · 4. *petersiana*
– Leaflets 3–5; calyx almost glabrous, not gland-dotted · · · · · · · · · · · · · · 5. *mossambicensis*

7. Inflorescences very short, axillary or sometimes appearing terminal, 1–5 cm long on
peduncles 1–1.4 cm long; flowers actinomorphic, c. 5 mm long, corollas 4-lobed; fruit c. 6
× 5 mm; leaflets scabridulous on upper surface, the lateral ones usually subsessile · · · · · ·
· 2. *schliebenii*

– Inflorescences usually much longer or if not then peduncles longer; flowers zygomorphic,
6–23 mm long, corollas 5-lobed; leaflets glabrous, pubescent or scabridulous on upper
surface, sessile or petiolulate · 8

8. Leaflet upper surface glabrous, shiny and dark green above drying dark olive-green or
brown; twigs and young stems ± ridged; median leaflet 1.5–5 cm long (Mozambique) · · ·
· 3. *harveyana*

– Leaflet upper surface hairy or glabrous, matt, drying yellow-brown; twigs and young stems
terete or ± ribbed; median leaflets longer (all areas) · 9

9. Leaflets sessile or subsessile · 10
– Leaflets with short or long petiolules · 11

10. Flowers less than 9 mm long, corolla tube 3–4 mm long; bracteoles linear; stamens included
in corolla tube; fruit c. 2–3 × 2 cm · 11. *payos*
– Flowers larger, corolla tube c. 10 mm long; bracteoles narrowly oblong-oblanceolate;
stamens and style exserted; fruit c. 4 × 3 cm · 12. *mombassae*

11. Leaflets glabrous, sometimes puberulous on midrib beneath, coriaceous; bracteoles
caducous; median leaflet 7–23 cm long · 10. *doniana*
– Leaflets hairy at least beneath, sometimes glabrous, not coriaceous; bracteoles persistent;
median leaflets smaller · 12

12. Leaflets narrowly to broadly lanceolate, usually long-acuminate; ovary lanate at apex, not
gland-dotted · 7. *ferruginea*
– Leaflets elliptic to obovate, rounded to acuminate at the apex; ovary glabrous, often gland-
dotted · 13

13. Large tree of evergreen rain-forest; leaves drying dark; median petiolule 2–3 cm long · · ·
· 6. *amaniensis*
– Shrubs or trees of woodlands and wooded grasslands or coastal forest; leaves drying pale
green; median petiolule less than 2 cm long · 14

14. Leaflets 3 rarely 5, upper surfaces scabridulous; fruit c. 1 × 0.8 cm · · · · · · · · · · 13. *patula*
– Leaflets 3–5(7), upper surfaces glabrous or ± pubescent; fruit 1.3–4 × 0.7–3 cm · · · · · · 15

15. Leaflet nerves numerous and ± closely spaced; flowers 8–10 mm long; stamens only slightly
exserted; fruit c. 1.3–0.7 cm; calycine cup not gland-dotted · · · · · · · · · · · · · · 8. *fischeri*
– Leaflet nerves fewer, not as closely spaced; flowers 8–20 mm long; stamens scarcely or
clearly exserted; fruit 2–4 × 1.5–3 cm; calycine cup gland-dotted or not · · · · · · · · · · · · 16

16. Leaflets mostly widest about the middle and rounded at the base, reticulation usually ±
sunken on upper surface; bracteoles narrowly oblong-oblanceolate; flowers up to c. 23 mm
long with stamens clearly exserted from corolla tube; fruit c. 4 × 3 cm · · · · 12. *mombassae*
– Leaflets mostly widest above the middle and cuneate at the base, reticulation on upper
surface not sunken; bracteoles linear; flowers 8–12 mm long with stamens scarcely exserted;
fruit c. 2 × 1.5 cm · 9. *madiensis*

Subgenus **Vitex**

Calyx shallowly 5-lobed; fruiting calyx cupular or saucer-shaped. Fruit a drupe, up
to 4 cm long and 3.5 cm wide but usually smaller, ellipsoid obovoid or globose, sitting
on the enlarged calyx, sometimes edible.

1. **Vitex buchananii** Baker ex Gürke in Engler, Pflanzenw. Ost-Afrikas **C**: 339 1895). —J.G.
Baker in F.T.A. **5**: 319 (1900). —Pieper in Bot. Jahrb. Syst. **62**, Beibl. 141: 53 (1928). —O.B.
Miller in J. S. African Bot. **18**: 76 (1952). —Garcia in Estud. Ensaios Doc. Junta Invest. Ci.
Ultramar [in Mendonça, Contrib. Conhec. Fl. Moçamb. II]: 170 (1954). —White, For. Fl.
N. Rhod.: 371 (1962). —Drummond in Kirkia **10**: 272 (1975). —K. Coates Palgrave, Trees

South. Africa: 807 (1977). —Verdcourt in F.T.E.A., Verbenaceae: 54, 55 (1992). —Sales in Kew Bull. **56**: 191 (2001). —M. Coates Palgrave, ed. 3 of K. Coates Palgrave, Trees South. Africa: 981 (2002). Type: Malawi, *Buchanan* 782 (B†, holotype; K; PRE).

    Vitex volkensii Gürke in Engler, Pflanzenw. Ost-Afrikas **C**: 339 (1895). —J.G. Baker in F.T.A. **5**: 318 (1900). —Brenan, Check-list For. Trees Shrubs Tang. Terr.: 642 (1949). Type from Tanzania.

    Vitex quadrangula Gürke in Bot. Jahrb. Syst. **28**: 463 (1900). Type from Tanzania.

    Vitex radula W. Piep. in Bot. Jahrb. Syst. **62**, Beibl. 141: 42, 55 (1928) (in key); Repert. Spec. Nov. Regni Veg. **26**: 161 (1929). —Brenan, Check-list For. Trees Shrubs Tang. Terr.: 643 (1949). —Garcia in Estud. Ensaios Doc. Junta Invest. Ci. Ultramar [in Mendonça, Contrib. Conhec. Fl. Moçamb. II]: 170 (1954). Lectotype from Tanzania.

    Vitex buchananii Gürke var. *quadrangula* (Gürke) W. Piep. in Bot. Jahrb. Syst. **62**, Beibl. 141: 54 (1928). —Brenan, Check-list For. Trees Shrubs Tang. Terr.: 642 (1949).

    Vitex robynsi De Wild., Pl. Bequaert. **5**: 13 (1929). Type from Dem. Rep. Congo.

    Vitex thyrsiflora sensu Sim, For. Fl. Port. E. Africa: 94 (1909). —sensu Moldenke in Phytologia **6**: 153 (1958) non Baker in Bull. Misc. Inform., Kew **1895**: 152 (1895).

Shrub 2–4.5 m tall, sometimes scandent or frequently a strong liane, occasionally a small tree to 6 m tall; bark fissured, dark grey-brown; older stems grey-brown tomentose; young stems conspicuously 4-angled, pale yellow tomentose. Leaves aromatic when crushed, (3)5-foliolate; petioles 4–14 cm long, terete; median petiolules 1–2.5 cm long; leaflets 5–16 × 2.5–7 cm, median leaflet largest, oblong-elliptic to obovate, long-acuminate at the apex, ± asymmetric and cuneate at the base, entire to occasionally serrate, ± discolorous, scabrid above with short hairs or glabrescent, sparsely to densely pubescent to subtomentose beneath and densely yellow gland-dotted, nerves ± sunken above and prominent beneath. Inflorescence terminal, paniculate, (4)9–35 cm long, white-tomentose; bracteoles 1–5 mm long, linear. Calyx obconical, shallowly 5-toothed, gland-dotted, pubescent sometimes lanate, strongly accrescent in fruit; tube 1.5–2 mm long; lobes 0.4–0.8 mm long, erect. Corolla white or mauve; tube 4–5 mm long, straight, glandular and pubescent outside; upper lip 2-lobed with lobes at an obtuse angle to tube; lower lip 3-lobed, 0.7–2 × 1.4–1.8 mm. Stamens scarcely exserted from tube; filaments glabrous. Ovary obovoid, gland-dotted at apex, glabrous; apex almost truncate; style 0.4–0.5 mm long. Fruit a globose drupe, c. 6–8 mm in diameter, enveloped in a distinctive bright orange-yellow papery flaking layer of glands, the fruit whitish beneath, the lower c. $^1/_3$ enclosed by the enlarged calyx; fruiting calyx cup-shaped, pubescent, gland-dotted.

    Zambia. N: Mbala Distr., Kalambo Falls, 15.ii.1964, *Richards* 19024 (K). **Zimbabwe**. E: Chipinge Distr., 4 km north of confluence of Musirizwi and Bwazi Rivers, 600 m, 28.ii.1975, *Gibbs Russell* 2613 (K). **Malawi**. N: Nkhata Bay Distr., Kabunduli road, 22.ii.1976, *Pawek* 10871 (K; MO). C: Dedza Distr., Mua-Livulezi Forest Reserve, 18.v.1960, *Adlard* 369 (K; SRGH). S: Mulanje Mt., foot of Chambe Peak, 1250 m, 16.ii.1986, *J.D. & E.G. Chapman* 7056 (E; MO). **Mozambique**. N: Nampula, 4.ii.1937, *Torre* 1268 (COI; K; LISC; LMA). Z: Maganja da Costa, 27.ix.1949, *Barbosa & Carvalho* 4224 (K; LMA). MS: Búzi Distr., Boca, Baixo Búzi River, 30 m, 20.xii.1906, *Swynnerton* 1053 (BM; K).

    Also in Tanzania and Dem. Rep. Congo. Forest margins in grassland with *Cussonia*, in riverine vegetation, wooded grassland and mixed deciduous woodlands with *Bauhinia*, *Brachystegia* and *Combretum*; 10–1800 m.

    Readily identified by the small spherical fruits covered with a characteristic bright orange-yellow papery glandular coat, but white beneath it. The scabrid adaxial surface of the leaflets and the abundant yellow oil-dots, especially abaxially, are also characteristic and apparent even on seedlings. Yellow oil-dots are also present on calyx and corolla.

2. **Vitex schliebenii** Moldenke in Phytologia **7**: 85 (1959). —Verdcourt in F.T.E.A., Verbenaceae: 57 (1992). —Sales in Kew Bull. **56**: 202 (2001). —M. Coates Palgrave, ed. 3 of K. Coates Palgrave, Trees South. Africa: 987 (2002). Type from Tanzania.

    Vitex amboniensis sensu Gomes e Sousa, Pl. Menyharth.: 32 (1936).

    Vitex sp. no. 2 of Drummond in Kirkia **10**: 272 (1975).

    Vitex sp. no. 2 of K. Coates Palgrave, Trees South. Africa: 813 (1977).

Straggling shrub to 3 m or tree to 12 m tall with upper branches scandent; bark fissured, grey-brown; older stems grey-brown tomentose; young stems 4-angled toward the apex, pale yellow tomentose. Leaves (3)5 foliolate; petioles 4–12 cm long, terete; median petiolules 0.3–3 cm long; leaflets 4–14 × 2.5–5.5 cm, median leaflet largest, elliptic to oblong-elliptic or obovate, acuminate or occasionally rounded at the apex, cuneate at the base, entire to occasionally dentate toward the apex; upper surface ± scabridulous with reticulation slightly sunken; lower surface paler, pubescent with hairs longer on the nerves, densely gland-dotted. Inflorescences axillary or sometimes appearing terminal, of short slender few-flowered panicles, 1–5 cm long, creamy tomentose; peduncles spreading, sometimes horizontally, 1–1.4 cm long, pedicels filiform, 1.5–6 mm long; bracteoles 1–5 mm long, linear. Flowers actinomorphic, c. 5 mm long, aromatic. Calyx campanulate, 4-toothed, gland-dotted, pubescent, accrescent in fruit; tube c. 1.5 mm long; lobes 0.3 mm long, erect. Corolla 4-lobed, ± actinomorphic, 3–5 mm long, pubescent and gland-dotted outside, hairy in the throat; tube white or mauve, 2.5 mm long, straight; lobes yellow or mauve, spreading to reflexed, the 2 lower lobes 2 × 1.5–2 mm, the 2 upper lobes somewhat smaller. Stamens clearly exserted from corolla tube; filaments glabrous. Ovary globose, gland-dotted at rounded apex, glabrous; style c. 2 mm long. Drupe 0.5–0.7 × 0.4–0.6 cm, obovoid-spherical, black at maturity; fruiting calyx thinly saucer-shaped, pubescent, gland-dotted.

Zimbabwe. W: Tsholotsho Distr., Gwayi River (Gwaai), 1933, *Eyles* 7562 (K). E: Chipinge Distr., Mwangazi (Mangaze), 610 m, ii.1966, *Goldsmith* 19/67 (BR; K; LISC; SRGH). **Malawi**. N: Chitipa Distr., Mafinga Mts., 1675 m, 3.iii.1982, *Brummitt, Polhill & Banda* 16282 (K). S: Mangochi Distr., 2 km SW of Monkey Bay, 530 m, 20.ii.1982, *Brummitt, Polhill & Patel* 16005 (K). **Mozambique**. T: Changara Distr., Boroma, ii.1891, *Menyharth* 762 (K; Z). MS: Cahora Bassa Distr., Tete to Chicoa, 25.vi.1949, *Barbosa & Carvalho* 3278 (K; LMA).

Also in Kenya and Tanzania. Mixed deciduous woodlands with *Combretum, Commiphora, Colophospermum mopane* and *Terminalia spp.*, at low altitudes, and in riverine thickets and woodland, often on termite mounds, on alluvium and sandy soils; 200–1500 m.

V. schliebenii is the only species in the Flora Zambesiaca region in which the calyx is 4-toothed, the corolla 4-lobed and the style distinctly bifid.

3. **Vitex harveyana** H. Pearson in F.C. **5**: 212 (1901). —Pieper in Bot. Jahrb. Syst. **62**, Beibl. 141: 56 (1928). —Garcia in Estud. Ensaios Doc. Junta Invest. Ci. Ultramar [in Mendonça, Contrib. Conhec. Fl. Moçamb. II]: 171 (1954). —Moldenke in Phytologia **8**: 40 (1961). — Ross, Fl. Natal: 300 (1972). —K. Coates Palgrave, Trees South. Africa: 808 (1977). — Bredenkamp & Botha in S. African J. Bot. **59**: 613 (1993); in Bothalia **26**: 142 (1996). — Sales in Kew Bull. **56**: 195 (2001). —M. Coates Palgrave, ed. 3 of K. Coates Palgrave, Trees South. Africa: 982 (2002). Lectotype from South Africa (chosen by Bredenkamp & Botha, 1993).

Vitex schlechteri Gürke in Bot. Jahrb. Syst. **33**: 299 (1904). —Pieper in Bot. Jahrb. Syst. **62**, Beibl. 141: 56 (1928). Type from South Africa.

Straggling shrub or small tree 2–4(7) m tall; bark finely fissured, dark grey-brown; older stems red-brown tomentose; young stems yellow-brown tomentose, angular, often ribbed. Leaves 3(5)-foliolate; petioles 0.5–3 cm long, grooved; median petiolules 0.1–0.9 cm long; leaflets 1.5–5 × 1–2.4 cm, median leaflet largest, elliptic to narrowly obovate, ± rounded or bluntly acuminate at the apex, cuneate at the base, entire or widely toothed or crenate toward the apex, somewhat coriaceous, shiny dark green and glabrous on upper surface, sparsely pubescent beneath, nerves sunken above with reticulation ± raised beneath, often with tufts of hairs in the axils of the lateral nerves beneath. Inflorescences axillary, of ± slender simple few-flowered dichasia 3.5–6 cm long, with scattered yellow hairs; bracteoles linear. Calyx obconical, 5-toothed, gland-dotted, pubescent, somewhat accrescent in fruit; tube 4–4.5 mm long. Corolla 16–18 mm long, usually gland-dotted, pubescent, glandular hairs present, sweet-scented; tube creamy-white, 7–10 mm long, straight; lobes bright mauve or violet, spreading to reflexed, middle lower lobe largest, 8–12 × 9–12 mm. Stamens clearly exserted from tube; filaments with glandular hairs. Ovary globose, truncate apex, not gland-dotted, pubescent at apex; style c. 15 mm long. Drupe (5)10 × 7–10 mm, obovoid, rounded at apex, not enclosed in the enlarged calyx, black at maturity; fruiting calyx cup-shaped, pubescent, not gland-dotted.

Mozambique. M: Namaacha Distr., Goba, 22.xii.1944, *Mendonça* 3440 (LISC; LMA; LMU). Also in South Africa (Mpumalanga, KwaZulu-Natal, Eastern Cape) and Swaziland. Locally frequent, usually in thickets, on rocky outcrops and in seasonally dry drainage lines; 40–380 m. This species may be distinguished by its rather coriaceous leaflets, glabrous above and entire or widely dentate on the margins, and by its large (16–18 mm long) fragrant flowers and clearly exserted stamens.

4. **Vitex petersiana** Klotzsch in Peters, Naturw. Reise Mossambique: 264 (1861). —Gürke in Engler, Pflanzenw. Ost-Afrikas C: 340 (1895). —J.G. Baker in F.T.A. **5**: 320 (1900). —Sim, For. Fl. Port. E. Africa: 94 (1909). —Eyles in Trans. Roy. Soc. S. Africa **5**: 459 (1916). — Pieper in Bot. Jahrb. Syst. **62**, Beibl. 141: 57 (1928). —Garcia in Estud. Ensaios Doc. Junta Invest. Ci. Ultramar [in Mendonça, Contrib. Conhec. Fl. Moçamb. II]: 171 (1954). — White, For. Fl. N. Rhod.: 371 (1962). —Drummond in Kirkia **10**: 272 (1975). —Sales in Kew Bull. **56**: 201 (2001). —M. Coates Palgrave, ed. 3 of K. Coates Palgrave, Trees South. Africa: 985 (2002). TAB. **12**. Type: Mozambique, Tete, *Peters* (B†, holotype); Tete at km 25 on road to Chicoa, *Torre & Correia* 13836 (LISC, neotype chosen by Sales, 2001).

   Vitex tettensis Klotzsch in Peters, Naturw. Reise Mossambique: 264 (1861). —Gürke in Engler, Pflanzenw. Ost-Afrikas C: 340 (1895). —J.G. Baker in F.T.A. **5**: 320 (1900). —Sim, For. Fl. Port. E. Africa: 94 (1909). —Moldenke in Phytologia **5**: 314 (1955). Type: Mozambique, Tete, *Peters* (B†, holotype).

   ?*Vitex dentata* Klotzch in Peters, Naturw. Reise Mossambique: 265 (1861). —Sim, For. Fl. Port. E. Africa: 94 (1909). Type: Mozambique, Tete, *Peters* (B†, holotype)*.

   Vitex kirkii Baker in F.T.A. **5**: 321 (1900). —Sim, For. Fl. Port. E. Africa: 94 (1909). —Eyles in Trans. Roy. Soc. S. Africa **5**: 459 (1916). —Gomes e Sousa, Pl. Menyharth.: 32 (1936). Type: Mozambique, Tete, (Livingstone's Zambezi Expedition), xi.1856, *Kirk s.n.* (K).

   Vitex petersiana var. *tettensis* (Klotzsch) W. Piep. in Bot. Jahrb. Syst. **62**, Beibl. 141: 58 (1928).

Spreading shrub, often much branched from ground level, 1–8 m tall; bark smooth or fissured, flaking, dark or light grey-brown; older stems grey-brown pubescent and young stems orange-tawny tomentellous, shortly tomentose on buds. Leaves 3-foliolate; petioles 0.5–4 cm long, terete; median petiolules longest, 0.1–1 cm long, lateral petiolules short or absent; leaflets 1.3–5 × 1–3 cm, median leaflet largest, ovate-lanceolate to broad elliptic, rounded to acute at apex, rounded at the base, subentire to irregularly deeply serrate toward the apex, somewhat membranous, glabrous or with a few hairs mainly on the midrib on upper surface, puberulous especially on the midrib and paler beneath, nerves sunken above and ± raised beneath, reticulation delicate. Inflorescences axillary, of ± slender simple 1–7-flowered dichasia 3–5 cm long, with orange-tawny hairs; bracteoles linear, up to 1.5 mm long; pedicels to 10 mm long. Calyx obconical, 5-toothed, pubescent, gland-dotted, accrescent in fruit; tube c. 2 mm long; teeth 0.3–0.5 mm long, erect. Corolla c. 14 mm long, gland-dotted, pubescent, with glandular hairs, flowers sweet-scented; tube white, 5–6.5 mm long, straight; lobes mauve to violet, spreading to reflexed, middle lower lobe largest, 7–8 × 5 mm. Stamens clearly exserted from tube; filaments with scattered glandular hairs. Ovary globose, with a few apical gland-dots, glabrous, rounded at apex; style c. 6.5–9 mm long. Drupe c. 1.2 × 0.8–1 cm, obovoid, rounded at apex, scarcely enclosed in the enlarged calyx, purple-black at maturity; fruiting calyx saucer-shaped, delicate, pubescent, not gland-dotted.

   Zambia. N: Mpika Distr., North Luangwa National Park, 11°48'S, 32°17'E, 650 m, 17.iii.1994, *P.P. Smith* 1456 (K). E: Chipata Distr., Luangwa Valley, near Jumbe (Njumbe), 30.xii.1965, *Astle* 4238 (K; SRGH). S: Livingstone Distr., Victoria Falls, 6.iii.1960, *Fanshawe* 5390 (K). **Zimbabwe**. N: Gokwe South Distr., near Tari (Tare) River, 19.iv.1963, *Bingham* 647 (K; SRGH). W: Hwange Distr., Zambezi R., between Matetsi and Deka Rivers, 28.iii.1963, *Wild* 6047 (LISC; SRGH). E: Chipinge Distr., Mutungawuna, c. 200 m, 24.i.1957, *Phipps* 155 (SRGH). **Mozambique**. T: Cahora Bassa Distr., c. 13 km from Chicoa to Chinhanda, Chianga, 27.ii.1972, *Macêdo* 4936 (K; LISC; LMU). MS: Báruè Distr., Changara to Catandica (Vila Gouveia), 29.v.1971, *Torre & Correia* 18707 (LISC; LMA).

   Known only from the Flora Zambesiaca area. Hot, low-rainfall areas of the main river valleys, often in mopane woodland and scrub, and in mixed deciduous woodlands with *Terminalia, Julbernardia, Combretum* and *Pterocarpus spp.*, also on termite mounds; 200–900 m.

   This is a distinctive species with delicate branches, peduncles and pedicels. The 3 leaflets are often irregularly serrate and covered with a whitish bloom. The flowers are large and few and borne on long pedicels in lax inflorescences; the stamens are clearly exserted. For its affinities, see *V. mossambicensis*.

---

*From its description and locality it is merely a variant of *V. petersiana* with clearly dentate leaves but it could also be a variant of *V. mossambicensis*.

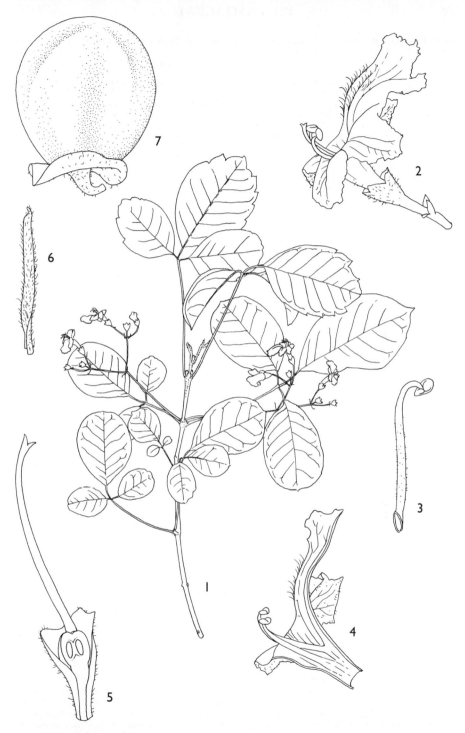

Tab. 12. VITEX PETERSIANA. 1, apical portion of flowering branch (× ²/₃), from *Rogers* 5377; 2, flower (× 4); 3, stamen (× 8); 4, half corolla (× 4); 5, half calyx with ovary and style (× 8); 6, bract (× 18), 2–6 from *Raymond* 232; 7, fruit (× 4), from *Raymond* 245. Drawn by Deborah Lambkin.

5. **Vitex mossambicensis** Gürke in Engler, Pflanzenw. Ost-Afrikas C: 340 (1895). —J.G. Baker in F.T.A. **5**: 329 (1900). —Sim, For. Fl. Port. E. Africa: 94 (1909). —Pieper in Bot. Jahrb. Syst. **62**, Beibl. 141: 57 (1928). —Moldenke in Phytologia **5**: 485 (1957). —Verdcourt in F.T.E.A., Verbenaceae: 57 (1992). —Sales in Kew Bull. **56**: 198 (2001). Type: Mozambique, Ilha Distr., Mussoril e Cabeceira, *M.R.P. Carvalho* 1884–1885 (B†, holotype; COI, neotype).
     Vitex carvalhi Gürke in Engler, Pflanzenw. Ost-Afrikas C: 339 (1895). —J.G. Baker in F.T.A. **5**: 326 (1900). —Sim, For. Fl. Port. E. Africa: 94 (1909). —Pieper in Bot. Jahrb. Syst. **62**, Beibl. 141: 57 (1928). Type: Mozambique, Ilha Distr., terrenos mais altos de Mussoril e Cabeceira, *M.R.P. Carvalho* 1884–1885 (B†, holotype; COI, isotype).
     Vitex oligantha Baker in F.T.A. **5**: 327 (1900). Type from Tanzania.
     Vitex mossambicensis var. *oligantha* (Baker) W. Piep. in Bot. Jahrb. Syst. **62**, Beibl. 141: 57 (1928). —Brenan, Check-list For. Trees Shrubs Tang. Terr.: 643 (1949).

Shrub or small tree up to c. 10 m tall; bark fissured, greyish-cream in colour; older stems grey-cream glabrous; young stems with scattered orange-tawny hairs, shortly tomentose only on buds. Leaves 3–5-foliolate; petioles 3–9 cm long, terete; median petiolules 0.3–1 cm long; leaflets 4–7 × 2–3.5 cm, median leaflet largest, obovate, curved downwards, acuminate at the apex, the tip acute or obtuse, rounded or cuneate at the base, entire or deeply toothed in the apical half, glabrous above, ± densely pubescent and sparsely gland-dotted beneath or pubescent only on the midrib and in the axils of nerves on the lower surface, nerves ± raised on both surfaces. Inflorescence axillary, of slender simple few-flowered dichasia, 6–15 cm long, glabrous; bracteoles 1–5 mm long, linear. Flowers fragrant. Calyx obconical, 5-toothed or truncate, with scattered hairs, not gland-dotted, accrescent in fruit; tube 2–2.5 mm long. Corolla 12–13 mm long, not gland-dotted, pubescent; tube white, 3–5 mm long, straight; lobes mauve, spreading, middle lower lobe largest, 6–9 × 6–10 mm, blue. Stamens clearly exserted from tube; filaments with glandular hairs. Ovary obovoid, not gland-dotted, with some hairs at apex; truncate at the apex; style c. 3.5 mm long. Drupe c. 1 × 0.5 cm, obovoid, rounded at apex, purplish-black at maturity, up to ± half enclosed in the enlarged calyx; fruiting calyx cup-shaped, pubescent, not gland-dotted.

Mozambique. N: Mossuril Distr., Floresta de Mossuril, c. 100 m, 17.ii.1984, *Groenendijk, de Koning & Dungo* 1154 (K).
     Also in Tanzania. Coastal scrub and thickets; c. 10–500 m.
     The few specimens seen from the Flora Zambesiaca area are without flowers and fruit; and the features and measurements given above are from Tanzanian material. Although there are some distinct similarities between *V. mossambicensis* and *V. petersiana*, the slender habit, long petiolules, leaflet size and consistency and the lax cymose inflorescence, they can usually be separated by the 3–5 leaflets in *V. mossambicensis* and 3 in *V. petersiana*.

6. **Vitex amaniensis** W. Piep. in Bot. Jahrb. Syst. **62**, Beibl. 141: 46, 60, (1928); in Repert. Spec. Nov. Regni Veg. **26**: 164 (1929). —Brenan, Check-list For. Trees Shrubs Tang. Terr: 643 (1949). —Verdcourt in F.T.E.A., Verbenaceae: 58, 66 (1992). Type from Tanzania.
     Vitex lokundjensis sensu Moldenke in Phytologia **5**: 444 (1956) pro parte non W. Pieper (1929).
     Vitex sp. no. 1 of Drummond in Kirkia **10**: 272 (1975).

Tree c. 15–18 m tall; bark white or pale brown, smooth; older branches dark red-brown; young branches crisply pilose. Leaves drying dark, 3–5-foliolate; petioles 9–15 cm long, terete; median petiolules 0.5–2.5 cm long; leaflets 10–20 × 5–6.5 cm, median leaflet largest, elliptic to broad obovate, rounded or apiculate to shortly acuminate at the apex, cuneate to ± attenuate at the base, entire to remotely serrulate, glabrous on both surfaces with a few short hairs on the nerves beneath, nerves ± raised on both surfaces. Inflorescences axillary, 7–11 cm long, of ± slender, few-flowered compound dichasia, shortly crisped-pubescent; bracteoles 0.4–0.5 cm long, linear. Calyx obconical, 5-toothed, not gland-dotted, densely pubescent, accrescent in fruit; tube c. 2 mm long; teeth c. 0.3 mm, erect. Corolla pale yellow, 10–11 mm long, with scattered gland-dots, densely pubescent; tube c. 5 mm long, straight; lobes at obtuse angle to tube, middle lower lobe 5.5 × 4.5 cm. Stamens scarcely exserted from tube; filaments pubescent at base, hairs eglandular. Ovary obovoid, sparsely gland-dotted at apex only, glabrous, truncate at apex; style 5 mm long. Drupe c. 1.4 × 1 cm, obovoid (?immature), rounded at apex, ± enclosed in the enlarged calyx; fruiting calyx cup-shaped, pubescent, not gland-dotted.

Zimbabwe. E: Chipinge Distr., Chirinda Forest, scattered trees up to 20 m high, 1100 m, i.1967, *Goldsmith* 14/67 (B; BR; K; LISC; SRGH).

Also in Tanzania. In the Flora Zambesiaca area known only from the evergreen rain forest at Chirinda in Zimbabwe; c. 1100 m.

Only four gatherings of this species from the Flora Zambesiaca region have been seen by this author – all from the Chirinda Forest in Zimbabwe. This rain forest species was apparently first collected there in 1948 (by *N.C. Chase* 435, Jan 1948, Chirinda Forest) and wrongly determined as *V. lokundjensis* W. Piep. by Moldenke (loc. cit. 1956). *V. amaniensis* and *V. lokundjensis* show some superficial similarities (both are tall rain forest trees with leaves drying dark, leaflets petiolulate (to 2.5 cm long) and flowers in lax few-flowered axillary compound dichasia). However, this mainly Cameroon species differs in many characters from our plant: in the terete petiole, longer bracteoles, lanate calyx, glabrous and gland-dotted ovary and the eglandular hairs on filaments. Our plant shows great uniformity and differs from many of the Tanzanian plants in the non-gland-dotted leaves, obsolete calyx teeth and median lower lobe of the corolla larger and yellow. Some Tanzanian specimens (e.g. Ulugurue, Tanana, Morogoro, *E.M. Bruce* 774 (K)) approach our plants, being clearly intermediates of a range of variation. Although there is a major disjunction between the Tanzanian and the Zimbabwean plants, both localities occur in similar montane habitats with a similar floristic composition.

7. **Vitex ferruginea** Schumach. & Thonn., Beskr. Guin. Pl.: 62 (1827). —J.G. Baker in F.T.A. **5**: 324 (1900). —Pieper in Bot. Jahrb. Syst.: **62**, Beibl. 141: 70 (1928). —Huber in F.W.T.A. ed. 2, **2**: 447 (1963). —Verdcourt in F.T.E.A., Verbenaceae: 66 (1992). —Sales in Kew Bull. **56**: 193 (2001). —M. Coates Palgrave, ed. 3 of K. Coates Palgrave, Trees South. Africa: 982 (2002). Type from Ghana.

Vitex rufescens Gürke in Bot. Jahrb. Syst. **18**: 169 (1893), non A. Juss., *nom. illegit.* Type from Angola.

Vitex tangensis Gürke in Engler, Pflanzenw. Ost-Afrikas **C**: 339 (1895). —J.G. Baker in F.T.A. **5**: 321 (1900). —Pieper in Bot. Jahrb. Syst. **62**, Beibl. 141: 68 (1928). —Brenan, Check-list For. Trees Shrubs Tang. Terr.: 642 (1949). —Garcia in Estud. Ensaios Doc. Junta Invest. Ci. Ultramar [in Mendonça, Contrib. Conhec. Fl. Moçamb. II]: 174 (1954). Type from Tanzania.

Vitex amboniensis Gürke in Engler, Pflanzenw. Ost-Afrikas **C**: 340 (1895). —J.G. Baker in F.T.A. **5**: 329 (1900). —Sim, For. Fl. Port. E. Africa: 94 (1909). —Brenan, Check-list For. Trees Shrubs Tang. Terr.: 642 (1949). —Garcia in Estud. Ensaios Doc. Junta Invest. Ci. Ultramar [in Mendonça, Contrib. Conhec. Fl. Moçamb. II]: 174 (1954). —White, For. Fl. N. Rhod.: 371 (1962). —Ross, Fl. Natal: 300 (1972). —Drummond in Kirkia **10**: 271 (1975). —K. Coates Palgrave, Trees South. Africa: 807 (1977). —Stirton in Gibbs Russell et al. in Mem. Bot. Surv. S. Africa, ed. 2, **2**: 169 (1987). Type from Tanzania.

Vitex guerkeana Hiern, Cat. Afr. Pl. Welw. **1**(4): 835 (1900). Type as for *V. rufescens.*

Vitex polyantha Baker in F.T.A. **5**: 321 (1900). Type from Kenya.

Vitex guerkeana H. Pearson in F.C. **5**: 217 (1901) non Hiern. Type: Mozambique, Baía de Maputo (Delagoa Bay), *Monteiro* 20 (K).

Vitex swynnertonii S. Moore in J. Linn. Soc., Bot. **40**: 168 (1911). —Pieper in Bot. Jahrb. Syst. **62**, Beibl. 141: 70 (1928). —Garcia in Estud. Ensaios Doc. Junta Invest. Ci. Ultramar [in Mendonça, Contrib. Conhec. Fl. Moçamb. II]: 175 (1954). Type: Mozambique, Madanda Forests, 122 m, 5.xii.1906, *Swynnerton* 1054 (BM; K).

Vitex dryadum S. Moore in J. Linn. Soc., Bot. **40**: 169 (1911). —Pieper in Bot. Jahrb. Syst. **62**, Beibl. 141: 66 (1928). Type: Mozambique, Beira, open woods, 15 m, 25.xii.1906, *Swynnerton* 1062 (BM; K).

Vitex amboniensis Gürke var. *amaniensis* W. Piep. in Bot. Jahrb. Syst. **62**, Beibl. 141: 69 (1928). —Brenan, Check-list For. Trees Shrubs Tang. Terr.: 642 (1949). Type from Tanzania.

Vitex amboniensis Gürke var. *schlechteri* W. Piep. in Bot. Jahrb. Syst. **62**, Beibl. 141: 69 (1928). Type: Mozambique, Maputo (Lourenço Marques), *Schlechter* 11715 (B†, holotype).

Vitex pearsonii W. Piep. in Bot. Jahrb. Syst. **62**, Beibl. 141: 75 (1928), as *nom. nov.* Type as for *V. guerkeana.*

Vitex oxycuspis Baker var. *mossambicensis* Moldenke in Bol. Soc. Brot., Sér. 2, **40**: 122 (1966). Type: Mozambique, Cheringoma, Dondo, *Torre* 6323 (LISC).

Vitex ferruginea Schumach. & Thonn. subsp. *amboniensis* (Gürke) Verdc. in F.T.E.A., Verbenaceae: 66 (1992). —Bredenkamp & Botha in S. African J. Bot. **59**: 615 (1993).

Vitex ferruginea subsp. *amboniensis* var. *amaniensis* (W. Piep.) Verdc. in F.T.E.A., Verbenaceae: 67 (1992).

Many stemmed shrub 1.5–4.5 m tall often scandent, or small tree to 6(15) m tall, sometimes a liane, often deciduous; bark fissured, grey to light brown; older stems brownish-yellow, red and grey; young stems orange lanate. Leaves (3)5(7)-foliolate; petioles 2.5–12 cm long, terete; median petiolules (0.2)1.0–2.0(3.5) cm long; leaflets 2.5–11 × 1.5–4 cm, median leaflet largest, narrow to broadly

lanceolate, acute to narrowly acuminate or sometimes short acuminate or rounded
at the apex, cuneate or rounded at the base, glabrous on upper surface or with
scattered small hairs, paler and densely finely hairy beneath, gland-dotted or not,
entire or occasionally deeply toothed toward the apex, nerves sunken above and
raised beneath.   Inflorescences axillary, few to many-flowered, of ± sturdy short
simple dichasia 2–4 cm long, orange-tawny tomentose; bracteoles 0.5–1.2 cm long,
narrowly lanceolate.   Calyx obconical, 5-lobed, ferruginous pubescent outside,
gland-dotted, accrescent in fruit; tube 2–3 mm long; lobes (1)1.5–3 mm long,
erect. Corolla creamy-white, c. 16 mm long, ferruginous pubescent outside, gland-
dotted; tube to c. 7 mm long, curved; lobes mauve, at acute angle to tube, middle
lower lobe 5–7 × 7–8 mm the others smaller.   Stamens exserted from tube;
filaments with glandular hairs.   Ovary narrowly ovoid, not gland-dotted, lanate
towards apex; apex long-acuminate; style c. 10 mm long.   Drupe c. 3 × 2.5 cm,
spherical-ellipsoid, rounded or sharp-tipped at the apex, not enclosed in the
enlarged calyx, purplish-black at maturity; fruiting calyx saucer-shaped, pubescent,
not gland-dotted.

Caprivi Strip — see below. **Zambia**. B: Scshckc, Nanga Forest, 19.xii.1952, *Angus* 963 (FHO,
K). **Zimbabwe**. E: Chipinge Distr., Nyamgambe R., 610 m, iii.1962, *Goldsmith* 79/62 (B; LISC;
S; SRGH). S: Chivi Distr., Madzivire Dip, 16.iii.1967, *Mavi* 209 (K; SRGH). **Mozambique**. N:
Meconta, 8.xii.1936, *Torre* 958 (COI; LISC; LMA). Z: Maganja da Costa Distr., Floresta de
Gobene, 12.ii.1966, *Torre & Correia* 14573 (LISC). T: Tambara Distr., Lupata to Tete, ii.1859,
Kirk 2 (K). MS: Dondo, 23.iii.1960, *Wild & Leach* 5203 (K; SRGH). GI: Vilankulo Distr.,
Vilankulo to Mapinhane, 17.xi.1941, *Torre* 3829 (BR; LISC). M: Inhaca Island, 23 miles east of
Maputo (Lourenço Marques), c. 200 m, vii.1956, *Mogg* 30855 (K; SRGH).
    Also in tropical Africa from Guinea-Bissau to S Nigeria, Dem. Rep. Congo and Angola, and
from Somalia to South Africa.   Riverine vegetation, semi-evergreen forest, mixed deciduous
woodlands, mutemwa thickets on Kalahari Sand and coastal forest, in sandy soils and coastal
dunes; sea level to 610 m.
    Recorded from the Caprivi Strip by Palmer & Pitman, Trees of South. Africa 3: 1951 (1972),
as *V. amboniensis*, and by Verdcourt in F.T.E.A., Verbenaceae 67 (1992), but no material from
there has been seen by this author.
    V. ferruginea can be confused with *V. mombassae* mainly in their similar inflorescences and
flowers: the peduncle in *V. ferruginea* is rather short (as it is in *V. mombassae*), the bracteoles are
also similar, the flowers of both are large with a prominent lower lip, and the stamens, style and
stigma are clearly exserted.   However, the two species can be readily separated on petiolule
length — absent or very small in *V. mombassae*, c. 1–2 cm long in *V. ferruginea*.   Other
distinguishing features are leaflet shape (narrow to broadly lanceolate in *V. ferruginea* and
elliptic to obovate in *V. mombassae*) and leaflet number ((3)5 in *V. ferruginea*, 3(5) in *V.
mombassae*).
    Some specimens with thin, deeply dentate leaves and long calyx teeth belong here, including:
the type of *V. oxycuspis* var. *mossambicensis*; the type of *V. dryadum*; *Torre & Correia* 17352
(Mozambique, Angoche, (LISC)) and *Wild* 7572 (Mozambique MS: 16.vi.1966, (K; SRGH)).
Although differing in leaf shape they belong to *V. ferruginea* on floral and inflorescence characters.
    It has been observed by field botanists that tropical savanna species go through a burst of
anomalous vegetative growth with the onset of the rains (pers. comm. David Harris).
Verdcourt, in F.T.E.A., Verbenaceae: 68 (1992), also refers to specimens of *Vitex* with such
anomalous leaves considering them to be dimorphic juvenile growth.   Such anomalous growth
would appear to be a characteristic of *Vitex* species.

8. **Vitex fischeri** Gürke in Bot. Jahrb. Syst. **18**: 171 (1893). —J.G. Baker in F.T.A. **5**: 330 (1900).
    —Pieper in Bot. Jahrb. Syst. **62**, Beibl. 141: 60 (1928). —Brenan, Check-list For. Trees
    Shrubs Tang. Terr.: 642 (1949). —Moldenke in Phytologia **5**: 335 (1955). —White, For. Fl.
    N. Rhod.: 371 (1962). —Verdcourt in F.T.E.A., Verbenaceae: 59 (1992). —Sales in Kew
    Bull. **56**: 195 (2001). Type from Tanzania.
        Vitex andongensis Baker in Hiern, Cat. Afr. Pl. **2**(4): 837 (1900); in F.T.A. **5**: 329 (1900).
    —Pieper in Bot. Jahrb. Syst. **62**, Beibl. 141: 60 (1928). Type from Angola.
        Vitex bequaertii De Wild. in Repert. Spec. Nov. Regni Veg. **13**: 142 (1914). —Pieper in Bot.
    Jahrb. Syst. **62**, Beibl. 141: 60 (1928). Syntypes from Dem. Rep. Congo (Kinshasa).

Tree up to 15 m tall or more; bark shallowly fissured, ± smooth, greyish to pale
yellow-brown; older branches yellow-brown; young branches and petioles orange-
tawny tomentose.   Leaves (4)5–7-foliolate; petioles (3)5–20 cm long; terete; median
petiolules 0.5–2 cm long. leaflets 8–19 × 3–9 cm, median leaflet largest, elliptic, ovate
to obovate, rounded to long acuminate at the apex, cuneate to rounded at the base,

± entire, somewhat scabrid-puberulous above to glabrescent, velvety-tomentose and sparsely to densely gland-dotted beneath and paler than the upper surface, nerves raised on both surfaces. Inflorescences axillary, of ± sturdy densely-flowered compound dichasia 5–16 cm long, pale yellow-tomentose; bracteoles 0.4–1 cm long, linear. Calyx obconical, 5-lobed, lanate, sparsely gland-dotted, accrescent in fruit; tube 2–2.3 mm long; lobes 0.5–1.0 mm long, erect. Corollas 8–10 mm long, not gland-dotted, lanate or pubescent; tubes white, 5–6 mm long, curved; lobes mauve, at an obtuse angle to tube, middle lower lobe 3–5 × 2–3 mm. Stamens scarcely exserted from the corolla tube; filaments glabrous. Ovary obovoid, gland-dotted and pubescent toward the apex; apex truncate; style c. 6 mm long. Drupe c. 1.3 × 0.7 cm, ellipsoid-obovoid, rounded at apex, the lower half enclosed by the enlarged calyx, purple-green to black with green spots at maturity; fruiting calyx cup-shaped, pubescent, not gland-dotted.

Zambia. B: Kabompo Distr., 16 km east of Kabompo Pontoon, 23.xi.1952, *Holmes* 1028 (FHO; K; SRGH). N: Samfya, 16.xi.1964, *Mutimushi* 1125 (K; NDO). W: Ndola, 8.xii.1953, *Fanshawe* 541 (BR; K; NDO; SRGH).

Also in Sudan, Uganda, Kenya, Tanzania, Dem. Rep. Congo and Angola. Miombo woodland, woodlands with *Entandrophragma*, *Erythrophleum*, *Newtonia* and *Pachystela*, wooded grassland and thickets, also on termite mounds; to 1220 m.

V. fischeri is similar to *V. madiensis* subsp. *milanjiensis* in its inflorescences (densely flowered and shorter than the peduncles) in its small flowers and in leaflet shape and size. However, it may be distinguished by stem colour (yellow-brown in *V. fischeri*, but dark purplish-brown in *V. madiensis* subsp. *milanjiensis*) and by the leaflet lower surface indumentum (usually of dense curly hairs in *V. fischeri* and rather straight scattered hairs in *V. madiensis*). *V. fischeri* is distinguished from *V. ferruginea*, with which it shares the characters of petiolulate, long acuminate leaflets and leaves sometimes 7-foliolate, by the flower size (much larger in *V. ferruginea*) and by the leaflet lower surface indumentum (scattered small hairs in *V. ferruginea*).

9. **Vitex madiensis** Oliv. in Trans. Linn. Soc., London **29**: 134, t. 131 (1875). —Gürke in Engler, Pflanzenw. Ost-Afrikas **C**: 339 (1895). —J.G. Baker in F.T.A. **5**: 322 (1900). —Pieper in Bot. Jahrb. Syst. **62**, Beibl. 141: 61 (1928). —Garcia in Estud. Ensaios Doc. Junta Invest. Ci. Ultramar [in Mendonça, Contrib. Conhec. Fl. Moçamb. II]: 173 (1954). —Huber in F.W.T.A. ed. 2, **2**: 447 (1963). —Drummond in Kirkia **10**: 272 (1975). —K. Coates Palgrave, Trees South. Africa: 809 (1977). —Verdcourt in F.T.E.A., Verbenaceae: 60 (1992). —Sales in Kew Bull. **56**: 196 (2001). —M. Coates Palgrave, ed. 3 of K. Coates Palgrave, Trees South. Africa: 983 (2002). Type from Uganda.

Shrub or small tree 3–8 m tall, or a suffrutex 0.3–1.5 m tall with annual stems from a massive underground woody rootstock; bark fissured, dark red-brown; older stems brownish-red, yellow and grey; young stems shortly tomentose with reddish-brown buds. Leaves aromatic when crushed, often 3-whorled, (3)5(6)-foliolate, often drying yellow-green; petioles 7–16 cm long, grooved; median petiolules 0.7–2.3 cm long; leaflets 6–17 × 4–9 cm, median leaflet largest, elliptic, narrow to broadly-obovate or oblanceolate, rounded or shortly acuminate at the apex, rounded, cordate or cuneate at the base, ± coriaceous, usually entire or sometimes coarsely shallowly crenate in upper half, often undulate, ± scabridulous or glabrous rather shining and not gland-dotted above, pubescent on the nerves beneath, conspicuously raised reticulate beneath. Inflorescences axillary, slightly aromatic when crushed, of sturdy usually few-flowered compound dichasia, 5–19(30) cm long, tomentose with stiff patent cream-coloured hairs; bracteoles (0.2)0.6–1.2 cm long, linear. Calyx deep purplish-violet, obconical, 5-toothed, gland-dotted, pubescent to tomentose, accrescent in fruit; tube 2.5–3.5 mm long; teeth c. 0.7–1.5 mm long, erect; lobes 0.5–1.0 mm long, erect. Corolla 8–12 mm long, gland-dotted, pubescent to tomentose; tube greenish-cream or greenish-mauve, 4.5–6.5 mm long, curved; lobes at obtuse angle to tube, dull pale mauve to lilac-violet or brilliant purple sometimes creamy with mauve or greenish marks, middle lower lobe 4–3.2 × (4.5)4–3.6 mm. Stamens scarcely exserted from corolla tube; filaments with glandular hairs. Ovary gland-dotted and ± truncate at apex, glabrous; style 4–5 mm long. Drupe 2 × 1.3–1.8 cm, ellipsoid, rounded at the apex, the base not enclosed by the enlarged calyx, purplish-black at maturity; fruiting calyx saucer-shaped, distinctly toothed, pubescent, not gland-dotted.

The variation, in this taxon, in leaflet number and shape and probably texture has led to the recognition of two species in the past. Plants in the north of its distribution range (subsp. *madiensis*) have leaves 1–3-foliolate with leaflets broadly-obovate, more leathery and green-brown on drying; while plants in the south (subsp. *milanjiensis*) have leaves 3–5-foliolate with leaflets narrowly-obovate, almost truncate apically and shortly acuminate at the apex, usually drying lemon-yellow. Plants from Zambia and Malawi exhibit features intermediate between these extremes with leaflets 5 and broad or 3–4 and narrow-obovate.

V. madiensis subsp. *milanjiensis* can sometimes appear to be very similar to *V. doniana*, especially in leaflet shape, but leaflets in *V. doniana* are totally glabrous on the upper surface and the bracteoles are extremely reduced. The reticulation in *V. madiensis* is darker than the leaflet blade, and not so in *V. doniana*. The two species are very distinct in their flowers.

Subsp. madiensis

Vitex simplicifolia Oliv. in Trans. Linn. Soc., London **29**: 133, t. 130 (1875). —Gürke in Engler, Pflanzenw. Ost-Afrikas **C**: 339 (1895). —J.G. Baker in F.T.A. **5**: 320 (1900). —Pieper in Bot. Jahrb. Syst. **62**, Beibl. 141: 64 (1928). —Huber in F.W.T.A. ed. 2, **2**: 447 (1963). Type from Sudan.

Vitex camporum Büttn. in Verh. Bot. Vereins Brandenburg **32**: 35 (1890). —J.G. Baker in F.T.A. **5**: 323 (1900); in F.W.T.A. **2**: 276 (1931). Types from Dem. Rep. Congo and Angola.

Vitex hockii De Wild. in Repert. Spec. Nov. Regni Veg. **13**: 143 (1914). —Pieper in Bot. Jahrb. Syst. **62**, Beibl. 141: 61 (1928). Type from Dem. Rep. Congo (Katanga).

Vitex madiensis var. *gossweileri* W. Piep. in Bot. Jahrb. Syst. **62**, Beibl. 141: 63 (1928). Syntypes from Angola, Uganda and Zambia.

Leaflets 1–3, broadly-obovate, ± leathery and green-brown on drying.

Zambia. N: Chinsali Distr., Mbesuma area, Kalunga River terrace, 1160 m, 18.x.1961, *Astle* 967 (K). C: Serenje Distr., 32 km SW of Serenje Corner, 15.vii.1930, *Hutchinson & Gillett* 3704 (K). **Zimbabwe**. E: Mutare Distr., Vumba Mts., Norseland Farm, 11.ii.1950, *Chase* 1965 (K; SRGH). **Malawi**. C: Salima to Balaka road, 12 km south of Chipoka, 4.v.1980, coll. ignot. (K). S: Machinga Distr., Chikala, 600 m, xi.1900, *Purves* 15 (K).
Also in Gambia, Senegal and Mali to N Nigeria, Cameroon, Gabon, Central African Republic, Sudan, Dem. Rep. Congo and Angola. Miombo and mixed deciduous woodlands with *Combretum* and *Diospyros*, floodplain grassland on termite mounds; 600–1450 m.

Subsp. milanjiensis (Britten) F. White, For. Fl. N. Rhod.: 372, 455 (1962). —Verdcourt in F.T.E.A., Verbenaceae: 61 (1992). —Sales in Kew Bull. **56**: 197 (2001). Syntypes from Malawi, Mulanje (Milanji), *Whyte* 138 (BM); Zomba, *Whyte* s.n. (BM).

Vitex milanjiensis Britten in Trans. Linn. Soc. London, Bot. **4**: 36 (1894). —J.G. Baker in F.T.A. **5**: 330 (1900).

Vitex golungensis Baker in F.T.A. **5**: 330 (1900). —Pieper in Bot. Jahrb. Syst. **62**, Beibl. 141: 64 (1928). Type from Angola.

Vitex epidictyodes Mildbr. ex W. Piep. in Bot. Jahrb. Syst. **62**, Beibl. 141: 48, 61 (1928). — Brenan, Check-list For. Trees Shrubs Tang. Terr.: 643 (1949). —Moldenke in Phytologia **5**: 332 (1955). Type from Tanzania.

Vitex madiensis var. *milanjiensis* (Britten) W. Piep. in Bot. Jahrb. Syst. **62**, Beibl. 141: 63 (1928). —Brenan, Check-list For. Trees Shrubs Tang. Terr.: 643 (1949).

?*Vitex seineri* Gürke ex W. Piep. in Bot. Jahrb. Syst, **62**, Beibl. 141: 60 (1928), *nomen*. — Pieper in Repert. Spec. Nov. Regni Veg. **26**: 164 (1929). Type: Zambia, Livingstone, *Seiner* 123 (B†, holotype).

Vitex grisea Baker var. *dekindtiana* sensu Norlindh & Weimarck in Bot. Not. **1940**: 67 (1940) non (Gürke) W. Piep.

Vitex madiensis subsp. *milanjiensis* var. *epidictyodes* (W. Piep.) Verdc. in F.T.E.A., Verbenaceae: 61 (1992).

Leaflets 3–5, narrow-obovate, ± truncate at the apex with a central shortly acuminate tip, drying lemon-yellow.

Zambia. B: Kabompo Distr., 35 km WSW of Kabompo to Zambezi (Balovale), 24.iii.1961, *Drummond & Rutherford-Smith* 7294 (K; SRGH). N: Nchelenge Distr., south of Chipampa River, 1.vi.1933, *Michelmore* 375 (K). W: Chingola, 14.x.1955, *Fanshawe* 2521 (K). C: Lusaka Distr., near Chongwe R., 16.iv.1952, *White* 2680 (FHO; K). S: 3 km south of Choma, 10.iv.1952, *White* 2634 (K). **Zimbabwe**. E: Mutare, Umtali Golf Course, 12.xii.1956, *Chase* 6261 (K; SRGH). **Malawi**. N: Mzimba Distr., 63 km north of Mzimba, 30.iv.1952, *White* 2532 (FHO; K). S: Zomba Distr., 14 km out of Zomba Town on road to Blantyre, 8.iv.1968, *Banda* 1024 (K). **Mozambique**. MS: Manica Distr., Macequesse, 3.xii.1943, *Torre* 6264 (BR; COI; LMA; LISC). M: Maputo R., 1.iii.1947, *R.M. Hornby* 2526 (LMA; P).

Also in Dem. Rep. Congo, Angola and Tanzania. Dense woodland on Kalahari Sand, miombo woodland on plateau and escarpments, mixed deciduous woodlands and wooded grassland and coastal bush on fixed dunes; 10–1540 m.

Moldenke (Phytologia **6**: 223 (1958)) cited *Vitex welwitschii* Gürke in Bot. Jahrb. Syst. **18**: 166 (1893) from Zimbabwe (*Chase* 4191) and Mozambique (*Hornby* 2526), but in my opinion both these specimens are *V. madiensis* subsp. *milanjiensis*. It is doubtful if *V. welwitschii* occurs in the Flora Zambesiaca area, but further gatherings from Angola are needed to assess its status.

A number of intermediate specimens that could not be assigned to either of the two subspecies were determined as *V. madiensis* sensu lat. Their distribution is:

Zambia. B: Zambezi (Balovale), 1.xi.1952, *Gilges* 258 (P). N: Mbala Distr., between Ndundu and Kawimbe, 1650 m, 4.xi.1965, *Richards* 20662 (K). S: Choma Distr., Siamambo Forest Reserve, 10.vi.1952, *White* 2945D (K). **Malawi**. N: Viphya Plateau, 1524 m, 12.xi.1977, *Pawek* 13225 (BR; COI; K; MO). C: Ntchisi (Nchisi) Mt., 14.iii.1967, *Salubeni* 613 (SRGH). S: Mulanje Mt., 860 m, 14.xi.1985, *J.D. & E.G. Chapman* 6808 (E).

10. **Vitex doniana** Sweet, Hort. Brit., ed. 1: 323 (1827). —J.G. Baker in F.T.A. **5**: 323 (1900). — Pieper in Bot. Jahrb. Syst. **62**, Beibl. 141: 64 (1928). —Brenan, Check-list For. Trees Shrubs Tang. Terr.: 642 (1949); in Mem. New York Bot. Gard. **9**: 38 (1954). —Moldenke in Phytologia **5**: 322 (1955). —Topham, Check List For. Trees Shrubs Nyasaland Prot.: 102 (1958). —White, For. Fl. N. Rhod.: 371 (1962). —Huber in F.W.T.A. ed. 2, **2**: 446, fig. 308 (1963). —Gomes e Sousa, Dendrol. Moçambique Estudo Geral **2**: 659 (1967). — Drummond in Kirkia **10**: 271 (1975). —K. Coates Palgrave, Trees South. Africa: 808 (1977). —Verdcourt in F.T.E.A., Verbenaceae: 62 (1992). —Sales in Kew Bull. **56**: 192 (2001). —White, Dowsett-Lemaire & Chapman, Evergr. For. Fl. Malawi: 588 (2001). —M. Coates Palgrave, ed. 3 of K. Coates Palgrave, Trees South. Africa: 981 (2002). Type from Sierra Leone.

Vitex cuneata Thonn. in Schumach. & Thonn., Beskr. Guin. Pl.: 62 (1827). —Gürke in Engler, Pflanzenw. Ost-Afrikas **C**: 339 (1895). —J.G. Baker in F.T.A. **5**: 328 (1900). —Eyles in Trans. Roy. Soc. S. Africa **5**: 459 (1916). —Pieper in Bot. Jahrb. Syst. **62**, Beibl. 141: 71 (1928). —Garcia in Estud. Ensaios Doc. Junta Invest. Ci. Ultramar [in Mendonça, Contrib. Conhec. Fl. Moçamb. II]: 175 (1954). —Moldenke in Phytologia **5**: 322 (1955). — Verdcourt in F.T.E.A., Verbenaceae: 62 (1992). Type from Ghana.

Vitex cienkowskii Kotschy & Peyr. in Kotschy & Peyr., Pl. Tinn.: 27, t. 12 (1867). —J.G. Baker in F.T.A. **5**: 328 (1900). —Sim, For. Fl. Port. É. Africa: 94 (1909). —Eyles in Trans. Roy. Soc. S. Africa **5**: 459 (1916). Type from Sudan.

Vitex paludosa Vatke in Linnaea **43**: 534 (1882). —Eyles in Trans. Roy. Soc. S. Africa **5**: 459 (1916). Type from Zanzibar.

Vitex pachyphylla Baker in F.T.A. **5**: 328 (1900). —Pieper in Bot. Jahrb. Syst. **62**, Beibl. 141: 71 (1928). Type from Gabon.

Vitex puberula Baker in F.T.A. **5**: 330 (1900). —Pieper in Bot. Jahrb. Syst. **62**, Beibl. 141: 56 (1928). Type from Angola.

Deciduous much branched tree 4–18 m tall; bark with vertical fissures and stringy ridges or smooth, pale grey-brown, sometimes reddish-brown; older branches brownish-red, yellow and grey; young branches shortly tomentose or glabrescent, with reddish-brown buds. Leaves (3)4–5(6)-foliolate; petiole 5–17 cm long, grooved; median petiolules 0.4–2 cm long; leaflets 7–23 × 3–9 cm, median leaflet largest, broadly-obovate or sometimes narrowly obovate, rounded or emarginate to cuspidate and shortly acuminate at the apex, ± asymmetric and cuneate at the base, coriaceous, entire, concolorous shiny and glabrous above, not gland-dotted, midrib sometimes puberulous beneath, nerves raised beneath. Inflorescences axillary, of sturdy ± densely-flowered compound dichasia, 2.5–10 cm long, orange-tawny tomentose; bracteoles 0.2–0.3 cm long, linear, caducous. Calyx conical, shallowly 5-toothed, rusty-brown lanate, not gland-dotted, accrescent in fruit; tube 2.5–4 mm long; teeth c. 0.5–1.5 mm long, erect. Corolla 8–12 mm long, not gland-dotted, densely brown velvety tomentose outside; tube white or violet, 6–8 mm long, curved; lobes mauve, at obtuse angle to tube, middle lower lobe c. 3 × 4 mm. Stamens scarcely exserted from corolla tube; filaments with glandular hairs. Ovary obovoid and almost truncate apically, not gland-dotted, crowned with erect rigid hairs; style c. 7 mm long. Fruit a shiny edible drupe, 2–2.7 × 1.5–2.5 cm, obovoid-ellipsoid, rounded at the apex, the lower half not enclosed in the enlarged calyx, purplish-black at maturity; fruiting calyx saucer-shaped, pubescent, not gland-dotted.

Zambia. N: Mbala Distr., Ndundu, 3.x.1959, *Richards* 11490 (K). W: Mwinilunga Distr., fringe of mushitu of the Kasombo R., 14.vi.1963, *Edwards* 769 (K; SRGH). C: Kabwe to Lukanga

Swamp, 5.vii.1964, *van Rensburg* 2943 (K; SRGH). E: Chadiza Distr., Mwangazi Valley, 750 m, 26.xi.1958, *Robson* 720 (BM; K; SRGH). **Zimbabwe**. E: Chimanimani Distr., 4 km east of Haroni/Lusitu confluence, 450 m, 8.i.1969, *Biegel* 2755 (K; SRGH). **Malawi**. N: Rumphi Distr., Lakeshore road, foot of Chombe Mt., 3.i.1973, *Pawek* 6321 (K; MO; UC). C: Ntcheu Distr., Livulezi Valley, 31.i.1968, *Jeke* 150 (K; SRGH). S: Mulanje Mt., Likhubula R. above Likhubula, 680 m, v.1980, *Blackmore & Brummitt* 1469 (K). **Mozambique**. N: Montepuez Distr., andados 5 km de Montepuez para Namuno, c. 430 m, 26.xii.1963, *Torre & Paiva* 9713 (K; LISC; LMA). Z: Gilé Distr., Alto Ligonha, Régulo to Namirué, 12.x.1949, *Barbosa & Carvalho* 4393 (K). T: Macanga Distr., Furancungo to Angónia, 29.ix.1942, *Mendonça* 501 (LISC). MS: Beira, 15 m, 25.xii.1906, *Swynnerton* 1058 (BM). GI: Vilankulo, 6.vi.1947, *Gomes Pedro* 3473 (LMA). M: wood of Namaacha, 28.ii.1950, *J.N. Galo* 5 (LMA).

Widespread throughout tropical Africa from Senegal to Cameroons, and in east Africa and Angola; also in Comores. Evergreen forest margins, swamp forest margins and riverine forest, in open *Brachystegia–Isoberlinia* woodland in high rainfall areas and in coastal forest, often on termite mounds; 5–1750 m.

Usually distinctive in its large (up to 23 cm long) glabrous smooth leathery leaves, pale-coloured on drying. Some atypical specimens approach *V. madiensis* subsp. *milanjiensis* in their general facies (see discussion under *V. madiensis*).

The type of *V. puberula* probably belongs here, although the whole plant is covered in a greyish bloom and leaflet lower surface is pubescent on the blade as well as on the veins.

11. **Vitex payos** (Lour.) Merr. in Trans. Amer. Phil. Soc., n. s. **24**, 2: 334 (1935). —Brenan, Check-list For. Trees Shrubs Tang. Terr.: 643 (1949). —Pardy in Rhodesia Agric. J. **50**: 186, photos (1953). —O.H. Coates Palgrave, Trees Central Africa: 430–3, plates (1956). — Moldenke in Phytologia **6**: 45 (1957). —White, For. Fl. N. Rhod.: 372 (1962). —Lopes & Oliveira in Mém. Inst. Invest. Agron. Moçambique **6**(1): 79–81 (1972). —Drummond in Kirkia **10**: 272 (1975). —K. Coates Palgrave, Trees South. Africa: 811 (1977). —Verdcourt in Taxon **38**: 155 (1989); in F.T.E.A., Verbenaceae: 63 (1992). —Sales in Kew Bull. **56**: 200 (2001). —M. Coates Palgrave, ed. 3 of K. Coates Palgrave, Trees South. Africa: 985, illust. 274 (2002). Conserved type from Tanzania.

    Allasia payos Lour., Fl. Cochinch.: 85 (1790). Type from Africa.

    Vitex hildebrandtii Vatke in Linnaea **43**: 534 (1882). —Gürke in Engler, Pflanzenw. Ost-Afrikas C: 339 (1895). —J.G. Baker in F.T.A. **5**: 326 (1900). —Pieper in Bot. Jahrb. Syst. **62**, Beibl. 141: 65 (1928). —Garcia in Estud. Ensaios Doc. Junta Invest. Ci. Ultramar [in Mendonça, Contrib. Conhec. Fl. Moçamb. II]: 172 (1954). Type from Tanzania.

    Vitex zambesiaca Baker in F.T.A. **5**: 322 (1900). —Sim, For. Fl. Port. E. Africa: 94 (1909). —Topham, Check-list For. Trees Shrubs Nyasaland Prot.: 102 (1958). Type: Mozambique, Lower Shire R., near the base of Morrumbala Mt., 15.i.1863, *Kirk*, s.n. (K).

    Vitex shirensis Baker in F.T.A. **5**: 326 (1900). Syntypes from Malawi: Mpemba (Impembe) Hill, *Kirk* 3; Shire Highlands, *Buchanan* 20; Zomba and vicinity, *Whyte*, excl. *Buchanan* 231 (= *V. mombassae*).

    Vitex iringensis Gürke in Bot. Jahrb. Syst. **28**: 464 (1900). —Pieper in Bot. Jahrb. Syst. **62**, Beibl. 141: 65 (1928). —Brenan, Check-list For. Trees Shrubs Tang. Terr.: 643 (1949). — Garcia in Estud. Ensaios Doc. Junta Invest. Ci. Ultramar [in Mendonça, Contrib. Conhec. Fl. Moçamb. II]: 172 (1954), as '*V. irvingensis*'. Type from Tanzania.

    Vitex isotjensis Gibbs in J. Linn. Soc., Bot. **37**: 463 (1906). —Eyles in Trans. Roy. Soc. S. Africa **5**: 459 (1916). —Pieper in Bot. Jahrb. Syst. **62**, Beibl. 141: 60 (1928). —Drummond in Kirkia **10**: 272 (1975). Type: Zimbabwe, Matopos Hills, 10.x.1905, *Gibbs* 236 (BM; K).

    Vitex eylesii S. Moore in J. Bot. **45**: 154 (1907); J. Linn. Soc., Bot. **40**: 168 (1911). —Eyles in Trans. Roy. Soc. S. Africa **5**: 459 (1916). Type: Zimbabwe, Bulawayo, *Eyles* 1201 (BM, holotype).

    Vitex villosa Sim in For. Fl. Port. E. Africa: 93, t. 78B (1909). —Pieper in Bot. Jahrb. Syst. **62**, Beibl. 141: 66 (1928). —Moldenke in Phytologia **6**: 217 (1958). —Sales in Kew Bull. **56**: 205 (2001). Type: Mozambique, Quelimane area, frequent, *Sim* 5572 (not traced – c.f. Sales, loc. cit.); lectotype: the illustration in Sim For. Fl. Port. E. Africa: t. 78B (1909), chosen by Sales 2001.

    Vitex hildebrandtii var. *glabrescens* W. Piep. in Bot. Jahrb. Syst. **62**, Beibl. 141: 66 (1928). Syntypes: Malawi, Blantyre, *Buchanan* 6970, Likoma Island, *Johnson* 17; Zimbabwe, Chirinda, *Swynnerton* 34, Mutare, *Engler* 3120; Mozambique, Kurumadzi R., *Swynnerton* 1059 (K), 1060, 1061, Madanda, *Johnson* 96, M'salu R. mouth, *Allen* 151; and from Tanzania.

    Vitex hildebrandtii Vatke var. *zambesiaca* (Baker) W. Piep. in Bot. Jahrb. Syst. **62**, Beibl. 141: 66 (1928).

    Vitex payos var. *glabrescens* (W. Piep.) Moldenke, Known Geogr. Distrib. Verbenaceae, ed. 2: 79 (1942); in Phytologia **6**: 47 (1957); in Phytologia **8**: 72 (1961).

    Shrub or small tree 4–10(13) m tall, with rounded crown; bark light brown to grey-brown, deeply vertically fissured; branches pale hairy becoming glabrous; young branches and twigs densely orange-tawny lanate. Leaves (3)5-foliolate;

petiole 3–16 cm, terete; petiolules ± absent; leaflets 8–16 × (2)3–9 cm, median leaflet largest, elliptic to obovate, rounded or obtuse at the apex or acute to shortly acuminate, rarely emarginate, cuneate at the base, entire, discolorous especially when young, ± roughly pubescent above, densely floccose beneath, venation impressed above, raised and reticulate beneath under the indumentum. Inflorescences axillary, of few–many-flowered, dense to somewhat lax compound dichasia, 3–18 cm long, orange-tawny tomentose or woolly when young; bracteoles 0.8–2.5 cm long, linear. Calyx obconical, 5-lobed, gland-dotted, lanate, strongly accrescent in fruit; tube 2.5–3 mm long; lobes 0.5–1 mm long, erect. Corolla fragrant, 6–7 mm long, gland-dotted, lanate; tube white-blue, mauve or purple, 3–4 mm long, curved, hairy outside; lobes at an obtuse angle to the tube, middle lower lobe mauve or whitish, 2–3 × 2–3 mm, oblong, ovate or ± round, other lobes white. Stamens enclosed within the corolla tube; filaments with glandular hairs. Ovary globose, apically rounded, not gland-dotted, lanate at apex; style c. 4 mm long. Fruit an edible shiny glabrous drupe, black at maturity, c. 2–3 × 2 cm, ellipsoid to obovoid, rounded at apex, the lower half enclosed in the enlarged calyx; fruiting calyx cup-shaped, 1–2 cm high, 1.5–2.5 cm wide, broadly crenate, pubescent, not gland-dotted.

Zambia. N: Nchelenge Distr., Lake Mweru, lake shore mushitu, 11.xi.1957, *Fanshawe* 3922 (K). W: Solwezi, 15.vi.1930, *Milne-Redhead* 516 (BR; K). C: Lusaka Distr., Mt. Makulu, Chilanga, 15.i.1958, *Angus* 1820 (BR; FHO; K; SRGH). S: Monze Distr., Pemba, 1200 m, xii.1933, *Trapnell* 1660 (K). **Zimbabwe**. N: Hurungwe National Park, Dokwa R., 18.xii.1956, *Phelps* 179 (K; SRGH). W: Hwange Distr., 3 km east of bridge over Inyantue River, 30.xii.1973, *Raymond* 237 (K; SRGH). C: Makoni Distr., Rusape, 3.xi.1957, *Norman* 12/ R (K). E: Chipinge Distr., near Chirinda, iv.1906, *Swynnerton* 34b (K). S: Chivi Distr., Madzivire Dip, 600 m, 29.xii.1962, *Moll* 399 (E; K; SRGH). **Malawi**. N: Rumphi Distr., Rumphi Waterworks road, 1070 m, 9.iii.1975, *Pawek* 9133 (K). C: Lilongwe Distr., 1 km SW of Malingunde, 1110 m, 24.ii.1982, *Brummitt, Polhill & Banda* 16059 (K). S: Zomba Distr., Magomero, 14.i.1981, *Salubeni* 2877 (E; SRGH). **Mozambique**. N: Namapa, Lúrio River, 8.iii.1960, *Lemos & Macuácua* 17 (K; LISC; SRGH). Z: Mocuba to Maganja, 8.ii.1966, *Torre & Correia* 14461 (LISC). T: Cahora Bassa Distr., Songo, 7.ii.1972, *Macêdo* 4813 (K; LISC; LMA; SRGH). MS: Mossurize Distr., Kurumadzi River, Jihu, 610 m, 26.xi.1906, *Swynnerton* 1059 (K). GI: Massinga, xii.1936, *Gomes e Sousa* 1932 (K).

Also in Uganda, Kenya, Tanzania and Angola. Open woodland, wooded grassland and on rocky outcrops at medium to low altitudes; 50–1500 m.

Characteristic features of this species are the sessile leaflets, the thick tawny woolly indumentum, especially on the young inflorescence, and the relatively small flowers. For comments on its affinities, see discussion under *V. mombassae*.

12. **Vitex mombassae** Vatke in Linnaea **43**: 533 (1882). —Gürke in Engler, Pflanzenw. Ost-Afrikas C: 339 (1895). —J.G. Baker in F.T.A. **5**: 326 (1900). —Pieper in Bot. Jahrb. Syst. **62**, Beibl. 141: 66 (1928). —Brenan, Check-list For. Trees Shrubs Tang. Terr.: 642 (1949). — O.B. Miller in J. S. African Bot. **18**: 76 (1952). —Garcia in Estud. Ensaios Doc. Junta Invest. Ci. Ultramar [in Mendonça, Contrib. Conhec. Fl. Moçamb. II]: 173 (1954). —Moldenke in Phytologia **5**: 481 (1955). —Burt Davy & Hoyle in Check-list For. Trees Shrubs Brit. Emp. 2, Nyasaland: 102 (1958). —White, For. Fl. N. Rhod.: 371 (1962). —Drummond in Kirkia **10**: 272 (1975). —K. Coates Palgrave, Trees South. Africa: 810 (1977). —Barnes & Turton, List Fl. Pl. Botswana at Nat. Mus., Sebele & Univ. Botswana: 27 (1986). — Verdcourt in F.T.E.A., Verbenaceae: 64, fig. 9 (1992). —Bredenkamp & Botha in S. African J. Bot. **59**: 617 (1993); in Bothalia **26**:146 (1996). —B.-E. van Wyk, Phot. Guide to Trees of Southern Africa: 317 (2000). —Sales in Kew Bull. **56**: 197 (2001). —M. Coates Palgrave, ed. 3 of K. Coates Palgrave, Trees South. Africa: 983 (2002). TAB. **13**. Type from Kenya.
Vitex flavescens Rolfe in Bol. Soc. Brot. **11**: 87 (1893). —Friedrich-Holzhammer in Merxmüller, Prodr. Fl. SW. Afrika, fam. no. 122: 10 (1962). Type from Angola.
Vitex shirensis sensu J.G. Baker in F.T.A. **5**: 326 (1900), quoad syntype *Buchanan* 231. — sensu Topham, Check-list For. Trees Shrubs Nyasaland Prot.: 102 (1958).
Vitex goetzei Gürke in Bot. Jahrb. Syst. **28**: 464 (1900). Type from Tanzania.
Vitex flavescens Rolfe var. *parviflora* Gibbs in J. Linn. Soc., Bot. **37**: 463 (1906). —Eyles in Trans. Roy. Soc. S. Africa **5**: 459 (1916). Type: Zimbabwe, Victoria Falls, *Gibbs* 135 (BM, photo.; B, photo.).
Vitex mombassae Vatke var. *parviflora* (Gibbs) W. Piep. in Bot. Jahrb. Syst. **62**, Beibl. 141: 68 (1928).
Vitex mombassae Vatke var. *erythrocarpa* Gürke ex W. Piep. in Bot. Jahrb. Syst. **62**, Beibl. 141: 68 (1928). Type: Zambia, Livingstone, *Seiner s.n.* (B†, holotype).
Vitex pooara sensu K. Coates Palgrave, Trees South. Africa: 810 (1977).

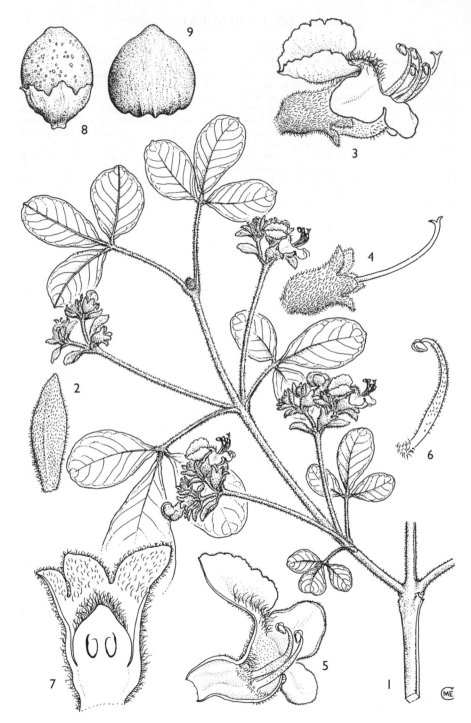

Tab. 13. VITEX MOMBASSAE. 1, portion of flowering branch (× ²/₃), from *Mgaza* 141; 2, inflorescence bract (× 2); 3, flower (× 3); 4, calyx and style (× 3); 5, longitudinal section of corolla (× 3); 6, stamen (× 4); 7, longitudinal section of ovary (× 6), 2–7 from *Tanner* 447; 8, fruit (× 1); 9, endocarp (× 2), 8 & 9 from *Milne-Redhead & Taylor* 8678. Drawn by M.E. Church. From F.T.E.A.

Much branched shrub or small deciduous tree up to 8(17) m tall; bark grey-brown, fissured becoming stringy; older branches grey-brown or red-brown pubescent; young branches and twigs densely orange-tawny lanate. Leaves 3–5-foliolate, aromatic when crushed; petiole 1.5–9 cm long, terete; petiolules absent or up to 0.6 cm long in median leaflet; leaflets 3–10 × 1.5–6 cm, elliptic to somewhat obovate, rounded or in apical leaves shortly acuminate at the apex, broadly cuneate to rounded at the base, entire, coriaceous, ± sparsely pubescent above, shortly lanate on the nerves and reticulation beneath or glabrescent, gland-dotted beneath, venation impressed above and the blade ± bullate, nerves and reticulation raised beneath. Inflorescences axillary, of few-flowered simple dichasia, 4–12 cm long, shortly orange-tawny villose or lanate with ± densely tangled pilose hairs; bracteoles 0.8–2 cm long, narrow-lanceolate or narrow-spathulate. Calyx obconical, 5-lobed, gland-dotted, lanate outside, strongly accrescent in fruit; tube 3–4 mm long; lobes 2–3 mm long, erect. Corolla c. 23 mm long, gland-dotted, villose with tangled pilose hairs and gland-tipped hairs outside, gland-dotted; tube creamy-white and mauve, 12–13 mm long, curved; lobes at acute angle to tube, middle lower lobe mauve, 6–7 × 6.5–7 mm, other lobes creamy-white smaller and rounded or triangular. Stamens exserted from corolla tube; filaments with glandular hairs. Ovary obovoid, truncate apex and lanate at the apex, gland-dotted; style c. 14–16 mm long, exserted. Fruit an edible spherical drupe, dark brown to purplish-black at maturity, c. 4 × 3 cm, globose, the lower end not or hardly enclosed in the enlarged calyx; fruiting calyx at first somewhat cup-shaped up to c. 1.5 cm long and c. 1 cm wide, later opening out becoming saucer-shaped, shortly toothed, puberulent, sometimes gland-dotted.

Caprivi Strip. c. 18 km south of northern border on firebreak, c. 29 km west of Katima Mulilo, 12.ii.1969, *de Winter* 9158 (K; PRE). **Botswana**. N: Chobe Distr., Kasane, x.1966, *Mutakela* 118 (GAB; SRGH). **Zambia**. B: Kalabo Distr., Luambimba River, c. 16 km north of Kalabo, 11.xi.1950, *Rea* 138 (K). N: Mbala Distr., Uningi Pans, 1.x.1960, *Richards* 13300 (K). W: Mwinilunga Distr., 27 km west of Mwinilunga on Matonchi road, 22.i.1975, *Brummitt, Chisumpa & Polhill* 13975 (K). C: Serenje Distr., South Luangwa National Park, near Mutinsase R., 11.x.1965, *Astle* 4074 (SRGH). E: Chipata Distr., Jumbe, Luangwa Valley, 30.xi.1966, *Mutimushi* 1594 (K; NDO). S: Namwala, 24.x.1963, *Lawton* 1136 (K). **Zimbabwe**. N: Gokwe, Copper Queen NPA, 7.xi.1963, *Bingham* 943 (SRGH). W: Lupane, gusu forest, 20.xii.1963, *D.H.M. Clarke* 339 (K; LMU; SRGH). C: c. 8 km NE of Gweru (Gwelo), xi.1957, *O.B. Miller* 4621 (SRGH). E: Mutare, 16.xi.1954, *Chase* 4328 (K). S: Chivi Distr., c. 72 km south of Masvingo (Fort Victoria), 15.iii.1967, *Mavi* 2000 (SRGH). **Malawi**. N: Karonga Distr., Vinthukhutu Forest Reserve, 550 m, i.1978, *Pawek* 13562 (K). C: Salima Distr., c. 3 km west of Lake Nyasa Hotel, 430 m, 28.vii.1951, *Chase* 3848 (K). S: Machinga Distr., Munde Hill, Mwekuwa Village, 27.i.1984, *Banda & Patel* 2112 (K). **Mozambique**. N: Ribáuè, 600 m, x.1931, *Gomes e Sousa* 848 (K). Z: Alto Mólocuè, road to Alto Ligonha, 29.xi.1967, *Torre & Correia* 16285 (LISC). T: Moatize Distr., Zóbuè, 800 m, 26.ix.1942, *Torre* 4562 (LISC; LMA). MS: Manica Distr., montes de Belas, 2.iii.1948, *Garcia* 456 (LISC). M: Maputo, Matutuíne (Bela Vista) towards Porto Henrique, 30.x.1980, *de Koning & Navunga* 8527 (K).

Also in Kenya, Tanzania, Burundi, Dem. Rep. Congo, Angola, Namibia and South Africa (Limpopo Province). Deciduous woodland and thickets on Kalahari Sand and on escarpments and rocky outcrops, and in wooded grasslands; 300–1800 m.

V. mombassae and *V. payos* can appear very similar – being similar in leaflet indumentum, inflorescence structure (though often much denser in *V. payos*), leaflets broadly lanceolate and often sessile and inflorescences often shorter than the leaves. However, *V. mombassae* usually has much more delicate branches, petioles and peduncles. The leaflet upper surface is bullate in *V. mombassae* and smooth or slightly bullate in *V. payos*, the flowers are much larger in *V. mombassae* (lower lip c. 6–7 mm long and 2–3 mm in *V. payos*) with clearly exserted stamens, the bracteoles are lanceolate to spathulate in *V. mombassae* and linear in *V. payos* and the calyx teeth c. 3 mm long in *V. mombassae* and c. 1.5 mm long in *V. payos*.

V. angolensis Gürke is cited as occurring in Caprivi Strip by Bredenkamp & Botha (Bothalia 26, 2: 146 (1996)), but no specimens from there have been seen.

V. pooara Corbishley was merged, wrongly in my opinion, with *V. mombassae* by K. Coates Palgrave in Trees South. Africa: 810 (1977). For further details on *V. pooara* see discussion under 'Doubtful Record' below.

13. **Vitex patula** E.A. Bruce in Bothalia **6**: 237 (1951). —Moldenke in Phytologia **6**: 44 (1957). —Bredenkamp & Botha in S. African J. Bot. **59**: 614 (1993); in Bothalia **26**, 2: 142, fig. 2, 143 (1996). —Sales in Kew Bull. **56**: 199 (2001). —M. Coates Palgrave, ed. 3 of K. Coates Palgrave, Trees South. Africa: 984 (2002). Type from South Africa (Limpopo Province).

Vitex isotjensis sensu K. Coates Palgrave, Trees South. Africa, ed. 1, second impression: 809 (1981) non Gibbs.

Many-stemmed, spreading shrub (1)2.5–4(5) m tall; bark fissured, grey to greyish-brown; young branches orange-tawny tomentellous to shortly lanate on new growth. Leaves 3- sometimes 5-foliolate; petiole (1)3–7 cm long, terete; median petiolule 0.2–1.0 cm long; leaflets (1.5)3–9 × (1)1.5–4 cm, oblong-elliptic to narrow obovate, rounded-obtuse or in apical leaves shortly acuminate to acute at the apex, occasionally emarginate, cuneate at the base, entire or sometimes margin ± crenate in upper half, somewhat chartaceous, drying a dark olive-green, paler beneath, ± scabridulous above with minute bristles, sparsely puberulous to shortly lanate in young leaves beneath with hairs mainly on the nerves, gland-dotted, nerves not or slightly raised above, raised beneath, the blade flat not bullate. Inflorescences axillary, patent spreading, of ± sturdy densely-flowered compound dichasia, 3–10 cm long, orange-tawny tomentose; bracteoles 0.5–1.5 cm long, usually linear. Calyx obconical, 5-lobed, gland-dotted, lanate outside with shortly tangled pilose hairs, accrescent in fruit; tube 2–3 mm long; lobes 0.5–1.3(2) mm long, erect. Corolla c. 8 mm long, shortly lanate outside, not gland-dotted; tube creamy-white, 4–5 mm long, curved; lobes mauve, at an obtuse angle to the tube, middle lower lobe 2–2.3 × 2–2.3 mm. Stamens not exserted from the corolla tube; filaments with glandular hairs. Ovary globose, rounded and pubescent at the apex; not gland-dotted; style 3.5–4.5 mm long. Fruit an ellipsoid-spherical drupe, 0.9–1.3 × 0.7–0.9 cm, apex rounded, the lower c. $^1/_3$ enclosed in the enlarged calyx, black at maturity; fruiting calyx cup-shaped, pubescent, not gland-dotted.

Zimbabwe. E: Chipinge Distr., c. 15 km north of Chisumbanje, 31.i.1975, *Gibbs Russell* 2711 (SRGH). S: Chiredzi Distr., west of Chipinda Pools, 12.x.1951, *McGregor* 72/51 (E). **Mozambique**. Z: Pebane, by the lighthouse, 8.iii.1966, *Torre & Correia* 15058 (LISC). GI: Chicualacuala Distr., between Dumela and Palfrey's Store, 30.iv.1961, *Drummond & Rutherford-Smith* 7642 (K). M: Matutuíne Distr., Santaca to Mazeminhama, 22.xi.1948, *Gomes e Sousa* 3885 (COI; K; LMA; PRE; S).

Also in South Africa (Limpopo Province). Deciduous woodland with *Erythroxylon* and *Rinorea*, and scrub, on sandy soil and often on rocky hillsides at low altitudes, and in coastal bush with *Carissa* and *Mimusops sp.*; 20–400 m.

V. patula has characteristically patent peduncles and small clusters of flowers, although this feature is not always clear in pressed specimens. This species can be distinguished from specimens of *V. mombassae* with 3-foliolate leaves by the colour of the third degree veins on the leaflet lower surface (×50 magnification required): the veins are darker than the epidermis in *V. patula* but are concolorous with epidermis in *V. mombassae*. 5-foliolate specimens of *V. patula*, e.g. *Pope, Biegel & Russell* 1513 from Chiredzi in Zimbabwe (BR; K; SRGH), may be distinguished from *V. ferruginea* by having leaflets cuneate at the base, shorter petiolules and smaller flowers.

Subgenus **Holmskioldiopsis** W. Piep.

Calyx broadly 5-lobed; fruiting calyx campanulate with spreading lobes. Fruit a small dry obovoid nut (mericarp) up to c. 3 mm long and 5 mm wide, ± contained within the enlarged persistent calyx tube.

14. **Vitex zeyheri** Sond. ex Schauer in A. de Candolle, Prodr. **11**: 693 (1847). —Pearson in F.C. **5**, 1: 216 (1901). —Pieper in Bot. Jahrb. Syst. **62**, Beibl. 141: 74 (1928). —O.B. Miller in J. S. African Bot. **18**: 76 (1952). —Moldenke in Phytologia **6**: 230 (1958). —K. Coates Palgrave, Trees South. Africa: 812 (1977). —Barnes & Turton, List Fl. Pl. Botswana at Nat. Mus., Sebele & Univ. Botswana: 27 (1986). —Bredenkamp & Botha in S. African J. Bot. **59**: 617 (1993); in Bothalia **26**, 2: 147 (1996). —Sales in Kew Bull. **56**: 203 (2001). —M. Coates Palgrave, ed. 3 of K. Coates Palgrave, Trees South. Africa: 987 (2002). Type from South Africa (Limpopo and North West Provinces).

Vitex zeyheri var. *brevipes* H. Pearson in F.C. **5**, 1: 216 (1901).

Shrub or tree (2.5)3–6 m tall; bark fissured, dark grey-brown; young branches and new growth densely whitish tomentose, becoming tomentellous to glabrescent on older branches. Leaves 3–5-foliolate; petiole 1–3 cm long, terete; median petiolules 0–4 mm long; leaflets 2–7 × 1–3.5 cm, elliptic to narrow-obovate, acute to rounded and sometimes shortly apiculate at the apex, cuneate to broadly cuneate at the base, entire, grey-green to silvery-white tomentose on both surfaces, the blades darker on the upper surface, not gland-dotted, nerves raised beneath. Inflorescences axillary, of ± sturdy compound dichasia, 5–9 cm long, densely white-tomentose; bracteoles 0.5–1.5 cm long, linear-spathulate. Calyx campanulate, 5-lobed, not gland-dotted, tomentose outside, accrescent in fruit; tube c. 4 mm long; lobes 1.5–2.5 mm long, spreading, tomentose outside and pubescent inside. Corolla sweet-scented, c. 10 mm long, not gland-dotted, tomentose outside; tube white-cream, 7–8 mm long, straight; lobes pale mauve, at an obtuse angle to tube, middle lower lobe 3 × 3.5 mm. Stamens scarcely exserted from corolla tube; filaments glabrous. Ovary obovoid, rounded at the apex, gland-dotted, tomentose except near the base; style c. 7 mm long. Fruit an obovoid nut, c. 4 × 5 mm, truncate at the apex and sharply tipped with remains of the style, almost fully enclosed in the enlarged calyx, brownish to black at maturity; fruiting calyx bell-shaped, tomentose, not gland-dotted.

Botswana. SE: Gaborone, near Notwani Dam, 11.xii.1961, *Yalala* 148 (K; LISC; SRGH).
Also in South Africa (Limpopo, Gauteng and North West Provinces). In wooded grassland with *Terminalia sericea*, on stony hill sides and in rocky valleys; c. 1000 m.
Characterized by the white tomentose indumentum on branches, leaves and inflorescences. It is related to the southern African *V. rehmannii* Gürke in which the median leaflets are narrowly oblong to lanceolate and both surfaces of the lamina are glabrous or with only a thin indumentum.

15. **Vitex obovata** E. Mey., Comment. Pl. Afr. Austr.: 273 (1835). —Pearson in F.C. **5**, 1: 214 (1901). —Pieper in Bot. Jahrb. Syst. **62**, Beibl. 141: 74 (1928). —Moldenke in Phytologia **6**: 22 (1957). —Ross, Fl. Natal: 300 (1972). —K. Coates Palgrave, Trees South. Africa: 810 (1977). —Bredenkamp & Botha in S. African J. Bot. **59**: 618 (1993); in Bothalia **26**, 2: 147, fig. 7, 148 (1996). —Sales in Kew Bull. **56**: 202 (2001). —M. Coates Palgrave, ed. 3 of K. Coates Palgrave, Trees South. Africa: 984 (2002). Type from South Africa.
    Vitex wilmsii Gürke in Notizbl. Königl. Bot. Gart. Berlin **3**: 76 (1900). —Pearson in F.C. **5**, 1: 216 (1901). —K. Coates Palgrave, Trees South. Africa: 812 (1977). —Bredenkamp & Botha in S. African J. Bot. **59**: 618 (1993). Lectotype from South Africa, chosen by Bredenkamp & Botha, 1993.
    Vitex reflexa H. Pearson in F.C. **5**: 215 (1901). Type from South Africa.
    Vitex wilmsii Gürke var. *reflexa* (H. Pearson) W. Piep. in Bot. Jahrb. Syst. **62**, Beibl. 141: 74 (1928). —Ross, Fl. Natal: 300 (1972).
    Vitex obovata subsp. *wilmsii* (Gürke) Bredenkamp & Botha in S. African J. Bot. **59**: 619 (1993); in Bothalia **26**, 2: 149, fig. 7 (1996). —M. Coates Palgrave, ed. 3 of K. Coates Palgrave, Trees South. Africa: 984 (2002).

Tree (2.5)3–7(9) m tall; bark fissured, light to dark grey; young branches and new growth densely puberulous to shortly pale yellow-tomentose. Leaves (4)5-foliolate, petiole 3–5 cm long, grooved at the base; median petiolules 0.3–1 mm long; leaflets 6–10 × 2.5–4 cm, elliptic, narrowly ovate to narrowly obovate, rounded and shortly acuminate to acute at the apex, cuneate at the base, entire, drying dark green and somewhat darker above, sparsely pubescent or glabrous and gland-dotted above, pubescent and gland-dotted beneath, nerves raised beneath. Inflorescences axillary of ± sturdy compound dichasia, 6–15 cm long, white-tomentose; bracteoles 0.5–0.8 cm long, linear-spathulate. Calyx campanulate, 5-lobed, not gland-dotted, tomentellous becoming puberulous or glabrescent outside, accrescent in fruit; tube 2.5–3.5 mm long; lobes 1–2 mm long, tomentellous outside and sparsely puberulous inside, spreading. Corolla 10–20 mm long, ± puberulous outside, gland-dotted; tube white, c.1.8 mm, funnel-shaped; lobes white or pale mauve, at obtuse angle to tube, middle lower lobe 6–9 mm long. Stamens clearly exserted from corolla tube; filaments with glandular hairs. Ovary obovoid, rounded at the apex, gland-dotted, puberulous except near the base; style c. 7–10 mm long. Fruit an obovoid nut, c. 3 × 4 cm, truncate at the apex and sharply tipped with remains of the style, almost fully enclosed in the enlarged calyx, brownish to black at maturity; fruiting calyx bell-shaped, pubescent, not gland-dotted.

Mozambique. M: Namaacha, 13.i.1948, *Torre* 7141 (COI; LISC).
Also in South Africa (Limpopo, Mpumalanga, KwaZulu-Natal and Eastern Cape Provinces), Transkei and Swaziland. Wooded grassland often on dry stony hillsides, seasonal water courses and riverine bush with *Adina microcephala* and mixed deciduous woodland; 200–600 m.
Characteristic features are the dark green leaves and the prominent chartaceous fruiting calyx enclosing the fruit.

DOUBTFUL RECORD

Vitex pooara Corbishley in Bull. Misc. Inform., Kew **1920**: 333 (1920). —Pieper in Bot. Jahrb. Syst. **62**, Beibl. 141: 60 (1928). —Moldenke in Phytologia **6**: 89 (1957). —Palmer & Pitman, Trees South. Africa **3**: 1953 (1974). —Bredenkamp & Botha in Bothalia **26**, 2: 144 (1996). —Sales in Kew Bull. **56**: 204 (2001). —M. Coates Palgrave, ed. 3 of K. Coates Palgrave, Trees South. Africa: 986 (2002). Type from South Africa (Limpopo Province).

Tree to 5 m tall with spreading branches. Leaves 5-foliolate; petioles c. 45 mm long; petiolules short or absent; leaflets c. 60 × 28 mm, broadly lanceolate, entire, velvety-pubescent on both surfaces. Inflorescences axillary, of few-flowered dichasia, c. 5.5 cm long. Calyx accrescent in fruit; tube 2 mm long; lobes 0.5 mm long. Corolla zygomorphic, white, curved. Stamens included. Ovary glabrous, sparsely pubescent at apex. Fruit a drupe, black at maturity, spherical or ellipsoidal, c. 15 × 13 mm; fruiting calyx cup-shaped.

Bredenkamp & Botha (1996) have recorded this South African species as "rare in North-West and Northern Provinces and Gauteng" and considered its range of distribution to extend into southern Zimbabwe. However, I have not seen material of it from the Flora Zambesiaca area.

The status and affinities of *V. pooara* have been variously treated by different authors: Corbishley (1920) treated it as related to *V. isotjensis* (= *V. payos*) while Palmer & Pitman (1972) as related to *V. harveyana*, and Coates Palgrave (1983) has merged it with *V. mombassae*. *V. pooara* is here considered to be a separate species distinguished by an indumentum of finer shorter hairs which give the leaflets a velvety feel, by leaflets broadly-lanceolate, thin and flexible in texture and very shortly petiolulate, by the few-flowered inflorescences and peduncles and pedicels thinner and more delicate than in *V. mombassae* and *V. payos*. Flower size appears to be intermediate between *V. payos* and *V. harveyana*, and much smaller than in *V. mombassae*. Also, the apex of the ovary in *V. pooara* is sparsely pubescent as opposed to lanate in *V. mombassae*.

Subfam. AJUGOIDEAE

3. CLERODENDRUM L.

Clerodendrum L., Sp. Pl.: 637 (1753); Gen. Pl., ed. 5: 285 (1754). —Thomas in Bot. Jahrb. Syst. **68**: 1–106 (1936). —Moldenke in Rev. Fl. Ceylon **4**: 407 (1983)*. —Harley et al. in Kadereit (ed.), Fam. Gen. Vasc. Pl. (Kubitzki, ed. in chief) **VII**: 199 (2004).
    Siphonanthus L., Sp. Pl.: 109 (1753); Gen. Pl., ed. 5: 47 (1754).
    Volkameria L., Sp. Pl.: 637 (1753); Gen. Pl., ed. 5: 284 (1754).
    Cornacchinia Savi in Mem. Soc. Ital. Moden. **21**: 184. t. 7. (1837).

Perennial herbs or suffrutices with woody rootstocks, shrubs (often with arching stems), lianes or trees. Leaves simple, opposite decussate or 3–4-verticillate, usually petiolate, caducous; lamina entire, undulate, dentate or crenate; petiole articulated, sometimes the lower part persisting as a woody spine. Flowers in lax or ± dense cymes grouped in axillary or terminal racemose, corymbose, paniculate or head-like ± compact inflorescences. Calyx usually regular, campanulate to tubular or cup-shaped, ± deeply 5-lobed or truncate, sometimes 5-angled, usually persistent and accrescent in fruit. Corolla symmetrical in bud or, if asymmetrical, usually expanding abruptly on upper side due to resupination, actinomorphic; tube ± long, narrowly cylindrical or funnel-shaped; limb of 5 equal or subequal spreading or reflexed lobes, anterior corolla lobe only slightly (if at all) larger than the others. Stamens 4(5?) didynamous inserted on the corolla tube, usually long exserted, erect and subparallel or ± diverging; anthers versatile. Carpels 2, each incompletely divided into 2 loculi; locules 1-ovulate. Style terminal, slender, ± equalling the filaments; stigma exserted, 2-fid, the lobes equal short, slender, acute. Fruit drupaceous, obovoid or globose, often 4-grooved, (2)4-lobed; mesocarp ± fleshy; endocarp bony or crustaceous, separated into four 1-seeded pyrenes or 2 pairs of pyrenes; seed without albumen.

---

* Moldenke has published exhaustive bibliographic notes on this genus in Phytologia, from **57**(2) (1985) to **64** (3) (1988). Only some references are given under individual species below.

A genus with about 400 species (584 taxa fide *Moldenke*), mainly in the Old World tropics and subtropics, but some in America and a few in temperate regions. Some species are cultivated, mainly as ornamentals.

Species introduced and cultivated as garden ornamentals in the Flora Zambesiaca area include the following, and are keyed out below.

Clerodendrum buchananii (Roxb.) Walp., Repert. Bot. Syst. **4**: 108 (1845). —White, For. Fl. N. Rhod.: 365 (1962). —Moldenke in Phytologia **58**: 283 (1985). Type from India.
    Volkameria buchananii Roxb., Fl. Ind., ed. 2, **3**: 60 (1832).

An erect (or climbing?) shrub to c. 4 m high. Leaves 7.8–30 × 7.3–20 cm, broadly ovate to suborbicular, shortly acuminate at the apex, deeply cordate at the base, entire or minutely remotely dentate or repand-dentate, hairy above, more densely so beneath; petiole 1–16 cm long, stout, densely hairy. Cymes 7–15-flowered, disposed in terminal, paniculate or corymbose inflorescences; peduncles of the lower cymes 2–6 cm long; pedicels of the dichotomies 6–11 mm long; rhachis, peduncles and pedicels puberulous, reddish. Calyx 3–5 mm long, campanulate or cylindric-campanulate; lobes 1.25–2.5 mm, triangular, acute. Corolla bright red, scarlet or carmine; tube 1.2–3 cm long; lobes 7–15 × 4–5(6–8) mm, ± reflexed. Stamens exserted portion of 2–2.5 cm long; anthers c. 3 mm long, reddish; style 3–4 cm long, pink-red. Fruit 8–9 mm wide, at first green or dark green, later orange or bright blue, blue-violet to black.

Zambia. C: Lusaka Forest Nursery, fl. 2.iii.1952, *White* 2296 (FHO). **Mozambique**. M: Jardim Tunduru (Jardim Vasco da Gama), fl. 24.xi.1971, *Balsinhas* 2279 (K; LMA).
Native to Java. It occurs in Africa as a cultivated garden ornamental.
According to Moldenke (op. cit.: 287), *C. buchananii* closely resembles *C. speciosissimum* C. Morren and it has been confused with it by many authors. The main distinctions are the different sizes of the floral parts: calyx 3–5 mm long in *C. buchananii* versus 7–12 mm long in *C. speciosissimum*; corolla tube 5–6-times as long as the calyx in the first versus 3–4-times as long as the calyx in the second; corolla lobes 7–10 × 4–5 mm, stamens 2–2.5 cm long and style 3–4 cm long in *C. buchananii* versus corolla lobes 10–25 × 10 mm long, stamens 4–6 cm long and style 6–7 cm long in *C. speciosissimum*.
After Backer & Bakhuizen van den Brink Jr. (Fl. Java **2**: 610 (1965), the calyx in *C. buchananii* is 5–8 mm long and has appressed lobes at most 1.5 mm wide, while in *C. speciosissimum* the calyx is 8–10 mm long and has erect-patent lobes, 2.5–3.5 mm wide.
In the Mozambican specimens seen, the calyx does not exceed 4 mm (total length), the lobes are 1.25–2 mm long, the corolla tube 1.6–2.2 cm long, the corolla lobes 9–14 × 6–8 mm and the stamens exceed the corolla tube for 20–22 mm. From these measurements, we think that they belong to *C. buchananii*, determination also reported by White (loc. cit.) for the Zambian specimen. However, in the specimen *Pedro* 4979 (LMA), cultivated in a garden in Maputo, the calyx is more deeply divided than in the other two specimens and its lobes are a little reflexed at the apex, as it is described for *C. speciosissimum*.
According to literature *C. speciosissimum* is cultivated in some countries of west and east Africa. We have seen São Tomé's specimens (*Espirito Santo* 40 and 3691; *Mendonça* 8 (COI) the first and the latter referred by Exell to *C. speciosissimum* (cf. Exell, Cat. Vasc. Pl. of S. Tomé: 265 (1944), for *Mendonça* 8; Suppl. Cat. Vasc. Pl. S. Tomé: 38 (1956) for *Espirito Santo* 40). However, by the size of their floral parts, they do not differ from the Mozambican ones.
A revision of these two taxa is perhaps necessary.

Clerodendrum costatum R. Br., Prodr. Fl. Nov. Holl. **1**: 511 (1810). —Bentham, Fl. Austr. **5**: 64 (1870). —Moldenke in Phytologia **59**: 113 (1986). —G.J. Leach in J. Adelaide Bot. Gard. **11**: 161, fig. 12 (1989). Type from Australia.
    Clerodendrum cunninghamii Benth., Fl. Austr. **5**: 64 (1870). —Backer & Bakhuizen van den Brink Jr., Fl. Java **2**: 607 (1965). —Moldenke in Phytologia **58**: 121 (1986). Type from Australia.

A shrub or small tree up to 5(7) m high. Leaves (4)6–18(25) × (2.5)4–10(15) cm, ± broadly ovate to elliptic-ovate, acute or acuminate at the apex, sub-cuneate or rounded to subcordate at the base, entire, somewhat membranous to thinly chartaceous, glabrous (except for the puberulous venation) above, ± pubescent beneath; petiole (1)2–5(8) cm long, shorter than the lamina, often collapsed at the base and apex on drying. Inflorescences showy, many-flowered, axillary corymbose cymes, up to 25 cm wide; peduncle, inflorescence branches and pedicels puberulous; primary branches (2.5)5–6(8) cm long; pedicels (3)5–10(15) mm long. Calyx 8–15 mm long, broadly campanulate, puberulous outside, glandular inside; lobes 3–7 mm long, 3–4 mm wide at the base. Corolla hypocrateriform, white, glabrous; tube (4.5)5–6(7) cm long, slender; lobes 5–10 (12) × 2–5(7) mm, elliptic, oblong or oblong-obovate. Stamens much exserted, with filaments 3.3–4 cm long and oblong anthers 2–3 mm long. Fruit 7–10 mm long and 6–12 mm in diameter at the top, purple-black or blue-black, surrounded by the enlarged, dark red, fruiting calyx.

Mozambique. M: Maputo (Lourenço Marques, cidade), Jardim Tunduru (Jardim Vasco da Gama), fl. 17.iii.1971, *Balsinhas* 1806 (K).

Originally from Australia and Papua New Guinea, it has been cultivated as an ornamental in British Guiana, Java, Mozambique and England.

Clerodendrum splendens G. Don in Edinburgh New Philos. J. **11**: 349 (1824). —J.G. Baker in F.T.A. **5**: 300 & 517 (1900). —R. Good in J. Bot. **68**, Suppl. 2 (Gamopet.): 141 (1930). — Hutchinson & Dalziel, F.W.T.A. **2**: 275 (1931). —Thomas in Bot. Jahrb. Syst. **68**: 56 (1936). —Huber in F.W.T.A., ed. 2, **2**: 444 (1963). —H. Moldenke & A. Moldenke in Rev. Fl. Ceylon **4**: 435 (1983). —Verdcourt in F.T.E.A., Verbenaceae: 87 (1992). —R. Fernandes in C.F.A., Verbenaceae. Type from Sierra Leone.

　　Clerodendrum aurantium G. Don in Edinburgh New Philos. J. **11**: 349 (1824). Type from Sierra Leone.

　　Sphinanthus splendens (G. Don) Hiern, Cat. Afr. Pl. Welw. **1**, 4: 841 (1900). Type as for *C. splendens.*

　　Clerodendrum botryodes sensu R. Good in J. Bot. **68**, Suppl. 2 (Gamopet.): 140 (1930) quoad *Gossweiler* 5232, non (Hiern) Baker.

　　Clerodendrum gilettii De Wild. & T. Durand in Compt. Rend. Soc. Bot. Belge, **38**: 113 (1899). Type from Dem. Rep. Congo.

　　Clerodendrum splendens var. *gilettii* (De Wild. & T. Durand) B. Thomas in Bot. Jahrb. Syst. **68**: 58 (1963).

Climber or sarmentose shrub up to 5 m high or a densely branched shrub; branchlets shortly pubescent or minutely puberulous; main branches long and flexuose. Leaves 1.7–18 × 1.5–12 cm, ovate-oblong to broadly ovate to suborbicular or lanceolate, shortly acuminate, cordate, truncate or rounded at the base, entire or with undulate margin, usually glabrous, membranous to subchartaceous, dark green above and paler or greyish-green and minutely glandular-punctate beneath, with nervation ± conspicuous to prominent and sometimes purplish beneath; petiole 0.3–2.7 cm long, the basal part persistent and sometimes ± spinescent. Inflorescences showy, many-flowered, axillary and terminal corymbose cymes; peduncles ± stout, up to 9 cm long; pedicels 3–13 mm long. Calyx 7–9 mm long, divided into broad lobes 3-4 mm long, purplish, accrescent in fruit. Corolla hypocrateriform, red to scarlet; tube c. 2 cm long, lobes 11–13 × 6–8 mm, broadly oblong or obovate, rounded, with sessile glands beneath. Stamens exserted for c. 3 cm. Drupes c. 1.5 cm long, ovoid-ellipsoid, smooth, black at maturity, glabrous.

Zambia. C: Chilanga, Mt. Makulu Research Station, fl. i.1979, *Critchett* 15/79 (K). **Mozambique**. M: Maputo (Lourenço Marques), Jardim Tunduru (Jardim Vasco da Gama), fl. 9.vi.1971, *Balsinhas* 1907 (LMA, cultivated).

Widespread in west and central Africa from Senegal to Angola and Dem. Rep. Congo. Cultivated as an ornamental. Recorded as naturalized in Dem. Rep. Congo and Angola.

Clerodendrum thomsoniae Balf. in Gard. Chron. **1862**: 70 (Jan 25, 1862); in Edinburgh New Philos. J., N.S. **15**: 232–235, t. II (April, 1862); in Trans. Bot. Soc. Edinburgh **7**: 264–267, t. 7 (1862, after July) & **7**: 580, t. 16 (1863), as "*thompsonae*". —J.G. Baker in F.T.A. **5**: 303 (1900). —Hutchinson & Dalziel, F.W.T.A. **2**: 271, fig. 272 (1931). —Gossweiler, Fl. Exot. Angola: 51 (1950). —Exell, Suppl. Cat. Vasc. Pl. S. Tomé: 38 (1956). —Huber in F.W.T.A., ed. 2, **2**: 442, fig. 307 (1963). —Backer & Bakhuizen van den Brink Jr., Fl. Java **2**: 611 (1965). —H. Moldenke & A. Moldenke in Rev. Fl. Ceylon **4**: 433 (1983). —Malaisse in Troupin, Fl. Rwanda **3**: 276, fig. 89/1 (1985). —Verdcourt in F.T.E.A., Verbenaceae: 87 (1992). Type a cultivated plant in Edinburgh Bot. Gard.

　　Clerodendrum scandens sensu Exell, Cat. Vasc. Pl. S. Tomé: 264 (1944), non P. Beauv.

Climber to 7 m or shrub 1–2 m tall; branchlets slender, obtusely tetragonous, greyish, puberulent and brown or purplish when very young. Leaves 6–18 × 3–10 cm, elliptic or ovate-elliptic, acuminate at the apex, rounded or shortly cuneate at the base, entire, subchartaceous or membranous, dark green above, lighter beneath, usually blackish on drying, glabrous on both faces or slightly puberulous on the nervation beneath; petiole 0.8–3.5 cm long, slender, puberulous. Cymes few-flowered, lax, axillary in the uppermost leaves or subtended by bracts, all forming a terminal, foliate, corymbose inflorescence, 5–9 cm long and 4–8.5 cm wide; pedicels of dichotomies up to 2.8 cm long. Calyx white at anthesis, turning yellow, pink or violet-red, 1.4–3 cm long, ± pentagonous and inflated; lobes 8–10 mm long, broadly ovate, apiculate or shortly acuminate, connivent at the apex, ± completely concealing the corolla tube. Corolla red, scarlet or crimson, sometimes deep rose; tube c. 2.5 cm long, slender, glandular-pubescent; lobes 6–10 × 3.5–4 mm. Stamens long-exserted. Fruit black, glossy.

Mozambique. MS: Gorongosa, fl. 12.xi.1946, *Pedro & Pedrógão* 149 (LMA). M: Maputo (Lourenço Marques), Jardim Tunduru (Jardim Vasco da Gama), fl. 4.xi.1971, *Balsinhas* 2251 (LMA).

A very ornamental plant originally from west tropical Africa and cultivated in Europe and in many tropical countries (cf. Moldenke, op. cit. **1**: 360 (1971).

1. Plants introduced, cultivated as ornamentals · 2
   - Plants native or naturalized · 5
2. Corolla white, the tube (4.5)5–6(7) cm long; leaf lamina (4)6–8(22) × (2.5)4–10(15) cm, broadly ovate or elliptic-ovate; a shrub or small tree, up to 5(7) m · · *Clerodendrum costatum*
   - Corolla red, scarlet, crimson or marked with crimson, the tube up to 3 cm long; shrubs, often sarmentose, or lianes · 3
3. Calyx 1.4–3 cm long, ± pentagonous or inflated, white at anthesis, turning yellow or pink on fading; lobes 8–10 mm long, broadly ovate, ± connivent, ± concealing the corolla tube; leaf lamina 6–18 × 3–10 cm, elliptic or ovate-elliptic, rounded or cuneate at the base; a woody or herbaceous liane, up to 7 m or more high · · · · · · · · · · *Clerodendrum thomsoniae*
   - Calyx 3–6 mm long, campanulate or cylindric-campanulate, red; lobes 1–5 mm long, triangular, not connivent, not concealing the corolla tube; leaf lamina 1.7–30 × 1.5–20 cm, broadly ovate to suborbicular, deeply cordate to rounded or truncate at the base; a shrub, often sarmentose, or climber · 4
4. Calyx lobes 1–3 mm long; leaves deeply cordate at the base; corolla tube 2–2.5 cm long · *Clerodendrum buchananii*
   - Calyx lobes 3–4 mm long; leaves rounded to shallowly cordate at the base; corolla tube 1.5–2 cm long · *Clerodendrum splendens*
5. Calyx cylindrical (sides parallel), 8–12(20) × 4 mm, lobes triangular or ovate, distinctly shorter than the tube; corolla tube more than 2.5 cm long, white; leaves broadly ovate or suborbicular, petiole relatively long; a glabrous shrub or tree · · · · · · · · · 10. *hildebrandtii*
   - Calyx ± narrowed to the base or constricted at the mouth (below the lobes), not cylindrical, if the tube cylindrical or subcylindrical then shorter than above; other characters not combined as above · 6
6. Calyx (tube + lobes) up to 10 mm long · 7
   - Calyx more than 10 mm long · 19
7. Perennial herbs or subshrubs up to 1(2.5) m high, from a woody rootstock; stems simple or few-branched; leaves sessile to indistinctly or shortly petiolate · · · · · · · · · · · · · · · · · · · 8
   - Trees, shrubs or lianes more than 1 m high, or if perennial herbs with annual stems then corolla tube less than 8 mm long (*C. formicarum*); leaves distinctly petiolate (at least the mid cauline ones) · 11
8. Corolla 1.5–4.5 cm long; cymes pedunculate, axillary in upper axils; leaves up to 10 × 2.2 cm, linear-lanceolate or oblanceolate, usually acute; rhizomatous subshrub up to 47(60) cm high · 19. *ternatum*
   - Corolla 5–11.5 cm long; cymes sessile or pedunculate, terminal in capitate or spicate inflorescences, or cymes terminal and in upper axils with some topping short lateral branchlets; subshrubs or perennial herbs · 9
9. A dwarf pyrophytic shrub, with short stems 1–15(25) cm long from a thick woody creeping rootstock; leaves up to 7.5(9) × 3.5 cm, entire, usually obtuse apiculate at the apex, broadly cuneate to rounded at the base; inflorescence ± capitate with cymes subsessile in the axils of clustered leaves on the much contracted upper portion of the stem · · · · · 20. *pusillum*
   - Subshrubs up to 1.5(2.5) m high, with erect stems; leaves larger, ± dentate, acute or acuminate at the apex, long attenuate towards the base; inflorescences spicate to racemose or with cymes terminal and in upper axils · 10
10. Calyx 4.5–5 mm long; cymes ± distinctly pedunculate in upper axils; inflorescences, pedicels and calyces not glandular hairy; leaves usually coarsely deeply incised toothed to pinnatilobed · 21. *incisum*
    - Calyx 8–10 mm long; cymes sessile or shortly pedunculate in the axils of bracts, forming spicate or narrowly racemose inflorescences; inflorescences, pedicels and calyces with ± sparse gland-tipped hairs; leaves with 1–5 coarse teeth on each side towards the apex or ± entire · 22. *lutambense*
11. Inflorescences cauline, mostly borne on leafless old wood, below the lowest leaves, composed of numerous 3–several-flowered cymes, lax; calyx 5–9.5 mm long, tube subcylindrical, slightly striate with triangular acute lobes 1–2 mm long; a liane climbing by means of persistent petiole-bases, glabrous except for the puberulous calyx · 11. *silvanum* var. *silvanum*
    - Inflorescences borne on young parts, terminal or axillary on leafy branches; other characters not as combined above · 12
12. Cymes racemose, mostly aggregated in terminal clusters 6(8) × 7 cm, with some peduncled cymes in axils below the clusters; calyx narrowly funnel-shaped, (4)5–7 mm long, lobes

triangular, acute, 1–2 mm long; petiole-bases persisting as hooks; a ± densely fulvous-pubescent or villose shrub · 12. *tanganyikense*
- Cymes not racemosely arranged; calyx usually campanulate · · · · · · · · · · · · · · · · · · · 13
13. Calyx lobes distinctly longer than the calyx tube, apically attenuate into a linear or filiform point · 14
- Calyx lobes usually shorter than or sub-equalling the tube, not ending in a long point · 15
14. Leaves densely woolly beneath, petiole up to 9 mm long; calyx hairy; corolla tube 9–11.5 mm long · 18. *tricholobum*
- Leaves sparsely appressed pubescent beneath; petiole up to 5.7 cm long; calyx glabrous; corolla tube 14–19 mm long, glabrous · 1. *pleiosciadium*
15. Leaves distinctly 3-nerved from the base; corolla tube (3)5–8 mm long, lobes 2–4 mm long; cymes borne on slender long spreading peduncles in terminal corymbiform inflorescences up to 16 cm wide; branches often hollow · 14. *formicarum*
- Leaves not distinctly 3-nerved at the base; cymes not as above, forming ± dense terminal or axillary inflorescences; branches not hollow · 16
16. Indumentum of young branches, petioles, inflorescence and calyx densely yellowish-orange pilose to bright red-ferruginous with shaggy hairs; shrub, tree or liane · · · · 15. *johnstonii*
- Young branches, petioles, inflorescence and calyx glabrous or if indumentum present then not as above; shrubs or small trees · 17
17. Inflorescence terminal paniculiform, subpyramidal, comprised of many-flowered cymes borne on opposite or 3–4-whorled spreading peduncles up to 4 cm long, sometimes panicles also axillary; leaves 6–19.5 × 3.2–10 cm, petioles up to 5 cm long; calyx lobes 1–2 mm long, triangular, acute; corolla tube 9.5–14 mm long, glabrous and eglandular · · · 13. *toxicarium*
- Inflorescences not as above; leaves smaller, less than 10(12–13) × 5(6–7.5) cm; petioles shorter, up to 2.7 cm long; corolla tube usually shorter, at least with glands · · · · · · · · · 18
18. Branchlets, petioles, inflorescence and calyx puberulous with short appressed antrorse hairs; leaves glabrous or with similar hairs scattered on the lamina and more dense on the main nerves; calyx tube 2.5–3.5 mm long, lobes (0.5)0.75–1.5 cm long, triangular at the base and narrowly acute at the apex; corolla tube (4)4.5–8(10) mm long · · · · ·16. *glabrum*
- Branchlets, petioles, inflorescence and calyx densely whitish pubescent with spreading hairs, leaves with similar hairs scattered on the lamina and more dense on the main nerves; calyx tube 2.5–2.75 mm long, lobes up to 3 mm long, linear-subulate; corolla tube (7)8–13 mm long · 17. *eriophyllum*
19. Inflorescences condensed, capitate, subglobose or subspicate; bracts large, coloured, similar to the calyx lobes; corolla tube 3.5–14 cm long · 20
- Inflorescences ± lax, not capitate, terminal with some cymes in the axils below; bracts not as above; corolla tube 7–11.5 cm long · 26
20. Perennial herbs or subshrubs with woody rootstocks; stems annual becoming ± woody, erect, simple or sparsely branched sometimes long and arching (up to 3 m), without spine-like modified petiole bases or abbreviated hook-like branchlets · · · · · · · · · · · · · · · · · 21
- Shrubs or lianes with spine-like modified petiole bases or abbreviated hook-like branchlets or, if spines or hooks absent, then corolla tube not more than 3.5 cm long · · · · · · · · · 23
21. Leaves up to (25)32 × 14 cm, usually cordate at the base; petiole up to 15.7 cm long; calyx lobes mostly light green, sometimes pale purplish-tipped, reticulation ± obscure; subshrubs with ± woody erect or scrambling stems ·7. *frutectorum*
- Leaves smaller, up to 15.5 × 8 cm, not cordate at the base; petiole up to 3.5 cm long; calyx lobes mostly ± purplish or purple-tipped, reticulation more evident; perennial herbs with woody rhizomes or rootstocks · 22
22. Corolla tube 5–7.5 cm long; calyx 15–20 mm long, with lobes 5.5–8 mm wide and ovate · 6. *buchneri*
- Corolla tube 8.5–14 cm long; calyx (15)19–28 mm long, with lobes 7–10 mm wide and ovate-lanceolate · 4. *robustum*
23. Calyx lobes ± ovate acuminate, densely to sparsely patent hispid-pilose outside or glabrous, sparsely to densely ciliate with pilose hairs 0.5–1.5 mm long on the margins (particularly toward the apex), sometimes ciliolate, reticulation evident or obscure; twigs ± densely hispid-pubescent with short erect or crisped hairs interspersed with scattered longer hairs · · · 24
- Calyx lobes ± lanceolate acute or acuminate, glabrous or sparsely puberulous with short hairs, the margins glabrous or obscurely ciliolate with shorter hairs, reticulation obscure; twigs shortly hispid puberulous · 25
24. Calyx lobe cilia mostly 1–1.5 mm long; corolla tube (4)6–8.5 mm long; inflorescences 3–5.5 cm in diameter (excluding corollas) · 8. *capitatum*

– Calyx lobe cilia short, scarcely 0.5 mm long; corolla tube c. 3.5 mm long; inflorescences smaller · 9. *cephalanthum* subsp. *montanum*
25. Calyx (8)10–13 mm long, with lobes 3–4.5 mm wide, usually glabrous, pale green (reddish in fruit), nerve reticulation not obvious; corolla tube 3.5–4.5(5) cm long; leaf lamina up to 17.5(21.5) × 9(11.5) cm, usually elliptic; petiole up to 4.2 cm long; a shrub or a liane climbing to 10 m or more, with spines up to 3.5 cm long · · · · · · · · · · · · 9. *cephalanthum*
– Calyx 17–21 mm long, with lobes 4–6 mm wide, sparsely hairy, reticulation slightly obvious; corolla tube 4.5–5.5 cm long; leaf lamina not more than 6.7(7.2) × 3.2(4) cm, obovate or suborbicular; petiole up to 7 mm long; shrub c. 1 m high, with spines up to 8 mm long
· 5. *abilioi*
26. Leaf lamina ovate-orbicular to broadly elliptic, up to 21 × 18 cm, cordate or truncate to rounded at the base, not tapering into the petiole; cymes few-flowered, subumbellately-arranged above and axillary below, aggregated into one lax terminal leafy inflorescence up to 17 cm long; calyx obconical at the base · 2. *rotundifolium*
– Leaf lamina deltoid or lanceolate, up to 11 × 4.8(5) cm, cuneate or attenuate at the base and tapering into the petiole; cymes 1–3-flowered, axillary, forming a terminal racemose inflorescence up to 20 cm long; calyx broadly rounded or truncate at the base · · 3. *baumii*

1. **Clerodendrum pleiosciadium** Gürke in Bot. Jahrb. Syst. **18**: 177 (1893); in Engler, Pflanzenw. Ost-Afrikas **C**: 341 (1895). —J.G. Baker in F.T.A. **5**: 303 (1900). —Thomas in Bot. Jahrb. Syst. **68**: 55 (1936). —Brenan, Check-list For. Trees Shrubs Tang. Terr.: 634 (1949). —Garcia in Estud. Ensaios Doc. Junta Invest. Ci. Ultramar [in Mendonça, Contrib. Conhec. Fl. Moçamb., II] **12**: 166 (1954). —Verdcourt in F.T.E.A., Verbenaceae: 95, fig. 12/1–3 (1992)*. Type from Tanzania.
    Clerodendrum syringifolium Baker in Bull. Misc. Inform., Kew **1898**: 160 (1898); in F.T.A. **5**: 300 (1900), as "*syringaefolium*". Type: Malawi, between Mpata and commencement of Tanganyika Plateau, *Whyte* s.n. (K).
    Clerodendrum congestum Gürke in Bot. Jahrb. Syst. **28**: 296 et 466 (1900). —J.G. Baker in F.T.A. **5**: 517 (1900). Type from Tanzania.

A woody herb, suffrutex or shrub, 0.65–3 m high. Branches 4-angled, flowering branches simple or sometimes with branchlets up to 11 cm long, the older branches with persistent spine-like or abbreviated hook-like petiole bases up to 1 cm long. Leaves opposite, subopposite or 3(4)-verticillate, petiolate; lamina up to 12 × 8 cm, ovate or ovate-oblong, abruptly acuminate or cuspidate or attenuate at the apex, truncate rounded or slightly cordate at the base, entire or sometimes few-dentate to irregularly dentate to repand-dentate, ± discolorous, membranous, sparsely and appressed pilose above and on the nerves beneath, densely glandular red-punctate beneath, ciliate, sometimes glabrescent; petiole up to 5.7 cm long, shorter in young leaves, slender, distinctly articulate, sparsely hairy. Cymes clustered in terminal, many-flowered, ± condensed, umbel-like inflorescences, 4–11 cm in diameter, sometimes also axillary at the 2–3 sub-apical nodes; peduncles up to 3 cm long, pubescent with short subappressed hairs. Calyx green, 6–8 mm long, campanulate, glabrous; lobes 4.5–5.5 mm long, longer than the calyx tube, c. 2 mm wide at the base, lanceolate, attenuate into a filiform point. Corolla white or cream, sweet-scented; tube 14–19 mm long, slender, glabrous, with sparse minute sessile glands, similar glands also in the lower part of the lobes; lobes 3.5–4 mm long, suborbicular. Stamens and style exserted for c. 9 mm. Fruit 8–10 mm long and wide, subglobose, deep red or purplish-black.

Zimbabwe. S: Chiredzi Distr., Gonakudzingwa, fl. 4.iv.1961, *R. Goodier* 1052 (K; LISC; SRGH). **Malawi.** N: Chitipa Distr., c. 72 km down Nthalire road from Nyika, c. 42 km south of Chisenga, c. 1750 m, fl. 18.iv.1975, *Pawek* 9351 (K). **Mozambique.** N: Malema Distr., Mutuáli, Estação Experimental do C.I.C.A., na margem do rio Nalume, fl. 27.iv.1961, *Balsinhas & Marrime* 462 (BM; COI; K; LISC; LMA). Z: Alto Molócuè, Dike, c. 160 m, fl. 8.vi.1949, *Gerstner* 7150 (PRE). MS: Manica Distr., Bandula, c. 735 m, fl. 24.iv.1958, *Chase* 6880 (COI; K; LISC; SRGH). GI: entre o Chibuto e a ponte de Cicacati, a 3 km do Chibuto, fl. & fr. 12.vi.1960, *Lemos & Balsinhas* 94 (BM; COI; K; LMA; SRGH).
    Also in Tanzania. Mixed forest and woodland, dry woodland, riverine forest, roadsides, thickets, on schist, deep sand, sandstone, etc.; 100–1750 m.

---

* For a more complete synonymy see Verdcourt in F.T.E.A., Verbenaceae: 95, fig. 12/1–3 (1992).

2. **Clerodendrum rotundifolium** Oliv. in Trans. Linn. Soc., London **29**: 132, t. 89 (1875). —
Gürke in Engler, Pflanzenw. Ost-Afrikas **C**: 341 (1895). —J.G. Baker in F.T.A. **5**: 308 (1900).
—Gürke & Loesener in Mildbraed, Wiss. Ergebn. Zweit. Deutsch. Zentr.-Afrika-Exped.,
Bot. 2: 282 (1914). —R.E. Fries, Wiss. Ergebn. Schwed. Rhod.-Kongo-Exped.: 275 (1916).
—Thomas in Bot. Jahrb. Syst. **68**: 61 (1936). —Robyns, Fl. Sperm. Parc Nat. Alb. **2**: 143
(1947). —Brenan, Check-list For. Trees Shrubs Tang. Terr.: 634 (1949); in Mem. New York
Bot. Gard. **9**: 36 (1954). —Jex-Blake, Gard. E. Africa, ed. 3: 206 (1950). —Dale &
Greenway, Kenya Trees & Shrubs: 585 (1961). —Blundell, Wild Fl. Kenya: 108, t. 3, fig. 15
(1982). —Verdcourt in F.T.E.A., Verbenaceae: 99, fig. 12/7–9 (1992). Type from Tanzania.
   Clerodendrum stuhlmannii Gürke in Bot. Jahrb. Syst. **18**: 173 (1893); in Engler, Pflanzenw.
Ost-Afrikas **C**: 341 (1895). —J.G. Baker in F.T.A. **5**: 309 (1900). Type from Tanzania.
   Clerodendrum guerkei Baker in F.T.A. **5**: 308 (1900). —Moldenke in Phytologia **60**: 136
(1986). Type from Tanzania.
   Clerodendrum zambesiacum Baker in F.T.A. **5**: 309 (1900). —Thomas in Bot. Jahrb. Syst. **68**:
62 (1936). Type: Malawi, Msapa, 1881, *Buchanan* 359 (K, lectotype).
   Siphonanthus rotundifolius (Oliv.) S. Moore in J. Linn. Soc., Bot. **37**: 198 (1905).
   Clerodendrum cavum De Wild. in Bull. Jard. Bot. État **7**: 164 (1920); Pl. Bequaert. **1**: 259
(1922). Type from Dem. Rep. Congo.
   Clerodendrum rotundifolium var. *keniense* T.C.E. Fr. in Notizbl. Bot. Gart. Berlin-Dahlem **8**:
701 (1924). —Thomas in Bot. Jahrb. Syst. **68**: 62 (1936). Type from Kenya.
   Clerodendrum rotundifolium var. *stuhlmannii* (Gürke) B. Thomas in Bot. Jahrb. Syst. **68**: 62
(1936). —Brenan, Check-list For. Trees Shrubs Tang. Terr.: 635 (1949). Type as for *C.
stuhlmannii.*
   Clerodendrum hildebrandtii var. *pubescens* Moldenke in Phytologia **53**: 197 (1983). Type
from Tanzania.

A soft-wooded somewhat virgate shrub 1–4 m high, sometimes scandent or a liane
up to c. 6 m or more, rarely a tree up to c. 8 m tall; flowering branches green-
brownish to dark brown, stout, cylindrical, hollow, hispid pubescent, glabrescent;
indumentum ± dense of short uniformly patent bristles, sometimes ± sparsely
intermixed with longer (up to 2 mm long) bristles, found on the upper part of
branches, petioles, the base and margins of inflorescence-leaves and the
inflorescence and calyx, sometimes reddish or reddish-yellow on the petioles.
Leaves opposite or 3-whorled; lamina 3.7–21 × 2.5–18 cm, ovate-orbicular to ovate
or broadly elliptic, mostly cordate (sometimes deeply so) or subtruncate
(sometimes the two base-types in the same plant) or rounded at the base, rounded
and shortly acuminate at the apex (the acumen sometimes cuspidate), usually
entire or rarely irregularly crenate, soft or somewhat rigid on drying, slightly
pubescent above, more densely so beneath (hairs soft, whitish, denser on the
nervation), sometimes nearly canescent or velvety beneath or on both faces; nerves
and reticulation somewhat raised beneath; petiole 1–16 cm long. Cymes few-
flowered, subumbellately arranged above and axillary below, aggregated into one
lax terminal leafy inflorescence up to 17 cm long; peduncles of cymes ascending,
those of the supra-axillary-cymes 5.5–8 cm long; pedicels long, those of the flowers
in the dichotomies up to 4 cm; bracteoles linear. Calyx 1.7–2.8 cm long, obconical
at the base; lobes up to 1.5 cm long, and 1 cm wide at the base, attenuate-cuspidate,
erect, sparsely puberulous. Corolla white or creamy-white, sweet-scented; tube 7–10
cm long, slender, glandular-pubescent on the outside; lobes 1.3–2 cm long and 5–9
mm wide, oblong or elliptic, obtuse. Stamens and style exserted for c. 3.5 cm. Fruit
1–1.5 cm long and 1.2–1.8 cm broad, black when dry.

Malawi. C: Dedza Distr., Dedza Mt., 1860 m, fl. 5.iv.1978, *Pawek* 14286 (K; SRGH). S: Mulanje
(M'lanje), Swazi Estate, fr. 20.vi.1949, *Faulkner* Kew 435 (BR; COI; S; SRGH). **Mozambique.** N:
Lago Distr., serra Jéci, prox. de Malolo (Malulo), a c. 60 km de Lichinga (Vila Cabral), c. 1800
m, fl. 3.iii.1964, *Torre & Paiva* 10992 (LISC). T: Angónia, environs de Colómuè, near the Malawi
frontier, fl. 11.x.1980, *Macuácua, Stefanesco & Mateus* 962 (LMA).
   Also in Sudan, Uganda, Burundi, Kenya, Tanzania and Dem. Rep. Congo. In evergreen
forest margins and clearings, dry montane forest, *Pinus* plantation, woodland, banks of streams;
1140–1860 m.
   Most specimens from Malawi and Mozambique have a simple indumentum of short hairs. In
some other specimens, however, this indumentum is intermixed with scattered longer bristles
(up to 2 mm long), as seen in *Buchanan* 359 (K, lectotype of the synonym *C. zambesiacum*),
Buchanan 332 (BM), *Chapman* 893 (FHO; SRGH) from Malawi; *Macêdo* 3409 and 3445 (LMA),
Kirk s.n. (K) from Mozambique; and in *Stolz* 614 (LD; S; Z) from Tanzania. Further
investigation may reveal that these represent a separate taxon.

3. **Clerodendrum baumii** Gürke in Warburg, Kunene-Samb.-Exped. Baum: 351 (1903). — Thomas in Bot. Jahrb. Syst. **68**: 62 (1936). —Moldenke in Phytologia **58**: 190 (1985). Type from Angola.

 Clerodendrum nyctaginifolium R. Good in J. Bot. **68**, Suppl. 2 (Gamopet.): 142 (1930). — Thomas in Bot. Jahrb. Syst. **68**: 62 (1936). —Moldenke in Phytologia **62**: 485 (1987). Type from Angola.

A suffrutex 24–150 cm high, with a woody rootstock. Stems annual, 1–numerous, erect, simple or sometimes branched below the main inflorescence, 4-angled, hollow, ± densely leafy, ± densely hispidulous with short patent hairs. Leaves opposite-decussate, petiolate; lamina up to 11 × 4.8(6.5) cm, smaller on the branchlets, deltoid or lanceolate, acute, entire, cuneate or attenuate at the base and decurrent on the petiole, hispidulous with indumentum similar to that on the stems and branches, with subspherical yellow glands on both surfaces but denser beneath; petiole up to 5.5 cm long, sometimes indistinct from the decurrent base of the leaves, articulate near the base. Inflorescence terminal, racemiform, up to 20 cm long (measurement excluding the corollas), formed by 1–3-flowered (or more) axillary erect cymes; inflorescence leaves similar to the cauline ones, but smaller; peduncles up to 1.5 cm long, solitary in each axil; pedicels 1.5–3 cm long, bibracteolate at the base; bracts very small. Calyx purplish-green, 20–25 mm long and 18–20 mm in diameter at the base of the lobes, broadly campanulate, broadly truncate at the base, hispidulous-puberulent, accrescent; lobes 10–12 mm long and 7–10 mm wide at the base, triangular, tapering to a ± narrowly acute apex, subauriculate. Corolla white or cream, scented; tube 8–11.5 cm long, hispidulous with short brownish hairs; lobes unequal, c. 10 mm long, oblong, obtuse. Style and stamens long exserted. Fruit black, surrounded by the somewhat accrescent reddish calyx.

Zambia. B: Kalabo, old garden, fl. 15.x.1953, *Fanshawe* 8080 (FHO; K).
Also in Angola and Dem. Rep. Congo. Kalahari Sand floodplain grassland and open woodland, and in old cultivation; c. 1000 m.

4. **Clerodendrum robustum** Klotzsch in Peters, Naturw. Reise Mossambique **6**, part 1: 259 (1861). —Verdcourt in F.T.E.A., Verbenaceae: 101 (1992). —R. Fernandes in Mem. Soc. Brot. **30**: 14 (1998). Type: Mozambique: Querimba Insel und Festland von Ilha de Moçambique (Mossambique), *Peters* s.n. (B†).

A bushy suffrutex with ± stout stems to c. 1 m tall from a woody rootstock, or a scandent shrub to c. 2(3) m; stems erect, cylindrical, simple or ± branched with internodes usually shorter than the leaves; younger part of the stems and branches shaggy with yellowish-brown hairs. Leaves opposite, petiolate; lamina 4.5–14.5 × 3–7(8) cm, usually up to 2.5 times longer than wide and widest in the upper one third, broadly obovate or oblong-obovate to elliptic, the lowermost smaller and suborbicular, rounded or obtuse and shortly apiculate or acuminate at the apex, rounded or somewhat narrowed at the base, usually entire or less often undulate-crenate, coriaceous, sparsely hispid above with scattered hairs, more densely so on the nervation beneath; nervation impressed above and ± strongly raised beneath; petiole 0.5–2.4(2.9) cm long, with indumentum similar to that on the branches. Inflorescence terminal, subglobose, up to c. 4.5 cm in diameter (measurement excluding the corollas), sometimes with shortly pedunculate cymes in the axils of upper leaves and then the inflorescence subspicate up to 5.5 × 3 cm; bracts green to purplish, ovate to lanceolate, acuminate and apiculate at the apex, ± hairy or glabrous on the back, ± ciliate at the margins. Calyx (15)19–25(28) mm long, hairy or villous on the oblong-obconical tube; lobes pale green or green at the base and purplish upwards, all similar, (12)16–21 × 7–9(10) mm, elliptic lanceolate, ± attenuate towards the acute apex, somewhat thick, ± conspicuously reticulate, ± glabrous, densely to sparsely ciliolate. Corolla usually white, or sometimes creamy-white, whitish-green or creamy-pinkish, scented; tube 8.5–13.5(14) cm long (usually more than 10 cm long), glabrous or ± glandular hairy and with sessile glands; lobes up to 17 × 8 mm, oblong, obtuse; terminal part of the bud 11–21 × 9–11 mm, obovoid; style and stamens exserted for c. 3.5 cm; anthers 2.5–3.5 mm long, brown. Fruit c. 10 mm long, black, shining.

Var. **robustum**

    Clerodendrum mossambicense Klotzsch in Peters, Naturw. Reise Mossambique **6**, part 1: 259 (1861). —R. Fernandes in Mem. Soc. Brot. **30**: 15 (1998), in adnot. Type: Mozambique, Festland von Ilha de Moçambique (Mossambique), *Peters* s.n. (B†).
    Clerodendrum fischeri Gürke in Bot. Jahrb. Syst. **18**: 172 (1893); in Engler, Pflanzenw. Ost-Afrikas **C**: 340 (1895). —J.G. Baker in F.T.A. **5**: 306 (1900). —Thomas in Bot. Jahrb. Syst. **68**: 64 (1936). —Brenan, Check-list For. Trees Shrubs Tang. Terr.: 635 (1949). —Dale & Greenway, Kenya Trees & Shrubs: 584 (1961). —Moldenke in Phytologia **59**: 412 (1986). Type from Tanzania (B†).
    Clerodendrum capitatum sensu J.G. Baker in F.T.A. **5**: 306 (1900) pro parte quoad syn. *C. robustum* et specim. mossambic. *Peters* s.n.; sensu Sim, For. Fl. Port. E. Africa: 92 (1909). — sensu Garcia in Estud. Ensaios Doc. Junta Invest. Ci. Ultramar [in Mendonça, Contrib. Conhec. Fl. Moçamb., II] **12**: 166–167 (1954) pro parte, non (Willd.) Schumach. & Thonn.
    Clerodendrum fischeri var. *robustum* (Klotzch) B. Thomas in Bot. Jahrb. Syst. **68**: 64 (1936). —Brenan, Check-list For. Trees Shrubs Tang. Terr.: 635 (1949). —Moldenke in Phytologia **59**: 414 (1986), *comb. illegit.* Type as for *C. robustum* Klotzsch.
    Clerodendrum capitatum var. *conglobatum* sensu Moldenke in Phytologia **58**: 439 (1985) quoad *Torre & Paiva* 10570 non (Baker) B. Thomas.
    Clerodendrum robustum var. *fischeri* (Gürke) Verdc. in F.T.E.A., Verbenaceae. 102, fig. 12/13–15 (1992). Type as for *C. fischeri*.
    Clerodendrum robustum var. *latilobum* Verdc. in F.T.E.A., Verbenaceae: 102 (1992). —R. Fernandes in Mem. Soc. Brot. **30**: 17 (1998), in adnot. Type from Tanzania.

Calyx (15)19–28 mm long with lobes 7–9(10) mm wide; leaves entire or rarely undulate-crenate.

    Zimbabwe. E: Chipinge Distr., Mwangazi Gap, c. 500 m, fl. 27.i.1975, *Pope, Biegel & Russell* 1371 (BR; K; SRGH). **Malawi**. S: Zomba Distr., Zomba Investig. Farm, fl. 10.i.1957, *Jackson* in *Banda* 339 (FHO; SRGH). **Mozambique**. N: Ribáuè Distr., Chinga Mt., in the valley between Chinga 1 and Chinga 2, coming from Mugasse R., c. 600 m, fl. 25.iv.1949, *Macêdo & Macuácua* 3053 (LMA). Z: Lugela Distr., Mocuba, Namagoa, fl. 11.vi.1949, *Faulkner* 469 (BM; BR; K; SRGH). MS: Beira, at forest edge at Dondo, fl. v.1960, *Pole Evans* 5892 (PRE). GI: Panda, fl. 25.ii.1955, *Exell, Mendonça & Wild* 608 (BM; LISC; SRGH).
    Also in Kenya and Tanzania. Sandveld deciduous woodland and wooded grassland, dry coastal woodland and forest, roadsides, usually on sandy soil; 100–1400 m.
    We agree with Verdcourt's interpretation of *C. robustum* Klotzsch, the type material of which was destroyed at B during World-War II. However, we do not distinguish the vars. *robustum* and *fischeri* (Gürke) Verdc., based only on the density of capitate-glandular hairs of the corolla tube (cf. R. Fernandes in Bol. Soc. Brot. **30**: 14 (1998).
    In herbaria *C. robustum* has been confused with *C. capitatum, C. cephalanthum* subsp. *swynnertonii* and *C. buchneri*. *C. robustum* is distinguished from *C. buchneri* mainly by the indumentum of its stems and branches, where the hairs are all of the same type in *C. robustum* (not intermixed as in *C. buchneri*); by its leaves usually broader and not narrowing towards the base, not so dense and not so upright; by its petiole stouter and shorter relative to the length of the lamina; by its longer calyx (19–28 mm long versus 15–20 mm in *C. buchneri*) with less apiculate and larger lobes (16–21 × 7–10 mm versus 13–17 × 5.5–8 mm in *C. buchneri*); by its broader and longer corolla tube, (8.5–13.5(14) mm long versus 5–7.5 cm long in *C. buchneri*), etc. These differences in characters are not always combined in the same specimen, as for example in *Farwell* 206 from Zimbabwe, and *Gomes e Sousa* 1880 from Panda, Mozambique. However, such specimens occur in the same areas where true *C. robustum* is found.
    Torre 6774 (LISC; SRGH) from Mozambique (Chibuto), without corollas, cited as *C. capitatum* by Garcia (loc. cit.) and as *C. fischeri* by Moldenke (loc. cit.), is a woody much branched shrub which approaches *C. buchneri* in leaf shape and the very attenuate calyx lobes, but is similar to *C. robustum* in the total length of the calyx (c. 25 mm).
    Correia & Marques 3315 (LMU), from Mozambique (Inhambane, Govuro, at 6 km from Tessolo to Jofane), without corollas, is a perennial herb c. 40 cm high with a stout stem; it approaches *C. robustum* in habit, rigid leaves, inflorescence and calyces, but differs from it by the shorter and less dense indumentum of stem and leaves and by the leaf lamina which is relatively narrower and attenuate towards both extremities. Perhaps it is a hybrid between *C. robustum* and *C. cephalanthum* subsp. *swynnertonii*.

Var. **macrocalyx** R. Fern. in Mem. Soc. Brot. **30**: 17 (1998). Type: Mozambique, Mossurize Distr., monte Sitatonga, Gogói (Nova Sintra) para Dombe, 4.iii.1966, *Pereira, Sarmento & Marques* 1669 (LMU, holotype).

Calyx 32–33 mm long with lobes 12–13 mm wide; leaves distinctly crenate-dentate.

    This is a perennial herb with a stout stem arising from a large woody rootstock; leaves up to 12 × 5 cm, distinctly dentate in the upper $\frac{1}{2}$–$\frac{1}{3}$; petiole relatively short (up to 1.8 cm long);

inflorescence terminal, not very dense. So far known only from the type specimen.

C. robustum var. *latilobum* Verdc. differs from var. *macrocalyx* in having a smaller calyx (c. 18 mm long with lobes 16 × 7–8 mm).

5. **Clerodendrum abilioi** R. Fern. in Mem. Soc. Brot. **30**: 7 (1998). Type: Mozambique, Eráti, Estação Experimental de Namapa, *Lemos & Macuácua* 50 (LISC, holotype; COI; K; LMA).
    Clerodendrum mossambicense sensu Moldenke in Phytologia **62**: 324 (1987) quoad specim. *Lemos & Macuácua* 50, non Klotzsch.

A suffrutex with stout stems to c. 1 m tall. Older stems erect, obscurely 4-angled, without leaves, glabrous with bark white, transversally fissured and somewhat lenticellate; old petiole-bases persistent and spine-like up to 8 mm long, patent, straight, also with white bark. Branchlets leafy, 4–25 cm long, erect or spreading, slender, brownish, sparsely pubescent with short whitish antrorse hairs, sparser towards the base; internodes shorter than the leaves. Leaves opposite or subopposite, shortly petiolate; lamina usually small, 1.1–6.7(7.2) × 0.7–3.2(4) cm, obovate or suborbicular, sometimes the pair at the same node very unequal, obtuse or rounded and acuminate-mucronate at the apex, broadly cuneate at the base, entire, somewhat rigid (but not coriaceous) on drying, sparsely and appressed-hairy above and on the nervation beneath; midrib and lateral nerves slightly impressed above, raised beneath, the reticulation raised beneath; petiole 0.2–0.7 cm long, 6–12-times shorter than the lamina, slender, pubescent with whitish appressed antrorse hairs. Inflorescences terminal on the branchlets, subcapitate, few–many-flowered, surrounded by the uppermost leaves. Calyx 17–21 mm long; tube obconical, slightly constricted above the ovary, sparsely hairy; lobes 4–6 mm wide at the base, erect, lanceolate, very attenuate to nearly filiform at the apex, sparsely hairy, ciliate with a few short cilia, nerves slightly raised. Corolla white; tube 4.5–5.5 cm long, slender, somewhat densely glandular hairy; head of buds ellipsoid, densely glandular hairy.

Mozambique. N: Namapa Distr., "Estação Experimental de Namapa, fl. 23.iii.1960, *Lemos & Macuácua* 50 (COI; K; LMA; LISC).

Not known elsewhere. Habitat unknown.

For comparison with *C. cephalanthum*, *C. robustum* and *C. mossambicence* (ex descript.), see R. Fernandes (op. cit.: 8 (1998).

6. **Clerodendrum buchneri** Gürke in Bot. Jahrb. Syst. **18**: 172 (1893); in Warburg, Kunene-Samb. Exped. Baum· 351 (1903). —J.G. Baker in F.T.A. **5**: 305 (1900). —Thomas in Bot. Jahrb. Syst. **68**: 64 (1936). —Hutchinson, Botanist South. Africa: 497 (1946). —Brenan, Check-list For. Trees Shrubs Tang. Terr.: 635 (1949). —White, For. Fl. N. Rhod.: 367 (1962). —Richards & Morony, Check List Fl. Mbala: 236 (1969). —Moldenke in Phytologia **58**: 300 et 330 (1985). —Verdcourt in F.T.E.A., Verbenaceae: 102, fig. 12/16–18 (1992). Type from Angola.
    Clerodendrum strictum Baker in F.T.A. **5**: 305 (1900). —R. Good in J. Bot. **68**, Suppl. 2 (Gamopet.): 141 (1930). Type from Angola.
    Clerodendrum cuneifolium Baker in F.T.A. **5**: 306 (1900). Type from Angola.
    Siphonanthus strictus Hiern, Cat. Afr. Pl. Welw. **1**, 4: 840 (1900), as "*stricta*". Type as for *C. strictum* Baker.
    Siphonanthus cuneifolius Hiern, Cat. Afr. Pl. Welw. **1**, 4: 840 (1900), as "*cuneifolia*". Type as for *C. cuneifolium* Baker.
    Clerodendrum hockii De Wild. in Bull. Jard. Bot. État **3**: 266 (1911). Type from Dem. Rep. Congo.
    Clerodendrum humile Chiov. in Nuov. Giorn. Bot. Ital., N.S. **29**: 117 (1923). Type from Dem. Rep. Congo.
    Clerodendrum impensum B. Thomas var. *buchneroides* B. Thomas in Bot. Jahrb. Syst. **68**: 66 (1936). —Brenan, Check-list For. Trees Shrubs Tang. Terr.: 636 (1949) *nom. inval.* Type from Tanzania.
    Clerodendrum capitatum sensu J.G. Baker in F.T.A. **5**: 306 (1900) quoad *Carson* s.n. (Tanganyika Plateau) et *Carson* 63. —sensu Richards & Morony, Check List Fl. Mbala: 236 (1969), non (Willd.) Schumach. & Thonn.

A suffrutex with a woody rhizomatous rootstock; stems herbaceous, 1–several, up to c. 1 m high (usually less than 30 cm), erect, usually simple, sometimes few-branched, often densely leafy in upper portion, with an indumentum of long spreading, ± densely hispid-pubescent, with short erect hairs interspersed with scattered or numerous long purplish to whitish pilose hairs. Leaves mostly opposite to subopposite, borne on ascending petioles; lamina of the median leaves 5–15.5 × 2.6–6.8 cm (1.5–3.5 times longer than wide, widest in the upper one third), lowermost leaves smallest, sometimes not more than 9 × 3 mm), oblong-obovate to obovate or ± broadly oblong, ± rounded and shortly acuminate or subtruncate and apiculate at the apex, cuneate or rounded at the base, usually entire or sometimes with a few broad teeth towards the apex, light to dark green, somewhat discolorous, sparsely setose-puberulous with short hairs above and beneath, densest on the nervation; midrib and lateral nerves slightly to much impressed above and prominent beneath, the leaves ± bullate; petiole 0.7–3.5 cm long, 3–10-times shorter than the lamina, slender, canaliculate; internodes shorter than the leaves. Inflorescence terminal, capitate, subglobose and 3.5–9 cm long or somewhat elongate with additional flower clusters in the axils beneath, often exceeded by the uppermost leaves (excluding the corollas); bracts maroon to purplish or purple, 1.2–3.5 × 0.5–1.5 cm, elliptic to narrowly lanceolate, reticulate-nerved, the lowermost subfoliaceous, pubescent outside, ciliate. Flowers ± pedicellate, the lowermost pedicels c. 6 mm long. Calyx 1.5–2 cm long, striate, glabrous or villose below; lobes dark purple or pale green margined with purple, 13–17 × 5.5–8 mm wide, ovate, acute or shortly apiculate (apiculum up to 1 mm long) or sometimes lanceolate and attenuate towards the apex, thin-textured, reticulate-nerved, glabrous to pilose outside, pale purple ciliate. Corolla sweet-scented ("hyacinth-scent"), white or creamy-white; tube 5–7.5(9) cm long, with glandular hairs or ± sessile glands; lobes 11–12 × 7 mm, rounded at the apex. Stamens exserted for c. 2.5 cm; anthers c. 3 mm long. Fruit c. 1.3 cm long and 0.8–1.3 cm wide.

Botswana. N: Ngamiland Distr., near the Movombi–Kwando road, 12.3 km west of its junction with the main Kwando riverine road at approx. 18°08'S, 23°16'E, fl. 28.i.1978, *P.A. Smith* 2303 (K; SRGH); near Gaenoma Village, on the Movombi–Gwiligwa track, 18°30'S, 22°23'E, fl. 16.ii.1983, *P.A.Smith* 4104 (COI). **Zambia.** B: c. 3 km west of Kalabo, foot of escarpment at edge of swamp bordering river, fl. 13.xi.1959, *Drummond & Cookson* 6442 (E; K; LISC; SRGH). N: Mbala Distr., 20 km from Mbala on road to Kalambo Falls, 08°44'S, 31°20'E, fl. 12.i.1975, *Brummitt & Polhill* 13736 (K). W: Solwezi Distr., c. 3 km east of Solwezi, fl. 17.iii.1961, *Drummond & Rutherford-Smith* 6948 (K; SRGH). C: Serenje Distr., Kundalila Falls, 1450 m, fl. 12.iii.1975, *Hooper & Townsend* 680 (K). S: Livingstone Distr., Simango, c. 1152 m., fl. 4.ii.1956, *Gilges* 573 (K; PRE). **Zimbabwe.** E: Chipinge Distr., c. 27 km southwest of Chirinda, c. 640 m, fl. iv.1963, *Goldsmith* 7/63 (K). S: Chiredzi Distr., Gonakudzingwa, fr. juv. 17.v.1962, *Savory* 842 (SRGH).

Also in Tanzania, Dem. Rep. Congo and Angola. Deciduous woodlands, mainly *Brachystegia*, and wooded grasslands, on Kalahari Sands and sandy soils often on anthills, dambo margins and sandy roadsides; 400–1700 m.

Sometimes confused with *C. capitatum* (Willd.) Schumach. & Thonn., but differs in habit (a rhizomatous suffrutex not a scrambling shrub or liane with spine-like modified petiole bases) and in its leaves (widest above the middle, with reticulation raised beneath).

7. **Clerodendrum frutectorum** S. Moore in J. Bot. **57**: 249 (1919). —Thomas in Bot. Jahrb. Syst. **68**: 63 (1936). —Moldenke in Phytologia **59**: 472 (1986). —Verdcourt in F.T.E.A., Verbenaceae: 103 (1992). —M. Coates Palgrave, ed. 3 of K. Coates Palgrave, Trees South. Africa: 989 (2002). Type from Dem. Rep. Congo.

Clerodendrum capitatum var. *rhodesiense* Moldenke in Phytologia **3**: 263 (1950); op. cit. **58**: 439 (1985). Type: Zambia, Mwinilunga Distr., south of Matonchi Farm, 24.i.1938, *Milne-Redhead* 4303 (BR; K).

Clerodendrum capitatum sensu White, For. Fl. N. Rhod.: 367 (1962), pro max. parte (excl. specim. *White* 3136b (FHO; K), non (Willd.) Schumach. & Thonn.

Clerodendrum capitatum var. *conglobatum* sensu Moldenke in Phytologia **58**: 439 (1985) pro parte quoad specim. *Bainbridge* 690 et 727 (SRGH), non (Baker) B. Thomas.

A weak-stemmed shrub to 3 m high or subshrub to 1.2 m high; stems dark grey, erect and sparsely branched or arching and scandent to c. 5 m, up to 5 cm in diameter at the base, lenticelled; branches green, hollow, young growth with a double indumentum of short densely patent hairs interspersed with longer spreading multicellular, pale to reddish hairs, older branches glabrescent, or only with short hairs; persistent spine-like or hooked petiole bases absent. Leaves decussate-opposite, large, somewhat clustered towards the ends of branchlets; lamina of the median leaves up to 25(32) × 14 cm, smaller immediately below the inflorescence (up to 10 × 8 cm), oblong-ovate to broadly ovate or oblong, shortly acuminate at the apex, ± cordate or sometimes rounded at the base, entire to remotely and irregularly obtusely dentate, coarsely crenate in the upper $^1/_2$–$^1/_3$, membranous, sparsely pubescent above, more densely pubescent beneath particularly on the venation, with 2–3 pairs of basal nerves, the reticulation slightly raised beneath; petiole 0.9–15.7 cm long, 1.7–3.5(4)-times shorter than the lamina in the larger leaves, indumentum as for the branches. Inflorescences terminal, subglobose or semi-globose, 2.5–5.5 × 3–5.5 cm (without the corollas), surrounded by the upper leaves; bracts 1.2–2.5 × 0.45–1.5 cm, lanceolate to elliptic, acuminate at the apex, suddenly narrowed into a stipe-like base, pubescent. Calyx light green, 18–24(26) mm long (total length), the tube obconical and slightly contracted above, uniformly shortly densely pilose; lobes pale green on drying (light green sometimes tipped with pink when fresh), 4–8.5 mm wide, oblong-lanceolate, attenuate towards the acute or sometimes cuspidate apex, erect, longitudinally nerved but without raised reticulation, shortly and densely ciliate at the margin, with cilia c. 0.5–1 mm long. Corolla white, or the tube sometimes greenish-white; tube 7.5–12.5 cm long, rather curved above, with short glandular hairs outside; lobes 13–18 × 5–8 mm, oblong; limb-bud 9–10 × 6–8 mm, ovoid, subdeflexed, with glandular hairs. Stamens exserted for 3.5–4 cm; style greenish. Fruit 12 mm long, nearly black (purple when fresh?).

Zambia. B: Sesheke Distr., Masese Forest Station, 1050 m, fl. 2.ii.1975, *Brummitt, Chisumpa & Polhill* 14235 (K; SRGH). N: Samfya Distr., above Samfya Beach, fl. 8.ii.1959, *Watmough* 225 (BR; K; LISC; S; SRGH). W: Kitwe, between Parklands and the slimes dam, 1250 m, fl. 2.iii.1961, *Linley* 84 (K; SRGH). C: Lusaka Distr., Makulu, stream near Chilanga, fr. juv. 10.vi.1957, *Angus* 1620 (BR; K). E: Petauke Distr., Minga Forest Reserve, 26.v.1952, *White* 2880B (BR; K). S: Kalomo Distr., Ngwezi (Ungwezi) R., fl. 15.i.1956, *Gilges* 549 (K; PRE; SRGH). **Zimbabwe**. N: Hurungwe Distr., Mana Pools Game Reserve, fl. juv. ii.1972, *Guy* 1953 (SRGH).

Also in Dem. Rep. Congo and Tanzania. Shade-loving shrub of riverine forest understorey and riverine thicket vegetation, dambo margin thickets, in lake shore vegetation, under shade trees on anthills, and in mixed woodland, on Kalahari Sand and alluvium; 360–1280 m.

C. frutectorum material has often been wrongly determined in herbaria as *C. capitatum*. However, *C. capitatum* may be distinguished by: its habit (a scrambler with ± strong spines on the older branches); by its indumentum (hairs appressed not patent and bristle-like); by its leaves (smaller, 1.5–17.5 × 1–12 cm, entire and not subreticulate beneath) and its shorter petiole (relative to lamina length); by its calyx (up to c. 20 mm long, 18–26 mm long in *C. frutectorum*), with lobes, like the bracts, often mauve or purple (not completely pale green or only tinged rose-purplish towards the apex); by its calyx lobes (lanceolate-elliptic and distinctly reticulately nerved and with scattered long cilia on the margins, not relatively wider with obscure reticulation and shorter and more numerous cilia as in *C. frutectorum*).

Verdcourt (loc. cit.) commented that *C. frutectorum* is perhaps a 'form or a state' of *C. poggei* Gürke (= *C. angolense* Gürke). We consider these as two separate species. In all Zambian material of *C. frutectorum* seen, there is a double indumentum consisting of a dense layer of short erect hairs interspersed with scattered long setiform hairs on the petioles and young branches, whereas in the Angolan specimens of *C. poggei* seen, the indumentum consists only of very short hairs (except on the petiole of a specimen of *Dundo*, where there are setiform hairs towards its apex); the leaves are smaller in *C. frutectorum* (up to 25(32) cm long and 13.7 cm wide versus 45 × 35 cm in *C. poggei*) and relatively narrower (1.5–2.1-times longer than wide versus 1–1.6-times in *C. poggei*) and not so deeply cordate at the base; the petiole, relative to the length of the lamina is shorter in *C. frutectorum* (lamina 1.7–3.9-times longer than the petiole in *C. frutectorum* versus 1.1–2.8-times in *C. poggei*); the inflorescences are globose or subglobose not exceeding 5.5 cm in diameter, whereas in *C. poggei* they are usually spiciform and conical up to 18 × 6.5 cm (at the base); the calyx is white-green on drying (like the bracts) and densely and shortly hairy in *C. frutectorum* and it is brown-purple (like the bracts) and only puberulous in *C. poggei*; the corolla tube is shorter in *C. frutectorum* (7.5–11(12.5) cm long versus (10–11)13–15 cm long in *C. poggei*); the habitat is different, *C. poggei* being a plant of dense forest, while *C. frutectorum* is a plant of woodland, thickets, savanna woodland, etc.

8. **Clerodendrum capitatum** (Willd.) Schumach. & Thonn., Beskr. Guin. Pl.: 61 (1827). — Schauer in A. de Candolle, Prodr. **11**: 673 (1847). —Bentham in Hooker, Niger Fl.: 486 (1849). —W.J. Hooker in Bot. Mag. **74**: t. 4355 (1848). —Gürke in Engler, Pflanzenw. Ost-Afrikas C: 340 (1895). —J.G. Baker in F.T.A. **5**: 305 (1900), pro parte. —T. & H. Durand, Syll. Fl. Congol.: 438 (1909). —Chevalier, Expl. Bot. Afr. Occ. Fr. **1**: 507 (1920). —De

Wildeman in Bull. Jard. Bot. État **7**: 165 (1920). —Broun & Massey, Fl. Sudan: 353 (1929). —Hutchinson & Dalziel, F.W.T.A. **2**: 275 (1931). —Thomas in Bot. Jahrb. Syst. **68**: 64 (1936). —Andrews, Fl. Pl. Sudan **3**: 195 (1956). —Watt & Breyer-Brandwijk, Medic. & Pois. Pl. S. & E. Africa: 1047 (1962). —Huber in F.W.T.A., ed. 2, **2**: 443 (1963). —Binns, First Check List Herb. Fl. Malawi: 103 (1968). —Moldenke in Phytologia **58**: 417, 432 (1985). — Malaisse in Troupin, Fl. Rwanda **3**: 274 (1985). —Verdcourt in F.T.E.A., Verbenaceae: 103, fig. 14/1–3 (1992). —R. Fernandes in Mem. Soc. Brot. **30**: 10 (1998). Type from Guinea.

Volkameria capitata Willd., Sp. Pl., ed. 4, **3**: 384 (1800).

Clerodendrum conglobatum Baker in F.T.A. **5**: 296 (1900). Type from Angola.

Siphonanthus conglobatus Hiern, Cat. Afr. Pl. Welw. **1**, 4: 840 (1900), as "*conglobata*". Type as for *C. conglobatum* Baker.

Siphonanthus capitatus (Willd.) S. Moore in J. Linn. Soc., Bot. **37**: 198 (1905). Type as for *C. capitatum* (Willd.) Schumach. & Thonn.

Clerodendrum talbotii Wernham in Cat. Talbot's Nigerian Pl.: 90 (1913).

Clerodendrum capitatum var. *conglobatum* (Baker) B. Thomas in Bot. Jahrb. Syst. **68**: 65 (1936). —Moldenke in Phytologia **58**: 436 (1985), pro parte. Type as for *C. conglobatum* Baker.

Clerodendrum capitatum var. *talbotii* (Wernham) B. Thomas in Bot. Jahrb. Syst. **68**: 65 (1936). —Moldenke in Phytologia **58**: 440 (1985).

A climbing or scrambling shrub up to 6 m high, the older stems and branches usually with stout straight or hook-like persistent petiole bases up to 2.5 cm long, or sometimes an erect or ± spreading shrub, 1.8–6 m high. Branches and twigs ± densely puberulous with an indumentum of short curved hairs interspersed with scattered, long (up to 6 mm), straight, spreading, purplish or whitish hairs, the latter often not persisting. Leaves opposite, subopposite or verticillate, petiolate; lamina 1.5–17.5(25) × 1–12 cm, elliptic, broadly-elliptic or ovate, wider at or below the middle, acuminate at the apex, rounded or subcordate or cuneate at the base, usually entire, membranous to chartaceous, sparsely pubescent above, sparsely to ± densely so beneath with longer and denser hairs on the venation; petiole 0.3–4(8.5) cm long, slender, hairy like the branches. Inflorescences 3–5.5 cm in diameter (without the corollas), subglobose or hemispherical, subsessile or on short peduncles, formed by congested cymes, surrounded by large bracts; bracts purplish, broadly ovate, abruptly acuminate, contracted at the base, reticulate-nerved, ciliate at the margin; bracteoles spathulate-lanceolate to linear, long-ciliate. Calyx 13–21 mm long; tube c. 2.5 mm long, cylindrical below, widened upwards; lobes pale green (whitish on drying), often mauve or purple, 10–18 × 3–8(10) mm, ovate or ovate-elliptic, attenuate or ± long acuminate at the apex into a slender acute point, long ciliate and ± pilose outside or sparsely hairy, papery, reticulately-nerved. Corolla white, sweetly scented; tube 4–8 cm long, cylindrical, very narrow, densely long hairy or with rather short gland-tipped hairs; lobes 11–16 × 6–7 mm, oblong. Anthers c. 2 mm long. Fruit c. 12 mm in diameter, green when young, black and shiny when ripe.

Zambia. N: Kawambwa, fl. 16.iv.1959, *Mutimushi* 4 (K); Kundabwika Falls on Kalungushi R., fl. 14.iv.1961, *Phipps & Vesey-FitzGerald* 3158 (K; SRGH). **Malawi.** N: Nkhata Bay Distr., Likoma Island, Lake Malawi (Nyasa), fl. s.d. *W.P. Johnson* s.n. (K). S: Zomba Distr., Zomba Mt. slopes, on Naisi (Naizi) road, fl. 8.iii.1976, *Banda* 1234 (SRGH).

Also in Uganda, Kenya, Tanzania, Rwanda, Sudan, Central African Republic, Dem. Rep. Congo and from Gambia and Senegal to Angola in West Africa. Mushitu and riverine forest; 475–1420 m.

All specimens included by White in For. Fl. N. Rhod: 367 (1962) under *C. capitatum* [except *White* 3136b (K) from Mansa (Fort Rosebery), without flowers and with stellate hairs on branches and leaves, which is not a *Clerodendrum sp.*], belong to *C. frutectorum* S. Moore.

Whyte s.n. (K), *Brummitt & Polhill* 16411 (K), *Buchanan* 1489 (BM; K) and *Jackson* 1651 (K), all from Malawi, approach *C. robustum* Klotzsch in the indumentum of the branches, but in features of the calyx (indumentum, shape, size and calyx lobes) and the shorter and more slender corolla tube, agree more with *C. capitatum.* Are they a hybrid between the two species?

9. **Clerodendrum cephalanthum** Oliv. in Hooker's Icon. Pl. **16**: t. 1559 (1887); in Gard. Chron., Ser. 3, **3**: 652 (1888). —Hemsley in Bot. Mag. **129**: t. 7922 (1903). —Thomas in Bot. Jahrb. Syst. **68**: 66 (1936) pro parte. —Brenan, Check-list For. Trees Shrubs Tang. Terr.: 635 (1949). —Verdcourt in F.T.E.A., Verbenaceae: 105 (1992). —M. Coates Palgrave, ed. 3 of K. Coates Palgrave, Trees South. Africa: 988 (2002). TAB. **14**. Type from Zanzibar.

A scandent or erect shrub up to 6 m high. Stems puberulous or pubescent when young, usually glabrous or glabrescent with age; stems and branches usually with stout straight or hook-like persistent petiole bases up to 3.5 cm long. Leaves

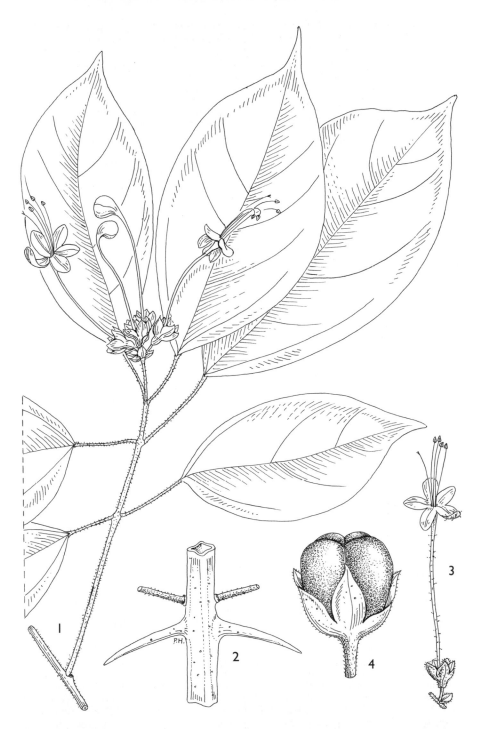

Tab. 14. CLERODENDRUM CEPHALANTHUM. 1, portion of flowering branch (× ²/₃),
from *Chase* 6927; 2, part of stem, with spines (× ²/₃), from *Müller & Gordon* 1354; 3, flower
(× ²/₃), from *Sands & Balaka* 2048; 4, fruit (× 2), from *Brass* 17831. Drawn by Pat Halliday.

opposite, petiolate; lamina 1.2–17.5(21.4) × 0.8–9(11.5) cm, narrowly- to broadly-elliptic, oblong-elliptic, oblong or ovate, acuminate at the apex, the acumen acute or obtuse, rounded or somewhat cuneate at the base, coriaceous or membranous, bullate or flat, with the venation ± impressed above and ± raised beneath, usually ± glabrous on both surfaces or sparsely hairy only on the venation beneath or sometimes ± pubescent on both faces to velvety beneath; petiole puberulous to pubescent, somewhat thick or ± slender.   Inflorescences capitate, subglobose or ovoid, few-flowered and lax or many-flowered and compact; bracts linear to elliptic. Calyx green, pink, purple or maroon; tube c. 3 mm long, not very constricted above the ovary; lobes erect or slightly spreading, acute to ± obtuse, somewhat thick (not membranous), with the venation usually obscure, glabrous or minutely puberulous, shortly and obscurely ciliate, sometimes glandular-punctate on the back.   Corolla white or whitish, scented; tube glabrous on the outside or with sessile glands and/or gland-tipped hairs.   Style and stamens ± exserted; anthers crimson.   Fruit black on drying, 4-lobed.

1. Leaf lamina ± pubescent on both surfaces, sometimes almost velvety beneath; petioles and ultimate twigs ± densely hispid-pubescent with short hairs interspersed with scattered longer hairs; bracts and calyx lobes puberulous and ciliate; corolla tube c. 3.5 mm long; leaves usually smaller than in the other subspecies · · · · · · · · · · · · · iii) subsp. *montanum*
  – Leaf lamina usually glabrous or ± pubescent or puberulous only on the venation beneath; petioles and ultimate twigs shortly hispid-puberulous or glabrous; bracts and calyx lobes puberulous or mostly glabrous; corolla tube c. 3.5–10 mm long; leaves usually ± large · · · 2
2. Corolla tube 6–10 cm long; calyx lobes 6–8 × (4)4.5–6 mm, suberect; inflorescences many-flowered, ± clustered, 3.5–5 cm in diameter; leaf lamina thick, sometimes coriaceous and sub-bullate; petioles relatively short, thick · · · · · · · · · · · · · · · · i) subsp. *cephalanthum*
  – Corolla tube 3.5–4.5(5) cm long; calyx lobes 5–7(8) × 3–4.5 mm, somewhat divergent; inflorescences usually few-flowered, lax, 1.5–3 cm in diameter; leaf lamina usually relatively narrower, membranous, flat; petioles slender · · · · · · · · · · · · · · · ii) subsp. *swynnertonii*

i) Subsp. **cephalanthum** —Verdcourt in F.T.E.A., Verbenaceae: 106, fig. 15 (1992).

Leaves often distinctly coriaceous, glabrous with venation impressed above. Calyx tube and pedicels finely puberulous; lobes glabrous or finely puberulous. Corolla tube 5.5–10 mm long, glabrous or with obscure sessile glands or rarely a few hairs. Usually scandent.

Lamina of the leaves usually large, up to 12.5 × 5.7 cm; petiole 0.8–2.5 cm long, stout, somewhat hirsute to glabrous; inflorescences condensed, many-flowered, up to 5 cm in diameter; calyx lobes c. 4.5 mm wide, erect · var. *cephalanthum*
Lamina of the leaves smaller, up to 7.3 × 3 cm; petiole shorter, up to 0.5 cm long, slender, appressed hairy; inflorescences small, 1.3–3 cm in diameter; calyx lobes c. 6 mm wide, somewhat divergent · var. *torrei*

Var. **cephalanthum**
      Clerodendrum capitatum var. *cephalanthum* (Oliv.) Baker in F.T.A. **5**: 306 (1900) quoad specim. Zanzibar. —Moldenke in Phytologia **58**: 432–436 (1985) pro parte.
      Clerodendrum swynnertonii sensu Thomas in Bot. Jahrb. Syst. **68**: 66 (1936) quoad specim. *Stuhlmann* 804, non S. Moore.

Usually scandent. Young branches pubescent, soon glabrescent or glabrous. Leaves relatively shortly petiolate; lamina up to 12.5 × 5.7 cm, elliptic or obovate-elliptic, acuminate or acute at the apex, rounded or cuneate at the base, usually coriaceous, with the venation impressed above, ± raised beneath (lamina sometimes sub-bullate), glabrous on both surfaces or shortly pubescent on midrib and nerves beneath, punctate beneath; petiole 0.8–2.5 cm long, glabrous or hirsute. Inflorescence 3.5–5 cm in diameter (without corollas), terminal, condensed, many-flowered, surrounded by the uppermost leaves or with leafy bracts at the base. Calyx 13.2–15.4 mm long; lobes 6–8 × 4.5 mm (at the base), ovate-lanceolate, acute, suberect, not reticulate when fresh (subreticulate with age).   Corolla tube (5.5)6–10 cm long, slender, glabrous or with sessile glands or few gland-tipped hairs.

No material has been seen from the Flora Zambesiaca area, but it would be expected to occur in northern Malawi and in the Niassa Province of Mozambique. In Kenya and Tanzania (incl. Zanzibar I.).

The above description of subsp. *cephalanthum* var. *cephalanthum* is based on Oliver's original description and that of Verdcourt (loc. cit.) and also on observation of photocopies of the specimens *Kirk* s.n. (K) (holotype of the taxon); and *Faulkner* 2223 (K), both from Zanzibar, and direct observation of the sheet *Stuhlmann* 804 (HBG), also from Zanzibar.

Barbosa 346 [Xai-Xai (Vila de João Belo), Lumane-Novo, fl. 25.vii.1947] closely approaches subsp. *cephalanthum* var. *cephalanthum*, as seen in the size and shape of the punctate, coriaceous leaf lamina, length of petiole, the compact inflorescence surrounded by the uppermost leaves and the length of the calyx. However, it is not here included in that taxon because the flowers in the specimen are too young to permit a precise conclusion, and Xai-Xai (the collecting locality) is separated by some distance from the natural area of the taxon. On the other hand, some specimens of subsp. *swynnertonii* (S. Moore) Verdc. (see below) were also collected near Xai-Xai.

Var. **torrei** R. Fern. in Mem. Soc. Brot. **30**: 10 (1998). Type: Mozambique, Palma Distr., Cabo Delgado, *Torre & Paiva* 12101 (LISC, holotype).
    Clerodendrum mossambicense sensu Moldenke in Phytologia **62**: 325 (1987) quoad *Torre & Paiva* 12101, non Klotzsch.

A scandent shrub to c. 2 m high with numerous erect or arching stems bearing many short, leafy, lateral branchlets; inflorescences terminal on the lateral branchlets. Corolla tube with sessile glands (only immature inflorescences seen, with corolla tubes still included within the calyx).

Mozambique. N: Palma Distr., andados 4 km do farol do Cabo Delgado para Palma, c. 10 m, fl. juv. 17.iv.1964, *Torre & Paiva* 12101 (LISC).

Not known elsewhere. Coastal bush, on coral sands.

In the size of the leaves and length of the petiole, this taxon is similar to the specimen *Carvalho* 1108 from Chibabava (Mozambique: MS), included in subsp. *swynnertonii*, but in the latter the calyx is smaller and the branchlets and petioles are only puberulous (see also the note on the same specimen in subsp. *swynnertonii*).

ii) Subsp. **swynnertonii** (S. Moore) Verdc. in F.T.E.A., Verbenaceae: 108 (1992). Type: Zimbabwe, Chirinda Forest, *Swynnerton* 85 (BM, lectotype; K).
    Clerodendrum capitatum sensu J.G. Baker in F.T.A. **5**: 306 (1900) quoad *Whyte* s.n. (K) in Masuku Plateau lectum; sensu Garcia in Estud. Ensaios Doc. Junta Invest. Ci. Ultramar [in Mendonça, Contrib. Conhec. Fl. Moçamb., II] **12**: 166–167 (1954) quoad specim mossambic. *Torre* 3008, 3018, 4341, 5832, 6785 et 8012, non Schumach. & Thonn.
    Clerodendrum swynnertonii S. Moore in J. Linn. Soc., Bot. **40**: 166 (1911). —Thomas in Bot. Jahrb. Syst. **68**: 66 (1936). —Brenan, Check-list For. Trees Shrubs Tang. Terr.: 636 (1949) quoad *Schlieben* 2579; in Mem. New York Bot. Gard. **9**: 37 (1954). —Drummond in Kirkia **10**: 272 (1975). Type as above.
    Clerodendrum schliebenii Mildbr. in Notizbl. Bot. Gart. Berlin-Dahlem **11**: 678 (1932). —Thomas in Bot. Jahrb. Syst. **68**: 66 (1936). —Brenan, Check-list For. Trees Shrubs Tang. Terr.: 636 (1949). Type from Tanzania.
    Clerodendrum cephalanthum sensu Thomas in Bot. Jahrb. Syst. **68**: 66 (1936) quoad *Whyte* s.n., non Oliv. sens. str.
    Clerodendrum cephalanthum subsp. *swynnertonii* var. *swynnertonii* —Verdcourt in F.T.E.A., Verbenaceae: 108 (1992).
    Clerodendrum cephalanthum subsp. *swynnertonii* var. *schliebenii* (Mildbr.) Verdc. in F.T.E.A., Verbenaceae: 109 (1992). Type as for *C. schliebenii*.
    Clerodendrum cephalanthum subsp. *mashariki* Verdc. in F.T.E.A., Verbenaceae: 108 (1992). Type from Tanzania.

A shrub, usually scandent, or a liane to 10 m or more. Stems arching, hollow, with stout straight or hook-like persistent woody petiole bases up to 3.5 cm long and 11 mm in diameter at the base; bark roughish, greyish, glabrous; branches lateral divaricate short hollow; branchlets short slender leafy, puberulous or very shortly pubescent, soon glabrous. Leaves opposite, petiolate, often unequal at the same node; lamina 1.2–17.5(21.5) × 0.8–9(11.5) cm, narrowly to broadly elliptic, oblong-elliptic, oblong or ovate, attenuate or acuminate at the apex, the acumen short to ± long, acute or obtuse, usually rounded or sometimes subcuneate at the base, mostly entire or sometimes with 1–3(4) teeth on each side, membranous to stiffly thinly chartaceous with age, nervation raised beneath, 3-nerved from the base with 3–4 pairs of arching lateral nerves, glabrous at both surfaces or with short sparse to dense hairs beneath on the midrib and sometimes on the lateral nerves; petiole 0.2–4.2 cm long, ± slender, glabrous or puberulous, rarely shortly pubescent. Cymes few, 1 or many-flowered, grouped in terminal inflorescences, 1–2 × 1.5–3 cm, subcapitate or subspicate; peduncles 1–2.7 cm long, shortly puberulous; bracts 4–7 × 3.5–4 mm, ovate to lanceolate, shortly acuminate or acute, pale green, tinged with purple upwards, glabrous. Calyx (8)10–13 mm long (up to 15 mm in fruit), puberulous at the base; lobes pale green, reddish or purplish upwards in fruit, somewhat divergent, 5–7(8) × 3–4.5 mm, ovate-lanceolate or ovate, acute or shortly apiculate, not reticulately nerved, glabrous or with minute, antrorse, appressed hairs on the back and margins, sometimes sparsely glandular-punctate. Corolla scented (jasmin-like scent), white; tube 3.5–4.5(5) cm long, slender, with minute sessile spherical glands or short gland-tipped hairs or glabrous; lobes 8–12.5 × 4–5 mm, ovate-spathulate to oblong, obtuse or rounded at the apex. Filaments white; anthers dark purple, 1.5–1.75 mm. Fruit 10–12 mm long, 4-lobed, dark glossy-green, black on drying.

Zimbabwe. E. Mutare Distr., NW base of Cross Hill, commonage, c. 1168 m, fl. 22.iii.1960, *Chase* 7313 (BM; K; LISC; SRGH). S: Masvingo Distr., Mutirikwi Recreational Park, Mtilikwe R., fl. 13.viii.1973, *Basera* in *GHS* 246827. **Malawi.** N: Chitipa Distr., Misuku Hills, Mugesse (Mughesse) rain forest, fl. 6.vi.1973, *Pawek* 7027 (K; SRGH). C: Dedza Distr., on top of Kalicelo (Kalichero), fl. 12.vi.1961, *Chapman* 1367 (FHO; SRGH). S: Mulanje Distr., Lichenya Crater, c. 1200 m, 4.ix.1970, *Müller* 1544 (K; SRGH). **Mozambique.** N: Serra Ribáuè, western-most mountain, fr. 16.viii.1968, *Macêdo* 3451 (LMA). Z: rio Malema, Maropo, c. 22 km from Gurué, c. 1140 m, veget. 1.viii.1979, *de Koning* 7518 (BR; K). MS: Gondola Distr., Mt. Zembe, eastern slopes, c. 850 m, fr. 20.vii.1970, *Müller & Gordon* 1354 (K; SRGH). GI: Xai Xai Distr., Floresta de Chirindzeni, fl. & fr. juv., s.d., *Junod* 358 (G; LISC; PRE).

Also in Tanzania. Evergreen forest and riverine woodland; 100–1820 m.

In subsp. *swynnertonii* we do not recognize varieties which are based only on the presence (var. *swynnertonii*) or absence (var. *schliebenii* (Mildbr.) Verdc.) of glands or of gland-tipped hairs on the corolla tube, because all transitions between total absence to ± dense gland-tipped hairs are found.

Most of the specimens of the Flora Zambesiaca area have relatively longer petioles than the type of the subspecific name. However, some specimens, e.g. *Whellan* 418 (SRGH) from Chirinda, *Torre* 4341 (LISC; LMA; LMU) from Manica, *Lemos & Balsinhas* 176 (BM; COI; K; LISC; LMA), from Inharrime (GI), have short petioles. The latter specimen, by virtue of its very short petioles (shorter than 1.7 cm and 6.9–21-times shorter than the lamina) and the relatively narrower lamina attenuate into a long acumen, deserves perhaps to be considered as a distinct variety or forma, because, in Zimbabwean plants, we found petioles 4.8–8.8(14) times shorter than the lamina (only completely developed leaves were considered).

The specimens *Torre* 6785 (LISC; LMA), collected between Xai-Xai (Vila de João Belo) and Bilene (GI), and *Chapman* 5837 (K) from Ntchisi Mt., Malawi, have calyces up to 13.5 mm long and many-flowered inflorescences more condensed than usual, and in these characters approach subsp. *cephalanthum*. However, the corolla tube length (4.5 cm long) and the leaf characters place these in subsp. *swynnertonii*.

E. Phillips 3509 (K), from Nkhata Bay Distr. (3 miles south of Chikangawa), and *Simão* 484 (LISC), from Manica (Dombe, Zomba Forest), have oblong inflorescences formed by additional pedunculate loose cymes below the terminal subcapitate inflorescence.

Chase 805 (SRGH) from Mutare, Bvumba (Vumba) Mt., Zimbabwe, and *Whyte* s.n. (K) from Masuku Plateau, Malawi, are similar to *C. cephalanthum* subsp. *marikishi* Verdc., a taxon which can probably be sunk in subsp. *swynnertonii*. *Chase* 6927, also from Bvumba Mt., has corolla tubes 5 cm long, which is the longest seen in subsp. *swynnertonii*. *Carvalho* 1108 (BR), from Mozambique (MS: Chibabava), approaches subsp. *montanum* in having small shortly petiolate leaves, but these lack the hairy indumentum characteristic for that subspecies.

iii) Subsp. **montanum** (B. Thomas) Verdc. in F.T.E.A., Verbenaceae: 109 (1992). Type from Tanzania.

    Clerodendrum montanum B. Thomas in Bot. Jahrb. Syst. **68**: 100, 67 (1936). —Brenan, Check-list For. Trees Shrubs Tang. Terr.: 636 (1949).

Leaves clustered towards the ends of the branchlets; laminas usually smaller and deeper green than in the other subspecies, usually pubescent on both surfaces, sometimes velvety beneath, the hairs somewhat long. Inflorescences small, condensed, terminal on the short branchlets. Calyx 12–13 mm long; lobes c. 6 × 4 mm, reticulation not evident, glabrous or pubescent. Corolla tube c. 3.5 cm long, with short gland-tipped hairs.

Malawi. N: Mzimba Distr., Viphya Plateau, c. 51 km SW of Mzuzu, 1730 m, fl. (corolla fallen), 23.vi.1974, *Pawek* 8754 (K; SRGH).
Also in Tanzania. Evergreen rainforest.
Characteristic subsp. *swynnertonii* has also been recorded from the same locality in Malawi.

10. **Clerodendrum hildebrandtii** Vatke in Linnaea **43**: 536 (1882). —Gürke in Engler, Pflanzenw. Ost-Afrikas **C**: 341 (1895). —J.G. Baker in F.T.A. **5**: 302 (1900). —Chiovenda, Fl. Somala **2**: 362 (1932). —Thomas in Bot. Jahrb. Syst. **68**: 67 (1936). —Brenan, Checklist For. Trees Shrubs Tang. Terr.: 637 (1949). —Garcia in Estud. Ensaios Doc. Junta Invest. Ci. Ultramar [in Mendonça, Contrib. Conhec. Fl. Moçamb., II] **12**: 167 (1954). —Dale & Greenway, Kenya Trees & Shrubs: 584 (1961). —Moldenke in Phytologia **60**: 191 (1986). —Verdcourt in F.T.E.A., Verbenaceae: 109, fig. 14/7–8 (1992). Type from Kenya.

A glabrous, rarely puberulous, much branched shrub 1.2–5 m high or a tree 3–9 m tall; branches ± pendulous, cylindrical, fissured, pale greyish or straw-coloured in old parts; branchlets 4-angled, dark, with characteristic raised persistent petiole-bases, lenticellate. Leaves opposite, petiolate; lamina 1.5–13 × 0.8–11.7 cm, broadly ovate to suborbicular, acuminate or rounded and apiculate at the apex, rounded subcordate or broadly cuneate at the base, the lowermost cordate, entire, slightly fleshy (but membranous when dry), densely and minutely punctate; petiole 0.7–7 cm long, slender. Inflorescences terminal panicles of lax dichasial cymes; panicles up to 6 cm long and 8 cm wide (in the Mozambican plants seen); cymes c. 9-flowered, and also some additional depauperate axillary cymes on short branchlets along the branches (in Mozambican plants); peduncles 1–3 cm long; pedicels up to 1.7 cm long; bracteoles setiform. Calyx dull purplish at base; tube 8–12(20?) × 4 mm, cylindrical, glabrous; lobes 2.5–9 mm long, triangular or ovate, distinctly shorter than the tube, becoming 12 mm long in fruit. Corolla white or creamy, pleasantly scented; tube 2.5–4 cm long; lobes 8–12.5 mm long, subequal. Stamens c. 2 cm long; filaments and style reddish or mauve above; anthers red-brownish. Fruit 1–1.7 cm long and wide, black, shiny.

Mozambique. N: Pemba (Porto Amélia), andados 2 km de Metuge para Quissanga, c. 20 m, fl. 19.xii.1963, *Torre & Paiva* 9613 (LISC).
Also in Kenya, Tanzania and Somalia. Wooded grassland, secondary bush and *Acacia* woodland, on sandy periodically inundated soils.
Var. *puberulum* Verdc., op. cit.: 110, from Tanzania, is distinguished by the presence of a shortly pubescent indumentum on the branchlets, the leaf lower surface and the flowers.
Only three of the specimens seen of typical *hildebrandtii* were collected in Mozambique. These were described as being shrubs and all have leaves up to 8.5 × 8 cm; petioles up to 5.5 cm long; calyces not more than 10.5 mm long with lobes 2–3.5 mm long; and corolla tubes c. 2.5 cm long. It is possible that future gatherings will match the maximum dimensions recorded for these organs by Verdcourt in F.T.E.A.

11. **Clerodendrum silvanum** Henriques in Bol. Soc. Brot. **10**: 148 (1892), as "*silvaeanum*". —J.G. Baker in F.T.A. **5**: 299 (1900). —Exell, Cat. Vasc. Pl. Sao Tomé: 264 (1944); in Bull. Brit. Mus. Bot. **4**: 386 (1973). —Verdcourt, F.T.E.A., Verbenaceae: 110 (1992). —R. Fernandes in Mem. Soc. Brot. **30**: 18 (1998). Type from S. Tomé.
Clerodendrum preussii Gürke in Bot. Jahrb. Syst. **18**: 175 (1893). —J.G. Baker in F.T.A. **5**: 302 (1900). —Thomas in Bot. Jahrb. Syst. **68**: 69 (1936). Type from Cameroon.
Clerodendrum buchholzii Gürke in Bot. Jahrb. Syst. **18**: 176 (1893). —J.G. Baker in F.T.A. **5**: 301 (1900). —Thomas in Bot. Jahrb. Syst. **68**: 69 (1936). —Irvine, Woody Pl. Ghana: 750, Pl. 32a (photo) (1961). —Huber in F.W.T.A., ed. 2, **2**: 443 (1963). —Richards & Morony, Check List Fl. Mbala: 236 (1969), excl. specim. *Bullock* 2618. —Moldenke in Phytologia **58**: 294 (1985). —Malaisse in Troupin, Fl. Rwanda **3**: 272, fig. 88/1 (1985). Type from Cameroon.
Clerodendrum thonneri Gürke in Bot. Jahrb. Syst. **28**: 292 (1900). —J.G. Baker in F.T.A. **5**: 517 (1900). —Thomas in Bot. Jahrb. Syst. **68**: 68 (1936). Type from Dem. Rep. Congo.
Siphonanthus costulatus Hiern, Cat. Afr. Pl. Welw. **1**, 4: 843 (1900), as "*costulata*". Syntypes from Angola.
Clerodendrum kentrocaule Baker in F.T.A. **5**: 296 et 515 (1905). Type from Angola.
Siphonanthus nuxioides S. Moore in J. Linn. Soc., Bot. **37**: 197 (1905). Type from Uganda.
Clerodendrum buchholzii var. *parviflorum* B. Thomas in Bot. Jahrb. Syst. **68**: 69 (1936). —Moldenke in Phytologia **58**: 300 (1985). Type from Angola.
Clerodendrum nuxioides (S. Moore) B. Thomas in Bot. Jahrb. Syst. **68**: 69 (1936). —Moldenke in Phytologia **62**: 484 (1987).

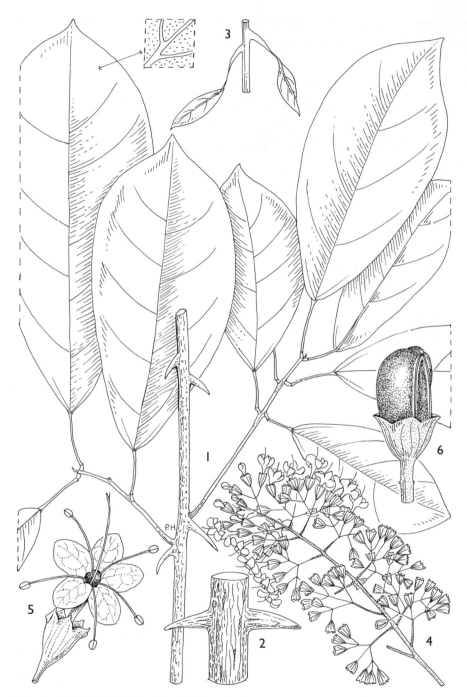

Tab. 15. CLERODENDRUM SILVANUM var. SILVANUM. 1, portion of branch, showing spines on older wood and articulated petioles on lateral branchlets (× 2/3); 2, spines on older branch (× 2/3); 3, pair of young leaves and developing spines (× 2/3); 4, cauliflorus inflorescence (× 2/3); 5, flower (× 2), 1–5 from *Richards* 9582; 6, fruit (× 2), from *Mutimushi* 3458. Drawn by Pat Halliday.

Clerodendrum preussii var. *silvanum* (Henriques) B. Thomas in Bot. Jahrb. Syst. **68**: 70 (1936), as "*silvaeanum*".
Clerodendrum silvanum var. *buchholzii* (Gürke) Verdc. in F.T.E.A., Verbenaceae: 110 (1992). Type as for *C. buchholzii*.

Var. **silvanum** TAB. **15**.

A woody liane often reaching the canopy 12–18 m high, or shrubs 3–4 m high with arching scandent stems, climbing or scandent by means of persistent spine-like and often retrorse petiole bases, the stems sometimes coiling on the ground and rooting; spines on stems up to 2.7 cm long, thick woody and straight, those of young shoots slender sharper and recurved; plants glabrous, except sometimes for the inflorescences and calyces which are minutely puberulous. Stems up to 10–15 cm in diameter with fissured corky grey bark, younger shoots pale grey-brown, with prominent lenticels and often with raised leaf scars or persistent petiole bases. Leaves opposite or subopposite, usually long-petiolate, usually borne toward the apex of the stems or on new shoots rising from old wood; lamina 5–22(30) × 1.8–11(13) cm, oblong-elliptic to oblong-obovate, distinctly acuminate at the apex, rounded at the base, membranous to thinly chartaceous; petiole (0.7)2–5 cm long, articulate and usually bent at nearly a right angle near the base, the lower part (later persistent) thicker. Cymes 2–several-flowered, in lax racemose or condensed subcapitate inflorescences; inflorescences usually borne on the old wood of main stems, near the ground far below the lowest leaves, or sometimes terminating lateral short leafy branchlets arising from old wood, or along somewhat thicker branchlets up to 35 cm long (or more?) with very small leaves, up to 1.5 × 0.6 cm, at the base; peduncles of racemes 2.5–20 cm long, with floral axis 4.5–25(30) cm long; peduncles of the cymes slender ± spreading, opposite or subopposite, the lower ones 1.5–2.7 cm long and 1.5–6.5 cm apart; pedicels of dichotomies 5–10 mm long, the others shorter and ± filiform. Calyx 5–9.5 mm long, cylindrical to subfunnel-shaped, finely striate on drying; lobes 1–2 mm long, triangular, acute, erect. Corolla white, sweet scented; tube (8)10–18(20–25) mm long, narrowly cylindrical, glabrous; lobes obovate, rounded at the top, spreading. Fruit up to 13 mm long and 10 mm wide, ovoid-ellipsoid, smooth, green when young, later black, surrounded by the cup-shaped fruiting calyx.

Zambia. N: Mbala Distr., Lunzua Gorge, fl. juv. 2.iv.1960, *Fanshawe* 5629 (BR; FHO; K). W: Mwinilunga Distr., Lunga R., Mwinilunga, fr. 17.v.1969, *Mutimushi* 3458 (B; K).
From Guineé to Gabon, Dem. Rep. Congo and Angola, and in Uganda. Evergreen forest in river gorge and in mushitu interior; 1000–1500 m.
Var. *brevitubum* R. Fern., an Angolan taxon, is distinguished by its corolla tube equalling or only a little longer than the calyx (c. 2–3 times longer than the calyx in var. *silvanum*).
See R. Fernandes in Mem. Soc. Brot. **30**: 18–23 (1998) for a treatment of this complex.

12. **Clerodendrum tanganyikense** Baker in Bull. Misc. Inform., Kew **1895**: 71 (1895); in F.T.A. **5**: 298 (1900). —Gürke in Engler, Pflanzenw. Ost-Afrikas **C**: 341 (1895). —Thomas in Bot. Jahrb. Syst. **68**: 68 (1936). —R.E. Fries, Wiss. Ergebn. Schwed. Rhod.-Kongo-Exped.: 275 (1916). —White, For. Fl. N. Rhod.: 367 (1962). —Hulstaert, Notes Bot. Mango: 185 (1966). —Richards & Morony, Check List Fl. Mbala: 237 (1969), pro parte. —Verdcourt in F.T.E.A., Verbenaceae: 111, fig. 14/12–14 (1992). Type: Zambia, Lake Tanganyika, Fwambo, *Carson* 52 (K, holotype).
Clerodendrum bequaertii De Wild. in Repert. Spec. Nov. Regni Veg. **13**: 144 (1914). —Thomas in Bot. Jahrb. Syst. **68**: 67 (1936). Type from Dem. Rep. Congo.
Clerodendrum dubium De Wild. in Repert. Spec. Nov. Regni Veg. **13**: 144 (1914). —Thomas in Bot. Jahrb. Syst. **68**: 67 (1936). Type from Dem. Rep. Congo.
Clerodendrum lupakense S. Moore in J. Bot. **57**: 247 (1919). Type from Dem. Rep. Congo.
Clerodendrum consors S. Moore in J. Bot. **57**: 248 (1919). —Moldenke in Phytologia **59**: 107 (1986). Type from Dem. Rep. Congo.
Clerodendrum tanganyikense var. *bequaertii* (De Wild.) Moldenke in Phytologia **4**: 177 (1953). Type as for *C. bequaertii* De Wild. (1914).
Clerodendrum tanganyikense var. *dubium* (De Wild.) Moldenke in Phytologia **4**: 177 (1953). Type as for *C. dubium* De Wild.
?*Clerodendrum tanganyikense* var. *microcalyx* Moldenke in Phytologia **4**: 177 (1953). Type from Dem. Rep. Congo.
Clerodendrum buchholzii sensu Richards & Morony, Check List Fl. Mbala: 236 (1969) quoad specim. *Bullock* 2618, non Gürke.

A low sparsely branched erect shrub to c. 1 m high or scandent shrub with arching stems to c. 6 m high; stems scandent by means of persistent hook-like and often retrorse petiole bases. Indumentum on young shoots and stems of low shrubs ± densely fulvous-pilose or villose consisting of simple many-celled soft patent hairs to c. 1 mm long, sometimes with some shorter ± crisped hairs intermixed; older stems becoming puberulous with mostly short hairs, becoming lenticellate and often with raised leaf scars or persistent petiole bases. Leaves opposite, subopposite or 3-whorled, petiolate; lamina 2.5–16 × 1.5–10.8 cm, broadly elliptic or suborbicular to oblong-obovate, abruptly and shortly acuminate at the apex, rounded to cuneate at the base, entire to repand-dentate or undulate in the margins, dull green and glabrous above, pale green and sparsely pubescent on the margin and venation beneath, with the reticulation ± raised beneath, sometimes chartaceous; petiole 0.2–5.5 cm long. Inflorescences terminal and paniculate on the young leafy branches, 8–11 cm long and up to 8 cm wide (including the opened flowers); cymes short, many-flowered sometimes 2-flowered, racemosely or subumbellately arranged on slender peduncles, sometimes a few in the upper axils or supra-axillary and these sometimes similar to the terminal inflorescence (but smaller), in some intermediates also borne on leafless lower stems but here the clusters are much shorter than in *C. silvanum* and more pubescent; peduncles of the inflorescences 3–6 cm long; peduncles of the cymes 1–3 cm long, the lowermost spreading; pedicels up to 5 mm long. Calyx (4)5–7 mm long and 2–3 mm in diameter, narrowly cylindrical-funnel-shaped, appressed pubescent or hairy with hairs similar to those of the inflorescence; lobes 1–2 mm long, triangular, acute, pale green, accrescent. Corolla white, sweet scented; tube 15–19 mm long, sparsely hairy to the base; lobes 2.5–4.5 × 2.5 mm, subequal, suborbicular to broadly ovate, obtuse. Anthers and stigma greenish. Style exserted 8 mm. Fruit c. 1.2 × 0.9 cm, oblong-ellipsoid, surrounded by the accrescent cupuliform strongly ribbed fruiting calyx c. 10 mm in diameter.

Zambia. B: Zambezi (Balovale), fl. 25.ii.1964, *Fanshawe* 8335 (BR; FHO; K). N: Mbala Distr., Lunzua (Unwin Moffat) Agricultural Station, fl. 13.ii.1964, *Angus* 3863 (FHO); Chinsali Distr., Mbesuma Ranch, Chambeshi R. floodplain, fl. 10.iv.1989, *Pope & Goyder* 2103 (BR; K; NDO). W: Mwinilunga Distr., c. 97 km south of Mwinilunga on Kabompo road, fr. 2.vi.1963, *Loveridge* 725 (BR; K; LISC; SRGH). C: Kabwe (Broken Hill), veg. vi.1920, *Rogers* 26150 (FHO; PRE). S: Mumbwa Distr., banks of the Kafue R. at Mswebe (Mswebi), fr. 16.v.1963, *van Rensburg* 2176 (K; SRGH).
    Also in Uganda, Tanzania, eastern Cameroon, eastern Dem. Rep. Congo, Burundi, Rwanda and Angola. *Brachystegia* woodland and wooded grassland, in thicket vegetation, dambo margins and lake shore woodland, often on termite mounds, sometimes in colonies; 1020–1720 m.
    This taxon approaches *C. silvanum* in the presence of cauliflorous inflorescences and a shorter and sometimes appressed indumentum on the inflorescences (*Bullock* 2618 (BR; K)).
    Moldenke (Fifth Summ. Verbenaceae: 249, 1971) recorded *C. tanganyikensis* from Malawi, but we have not seen any material from that country.

13. **Clerodendrum toxicarium** Baker in F.T.A. **5**: 298 (1900). —Gürke in Engler, Pflanzenw. Ost-Afrikas **C**: 341 (1895) *nomen*. —Thomas in Bot. Jahrb. Syst. **68**: 73 (1936). Type: Malawi, River Shire, fl. 1863, *Kirk* s.n. (K, lectotype).
        Clerodendrum kirkii Baker in F.T.A. **5**: 299 (1900). —Thomas in Bot. Jahrb. Syst. **68**: 73 (1936). —Moldenke in Phytologia **59**: 335 (1986); op. cit. **61**: 406 (1986). Type: Malawi, Upper Shire Valley, viii.1861, *Kirk* s.n. (K, holotype).

A subshrub with herbaceous stems to c. 1 m tall, or a scandent shrub to c. 2(9.5) m; herbaceous stems and young shoots longitudinally 4–6-ridged, sparsely pubescent with short crisped antrorse hairs; older stems becoming woody, arching and scandent with petiole bases persisting as short woody obliquely truncate projections, glabrescent, bark grey and lenticellate, lateral branches up to 30 cm long. Leaves usually ternate, rarely opposite, petiolate; lamina 6–19.5 × 3.2–10 cm, usually shorter than twice the width, widest about the middle, elliptic to broadly elliptic, elliptic-obovate or ovate, acuminate at the apex, rounded or broadly cuneate at the base, entire or sometimes ± irregularly dentate, sparsely setose-glabrous or sparsely puberulous above and more densely so beneath particularly on the nerves; petiole 1.2–5 cm long, slender. Inflorescence terminal on the stems and lateral branches, paniculate, subpyramidal or corymbose, formed by many-flowered cymes in 3(4)-whorls at the upper 3–8 nodes on peduncles up to 4 cm long, usually also with inflorescences in leaf axils of the 2–8 nodes below these and together forming a large leafy inflorescence up to 27 cm long and 24 cm wide; the axillary inflorescences subpyramidal and equalling or longer than the subtending leaf, on divaricate straight or slightly arched nude peduncles up to 9 cm long; pedicles up to 11 mm long; bracts filiform, shorter than the pedicels; peduncles and pedicels shortly puberulous as on the petioles and young branches. Calyx green, 2.5–4.5 mm long, subcampanulate, shortly appressed hairy; lobes 1–2 mm long, triangular, acute. Corolla white; tube 9.5–14 mm long, narrowly cylindrical, glabrous. Fruit 7 mm long.

Malawi. C: Dedza Distr., Mua-Livulezi Forest Reserve, Naminkokwe (Namkokwe) R., 640 m, fl. 19.iii.1955, *Exell, Mendonça & Wild* 1047 (LISC; SRGH). S: Mulanje Distr., road skirting Mulanje Mountain, c. 13 km south of Mulanje, 900 m, fl. 2.iv.1978, *Pawek* 14184 (BR; K). **Mozambique.** N: north of Mandimba, fl. juv. 14.iii.1942, *A.J.W. Hornby* 2573 (PRE). Z: Morrumbala Distr., Massingire, M'bobo, fl. 28.iv.1943, *Torre* 5222A (COI; LISC; LMU; PRE). MS: Marromeu Distr., between Lacerdónia and Chupanga Mission, banks of Zambeze R., fl. 8.v.1942, *Torre* 4098 (LISC).
Not known elsewhere. Low altitude open woodlands, lake shore plain and riverine vegetation, in sandy soils and sometimes on termitaria, also in old cultivation; 50–900 m.
Moldenke (Fifth Summ. Verbenaceae: 246, 1971) cites this species from Zambia, but we have not seen any specimen of this country.

14. **Clerodendrum formicarum** Gürke in Bot. Jahrb. Syst. **18**: 179 (1893); in Engler, Pflanzenw. Ost-Afrikas **C**: 341 (1895). —Henriques in Bol. Soc. Brot. **16**: 69 (1899). —J.G. Baker in F.T.A. **5**: 297 et 516 (1900). —De Wildeman in Bull. Jard. Bot. État **7**: 167 (1920). —T. & H. Durand, Syll. Fl. Congol.: 438 (1909). —R.E. Fries, Wiss. Ergebn. Schwed. Rhod.-Kongo-Exped.: 275 (1916). —Broun & Massey, Fl. Sudan: 353 (1929). —Mildbraed, Wiss. Ergebn. Zweit. Deutsch. Zentr.-Afrika-Exped. **2**: 62 (1922). —R. Good in J. Bot. **68**, Suppl. 2 (Gamopet.): 141 (1930). —Hutchinson & Dalziel, F.W.T.A. **2**: 274 (1931). —Thomas in Bot. Jahrb. Syst. **68**: 74 (1936). —Brenan, Check-list For. Trees Shrubs Tang. Terr.: 637 (1949). —F.W. Andrews, Fl. Pl. Anglo-Egypt. Sudan **3**: 194 (1956). —Irvine, Woody Pl. Ghana: 753 (1961). —Huber in F.W.T.A., ed. 2, **2**: 444 (1963). —Schnell & Beaufort in Bull. Inst. Fondam. Afrique Noire, no. 75: 41, t. 10 fig. F (1966). —Malaisse in Troupin, Fl. Rwanda **3**: 274, fig. 88/2 (1985). —Verdcourt in F.T.E.A., Verbenaceae: 116, fig. 17 (1992). TAB. 16. Type from Dem. Rep. Congo.
    Clerodendrum triplinerve Rolfe in Bol. Soc. Brot. **11**: 87 (1894). —White, For. Fl. N. Rhod.: 367 (1962). Type from Angola.
    Clerodendrum lujaei De Wild. & T. Durand in Compt. Rend. Soc. Bot. Belg. **38**: 213 (1899). Type from Dem. Rep. Congo.
    Siphonanthus formicarum (Gürke) Hicrn, Cat. Afr. Pl. Welw. **1**, 4: 843 (1900).
    ?*Clerodendrum yaundense* Gürke in Bot. Jahrb. Syst. **28**: 297 (1900). —J.G. Baker in F.T.A. **5**: 516 (1900). —Thomas in Bot. Jahrb. Syst. **68**: 73 (1936). Type from Cameroon.
    Clerodendrum oreadum S. Moore in J. Bot. **45**: 93 (1907). —Thomas in Bot. Jahrb. Syst. **68**: 69 (1936). —Moldenke in Phytologia **63**: 49 (1987). Type from Uganda.
    Clerodendrum formicarum var. *sulcatum* B. Thomas in Bot. Jahrb. Syst. **68**: 74 (1936). Type from Tanzania *nom. inval.*
    Clerodendrum triplinerve var. *sulcatum* (B. Thomas) Moldenke in Phytologia **3**: 409 (1951).
    ?*Clerodendrum triplinerve* var. *grandiflorum* Moldenke in Phytologia **4**: 177 (1953). Type from Dem. Rep. Congo.

    Erect tufted rhizomatous suffrutex 40–100 cm high from a woody rootstock with stems numerous, annual, herbaceous becoming woody (in East Africa a shrub with scandent branches to 6 m long or small tree with spreading crown to 8 m). Stems simple or few-branched towards the apex, petiole bases persisting as short woody obliquely truncate projections, puberulous on young growth, to glabrescent; internodes 3–7 cm long. Leaves 3–4-whorled; lamina 1.2–12.5(14) × 0.8–5.5 cm, ± narrowly oblong-elliptic to elliptic or lanceolate, acuminate to ± caudate at the apex, broadly cuneate to rounded at the base, entire, thin textured, dark green and glossy above, glabrous or puberulous only on the nerves beneath, distinctly 3-nerved from the base with first pair of lateral nerves inserted just at the lamina-base and somewhat stronger than the lateral nerves; petiole articulated where it rises from the woody basal part, 0.2–1.5 cm long, erect or ± spreading. Inflorescence a terminal loose umbel-like corymb with a relatively short axis, up to 10 cm long and 16 cm in diameter in fruit, shortly pubescent, formed by small

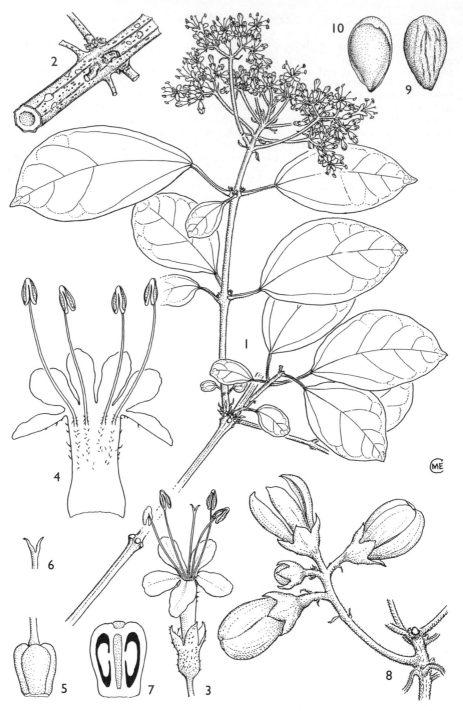

Tab. 16. CLERODENDRUM FORMICARUM. 1, portion of flowering branch (× ²/₃), from *Drummond & Hemsley* 4701; 2, part of branch, showing hollow centre and persistent petiole base (× ²/₃), from *Hansen* 904; 3, flower (× 4); 4, corolla, opened out (× 6); 5, ovary (× 12); 6, stigma (× 12); 7, longitudinal section of ovary (× 16), 3–7 from *Paulo* 537; 8, part of fruiting branchlet (× 2); 9, 10, pyrene, 2 views (× 2), 8–10 from *Snowden* 38. Drawn by M.E. Church. From F.T.E.A.

1–few-flowered cymes on slender peduncles, the lower peduncles longest up to 7 cm long, spreading and finely puberulous. Calyx campanulate, puberulous or glabrous, pale green, strongly accrescent in fruit; tube 1.5–2.5 mm long; lobes 1.5–2 mm long, ovate-deltoid or oblong. Corolla cream or cream-greenish, sweet-smelling, glabrous or puberulous outside; tube (3)5–8 mm long; lobes 2–4 × 1.5 mm, oblong. Fruit c. 10 mm high, black (green or greenish-black when unripe), with juicy red flesh.

Zambia. N: Nchelenge, Lake Mweru, 900 m, fl. 22.iv.1957, *Richards* 9408 (BR; K). W: Mwinilunga Distr., Matonchi Farm, fl. 4.ix.1930, *Milne-Redhead* 1054 (BR; K).
Also from Guinée to Angola and Dem. Rep. Congo, and in Egypt (fide F.W.T.A.), Sudan, Burundi, Rwanda, Uganda, Kenya and Tanzania. Chipya woodland, tall grass and woodland on lake shore and river bank alluvium, watershed grassland and forest thicket on Kalahari Sand and rocky outcrops; 950–1420 m.

15. **Clerodendrum johnstonii** Oliv. in Trans. Linn. Soc. London, Bot. **2**: 346 (1887). —Gürke in Engler, Pflanzenw. Ost-Afrikas **C**: 341 (1895). —J.G. Baker in F.T.A. **5**: 300 (1900). —Gürke & Loesener in Mildbraed, Wiss. Ergebn. Deutsch. Zentr.-Afrika-Exped., Bot. 2: 282 (1914). —De Wildeman in Bull. Jard. Bot. État **7**: 170 (1920); Pl. Bequaert. **1**: 262 (1922). —R.E. Fries in Notizbl. Bot. Gart. Berlin-Dahlem **8**: 701 (1924). —Thomas in Bot. Jahrb. Syst. **68**: 75 (1936). —Robyns, Fl. Sperm. Parc Nat. Alb. **2**: 142 et 146, t. 14 (1947). —Brenan, Check-list For. Trees Shrubs Tang. Terr.: 638 (1949). —Dale & Greenway, Kenya Trees & Shrubs: 584 (1961). —Watt & Breyer-Brandwijk, Medic. & Pois. Pl. S. & E. Africa: 1048 et 1372 (1962). —White, For. Fl. N. Rhod.: 367, fig. 65 (1962). —Moldenke in Phytologia **61**: 332 (1986). —Malaisse in Troupin, Fl. Rwanda **3**: 274, fig. 86/2 (1985). —Verdcourt in F.T.E.A., Verbenaceae: 118, fig. 16/7–9 (1992). Type from Tanzania.
Clerodendrum murigono Chiov., Racc. Bot. Miss. Consol. Kenya: 99 (1935). Syntypes from Kenya.
Clerodendrum johnstonii var. *rubrum* B. Thomas in Bot. Jahrb. Syst. **68**: 75 (1936). —Brenan, Check-list For. Trees Shrubs Tang. Terr.: 638 (1949). —Moldenke in Phytologia **61**: 336 (1986) *nom. inval.* Type from Tanzania.

An erect shrub 1.5–4.5 m high or a scrambling shrub with scandent stems up to 10.5 m long, or a tree up to 8 m tall. Stems with paired truncate projections or stout recurved hook-like persistent woody petiole bases; young shoots and branches densely tomentose-pubescent with soft crisped yellow-orange or ferruginous to bright red hairs, a similar indumentum also present on the petiole, inflorescences and calyces; older stems 4-angled, subglabrous. Leaves opposite or ternate, petiolate; lamina 3.3–19.5 × 3–14.5 cm, oblong-ovate, ovate or suborbicular, rounded and shortly cuspidate or acuminate at the apex, broadly cuneate to subtruncate or cordate at the base, entire or rarely irregularly crenate, discolorous, sparsely pubescent above with patent acicular hairs, ± pubescent-velutinous beneath with sessile spherical translucent to pale yellowish or reddish glands; petiole 0.8–10.5 cm long, articulate above the base, the basal part in stronger branches woody even before the fall of the leaf lamina. Inflorescences axillary and terminal forming dense corymbose panicles of cymes 5–20 cm in diameter, sometimes with cymes also in the axils of the upper leaves; cymes many-flowered; peduncles of cymes somewhat thick, spreading or erect-spreading, the lower peduncles 4.5–6 cm (or more?) long; pedicels 0–3 mm long; bracts 5–10 mm long, linear-spathulate. Flowers sweet-scented. Calyx campanulate, 3–4.5 mm long; lobes 1.5–2 mm long, triangular. Corolla white or creamy; tube 8–10 mm long, glabrous or nearly so; lobes 3–4 × 3–4 mm, elliptic-oblong, ciliate and pubescent outside with sessile spherical, reddish, sparse, minute glands on the outside. Style and stamens exserted for c. 8 mm. Fruit 10–12 mm long and 9 mm wide, black, surrounded by the persistent, obtusely lobed, sitting in a cup-like calyx c. 12 mm in diameter.

Zambia. E: Isoka Distr., Nyika Plateau, upper slopes of Kangampande (Kangampandi) Mt., 2240 m, fl. 3.v.1952, *White* 2567 (BR; FHO; K). **Malawi.** N: Chitipa Distr., Misuku Hills, Chisasu–Itera road, fl. 27.xii.1977, *Pawek* 13430 (BR; K; SRGH).
Also in Ethiopia, Rwanda, Burundi, Uganda, Kenya, Tanzania and Dem. Rep. Congo. Montane evergreen rain forest and forest margins; 1750–2100 m.
Subsp. *marsabitense* Verdc., occurring in Kenya, is distinguished by the leaves sparsely pubescent beneath and the corolla tube only 3.5–5 mm long.

16. **Clerodendrum glabrum** E. Mey., Comment. Pl. Afr. Austr.: 273 (1838) —Walpers, Repert,
Bot. Syst. **4**: 110 (1845). —Schauer in A. de Candolle, Prodr. **11**: 661 (1847). —Briquet in
Engler & Prantl, Nat. Pflanzenfam. **4**, 3a: 175 (1897). —Wood & Evans, Natal Pl.: t. 45
(1898). —Baker in F.T.A. **5**: 297 et 515 (1900). —H. Pearson in F.C. **5**, 1: 219 (1901). —
Sim, For. Fl. Port. E. Africa: 92 t. 98A (1909). —R. Good in J. Bot. **68**, Suppl. 2 (Gamopet.):
140 (1930). —Thomas in Bot. Jahrb. Syst. **68**: 76 (1936). —Brenan, Check-list For. Trees
Shrubs Tang. Terr.: 637 (1949). —Garcia in Estud. Ensaios Doc. Junta Invest. Ci. Ultramar
[in Mendonça, Contrib. Conhec. Fl. Moçamb., II] **12**: 167 (1954). —Mogg in Macnae &
Kalk, Nat. Hist. Inhaca Isl., Moçamb.: 152 (1958). —Dale & Greenway, Kenya Trees &
Shrubs: 584 (1961). —Huber in F.W.T.A., ed. 2, **2**: 444 (1963). —Friedrich-Holzhammer et
al. in Merxmüller, Prodr. Fl. SW. Afrika, fam. 122: 5 (1967). —van der Schijff &
Schoonraad in Bothalia **10**, 3: 496 (1971). —Ross, Fl. Natal: 300 (1972). —Palmer &
Pitman, Trees Southern Africa **3**: 1963, photo p. 1964, fig. p. 1965 (1973). —Fosberg in
Kew Bull. **33**: 143 (1978). —H. Moldenke & A. Moldenke in Dassanayake & Fosberg, Rev.
Fl. Ceylon **4**: 457 (1983). —Moldenke in Phytologia **59**: 486 (1986). —Verdcourt in
F.T.E.A., Verbenaceae: 119, fig. 16/10–12 (1992). —M. Coates Palgrave, ed. 3 of K. Coates
Palgrave, Trees South. Africa: 989 (2002). Type from South Africa (Cape Prov.).

   Clerodendrum glabrum var. *angustifolium* E. Mey., Comment. Pl. Afr. Austr.: 273 (1838). —
Walpers, Repert. Bot. Syst. **4**: 110 (1845). —Moldenke in Phytologia **59**: 496 (1986). Type
from South Africa (Cape Prov.).

   Clerodendrum ovale Klotzsch in Peters, Naturw. Reise Mossambique **6**, part 1: 257 (1861).
—Vatke in Linnaea **43**: 537 (1882). —Gürke in Engler, Pflanzenw. Ost-Afrikas **C**: 341
(1895). —Sim, For. Fl. Port. E. Africa: 92 (1909). Type: Mozambique, Rios de Sena, *Peters*
s.n. (B†, holotype; K).

   Clerodendrum ovalifolium Engl., Pflanzenw. Ost-Afrikas **A**: 124 (1895).

   Siphonanthus glaber (E. Mey.) Hiern, Cat. Afr. Pl. Welw. **1**, 4: 842 (1900), as "*glabra*". Type
as for *C. glabrum*.

   ?*Clerodendrum glabratum* Gürke in Bot. Jahrb. Syst. **28**: 295 (1900). —J.G. Baker in F.T.A.
5: 516 (1900). —Thomas in Bot. Jahrb. Syst. **68**: 77 (1936). —Brenan, Check-list For. Trees
Shrubs Tang. Terr.: 637 (1949). —Moldenke in Phytologia **59**: 485 (1986). Type from
Tanzania.

   Clerodendrum glabrum var. *ovale* (Klotzsch) H. Pearson in F.C. **5**, 1: 219 (1901) quoad
basion.

A shrub 1–4 m tall or a small tree up to 6.5 m. Stems erect, cylindrical, brown when
young eventually grey; branches straight, somewhat thick, lenticellate, glabrous or
finely imperceptibly puberulous, leaf scars ± raised with tomentellous buds in the axils.
Leaves opposite or 3-whorled; lamina 5–10 × 0.8–5.5 cm, elliptic to elliptic-ovate or
ovate, obtuse or acute or sometimes attenuate to a very acute apex, cuneate or rounded
at the base, entire with slightly revolute margins, ± fleshy, drying ± coriaceous and
somewhat wrinkled, very densely punctate, foetid, sometimes folded along the midrib,
rather caducous; petiole 1–2 cm long, slender, ascending or spreading. Inflorescences
terminal, corymbose, subspherical, loose to ± dense, 3–14 cm in diameter, sometimes
surrounded and exceeded by the uppermost leaves; smaller inflorescences in the axils
of the upper leaves sometimes also present; axes, pedicels and bracts ± puberulous;
pedicels 0.75–6 mm long; bracts and bracteoles linear, 1.5–2.5 mm long, ± equalling
the calyx. Calyx campanulate, 2.5–3.5 mm long; tube 2.5–3 mm in diameter,
glabrescent to ± densely pubescent with ± appressed, whitish hairs; lobes
(0.5)0.75–1.5(3) mm long, triangular and narrowing to the apex but not linear,
separated by rounded sinuses. Corolla usually white, sometimes mauve or white
suffused with mauve, cream or greenish, sweet-scented; tube (4)4.5–8(10) mm long,
slender, only with sessile glands or rarely also sparsely hairy above; lobes 2.5–4 × 2 mm.
Stamens mauve or lilac, exserted 5–7 mm; anthers 1–1.25 mm long, violet or brown.
Fruits marbled-white, but black when dry, 6–10 mm in diameter, strongly wrinkled and
± lobed when dry, sitting in a ribbed shallow, lobed fruiting calyx 8 mm wide; edible.

Branches and petioles glabrous or ± appressed hairy; leaves usually glabrous or sometimes hairy
    on the nerves beneath · forma *glabrum*
Branches and petioles ± densely puberulous, with hairs short, usually appressed, antrorse or
    subspreading; leaves often rather pubescent beneath and on the nerves above · · · · · · · ·
    · forma *pubescens*

Forma glabrum

Botswana. SE: Central Distr., Mahalapye, Shoshong, 1100 m, fl. 6.i.1961, *J. Denison in J.S. de
Beer* 915 (K; SRGH). **Mozambique**. N: Palma, farol do Cabo Delgado, 10 m, fl. 17.iv.1964, *Torre*

& *Paiva* 12124 (LISC; LMU). Z: c. 32 km north of Quelimane, fl. 10.viii.1962, *Wild* 5864 (K; LISC; LISJC). T: Marromeu Distr., Mungari R. mouth (West Luabo R.), 3.vi.1858, *Kirk* s.n. (K). MS: Búzi Distr., c. 19 km east on Búzi (Nova Lusitânia) road from Muda, fl. 23.iii.1960, *Wild & Leach* 5213 (K; SRGH). GI: entre o Chibuto e a ponte do Cicacata a 3 km do Chibuto, fl. 12.vi.1960, *Lemos & Balsinhas* 91 (BM; COI; K; LISC; LMA; SRGH). M: Maputo (Lourenço Marques, cidade), Polana, fl. 20.i.1960, *Lemos & Balsinhas* 17 (COI; K; LISC; LMA).

Also in Kenya, Tanzania, Angola, Namibia, South Africa (KwaZulu-Natal and Eastern Cape) and Swaziland; and in the Seychelles and Comores Is. Introduced in some tropical countries as a hedge plant. Coastal dunes and evergreen coastal thicket, and dry rocky hillsides; 0–200 m in Mozambique; 1100 m in Botswana.

Forma **pubescens** R. Fern. in Mem. Soc. Brot. **30**: 11 (1998). Type from Angola.
 Clerodendrum glabrum var. *pubescens* B. Thomas in Bot. Jahrb. Syst. **68**: 77 (1936) *nom. illegit.*
 Siphonanthus glaber var. *vagus* Hiern, Cat. Afr. Pl. Welw. **1**, 4: 842 (1900), as "*vaga*". Type from Angola.
 Siphonanthus glaber var. *incarnatus* Hiern, Cat. Afr. Pl. Welw. **1**, 4: 842 (1900), as "*incarnata*". Type from Angola.
 Clerodendrum rehmannii Gürke in Bot. Jahrb. Syst. **28**: 294 (1900). —H. Pearson in F.C. **5**, 1: 220 (1901). —Thomas in Bot. Jahrb. Syst. **68**: 77 (1936). —Sim, For. Fl. Port. E. Africa: 92 (1909). —Hutchinson, Botanist South. Africa: 400, fig. page 399 (1946). Type from South Africa (Transvaal).
 Clerodendrum glabrum var. *ovale* H. Pearson in F.C. **5**, 1: 219 (1901), non *C. ovale* Klotzsch.
 Clerodendrum glabrum var. *vagum* (Hiern) Moldenke in Phytologia **13**: 306 (1966); op. cit. **59**: 499 (1986), pro parte.

Mozambique. N: Nampula, fl. 3.ii.1937, *Gomes e Sousa* 1329 (COI). M: Baía de Maputo (Delagoa Bay), 1890, *Junod* 161 (BR) and 165 (Z).

Also in Angola, South Africa (Transvaal) (and elsewhere?). This form is also cited from Dem. Rep. Congo (cf. De Wildeman, Études Fl. Katanga: 121, (1903); T. & H. Durand, Syll. Fl. Congol.: 440 (1909), but we cannot confirm this because we have not seen material from this country.

Mahundu 74 (PRE) from Chobe National Park in Botswana, without flowers, and shortly hairy on the branchlets, probably belongs here.

The specimens from Zambia, Zimbabwe and Malawi, referred by some authors as *C. glabrum*, *C. glabrum* var. *ovale* or *C. glabrum* var. *vagum* belong to *C. eriophyllum*.

Var. *minutiflorum* (Baker) Fosberg [= *Clerodendrum minutiflorum* Baker], cited from Aldabra and neighbouring islands by Fosberg in Kew Bull. **33**: 143–144 (1978), is distinguished from the type mainly by the subtruncate or denticulate (not lobed) calyx.

17. **Clerodendrum eriophyllum** Gürke in Bot. Jahrb. Syst. **18**: 178 (1893); in Engler, Pflanzenw. Ost-Afrikas **C**: 341 (1895). —J.G. Baker in F.T.A. **5**: 299 (1900). —Thomas in Bot. Jahrb. Syst. **68**: 77 (1936). —Brenan, Check-list For. Trees Shrubs Tang. Terr.: 637 (1949). —Dale & Greenway, Kenya Trees & Shrubs: 584 (1961). —Richards & Morony, Check List Fl. Mbala: 237 (1969), as "*C. cryophytum*". —Verdcourt in F.T.E.A., Verbenaceae: 120, fig. 16/13–15 (1992). —M. Coates Palgrave, ed. 3 of K. Coates Palgrave, Trees South. Africa: 988, illust. 275 (2002). Type from Tanzania.
 Clerodendrum glabrum var. *ovale* sensu Wild in Clark, Victoria Falls Handb.: 158 (1952). —sensu Pardy in Rhodesia Agric. J. **52**: 414 (1955), non var. *ovale* H. Pearson neque *C. ovale* Klotzsch.
 Clerodendrum glabrum sensu auct. incl.: White, For. Fl. N. Rhod.: 367 (1962). —Richards & Morony, Check List Fl. Mbala: 237 (1969). —Jacobsen in Kirkia **9**: 172 (1973). —Drummond in Kirkia **10**: 272 (1975). —Moldenke in Phytologia **59**: 495 (1986), quoad specim. *Pole Evans* 3231 et *Stolz* 1166), non E. Mey.
 Clerodendrum eryophylloides Moldenke in Phytologia **9**: 183 (1963); op. cit. **59**: 350 (1986) quoad *Torre & Paiva* 10113.
 Cleodendrum glabrum var. *vagum* sensu Moldenke in Phytologia **13**: 306 (1966); op. cit. **59**: 499 (1986) et **60**: 59 (1986) pro parte, non *Siphonanthus glaber* var. *vagus* Hiern.

A few- to many-stemmed shrub 1–5.80 m tall, or small tree 3.5–9.5 m tall, densely branched with a ± rounded crown; bark grey or pale brown; branchlets as well as petioles and inflorescences finely ± densely greyish pubescent or velvety, leaf scars ± raised with tomentellous buds in the axils. Leaves aromatic, opposite or 3(4)-whorled; lamina (1)2–10.5(12–13) × (0.5)1–5(6–7.5) cm, ovate to elliptic or broadly elliptic, acute or distinctly acuminate at the apex, rounded or broadly cuneate at the base, pubescent above, more densely so beneath, ± discolorous, dull green above and paler beneath, entire and flat at the margin, thin textured to somewhat rigid on drying but usually not coriaceous, densely punctate with an unpleasant foetid scent when crushed; petiole (0.3)0.5–2.7 cm long. Inflorescences terminal with cymes ± densely to laxly corymbosely arranged, 4–6 cm long and 1–5.5(12) cm in diameter, sometimes with some axillary cymes beneath; bracts and bracteoles 3–5 mm long, linear-lanceolate, similar to the calyx lobes but longer; peduncle 0–7 cm long; secondary branches up to 1.3 cm long; pedicels 2–4 mm long. Calyx densely white-greyish pubescent; tube 2.5–2.75 mm long, campanulate, 2.5–2.75 mm in diameter below the lobes; lobes (1)2–3 mm long, linear-subulate from a ± triangular base, separated by rounded sinuses. Corolla white, rarely mauve or pinkish-white or cream-white, fragrant; tube (7)8–13 mm long, sparsely pubescent and also with small sessile or shortly stalked glands outside; lobes 3.5 mm long, oblong. Style and stamens exserted 8 mm. Filaments pink; anthers brown, (1)1.25–1.5 mm long. Fruits white to marble-coloured, 6–10 mm in diameter, strongly wrinkled and ± lobed when dry, sitting in a ribbed shallow, lobed fruiting calyx up to 8 mm wide.

Botswana. N: North East Distr., Francistown to Bisole (Bosoli), fl. & fr. juv. 17.iv.1931, *Pole Evans* 3231 (K; PRE). **Zambia**. N: Mbala Distr., road to Kaka Village, 1740 m, fl. 20.ii.1960, *Richards* 12517 (B; BR; K). W: Kitwe Distr., fl. 20.i.1956, *Fanshawe* 2752 (BR; K; LISC; SRGH). E: Lundazi to Mzimba 'mile 4', fr. 28.iv.1952, *White* 2499 (FHO; K). S: Monze Distr., Lochinvar National Park, anthill-zone, 1040 m, fl. 14.ii.1972, *van Lavieren, Sayer & Rees* 557 (SRGH). **Zimbabwe**. N: Makonde Distr., Whindale Farm, 1232 m, fl. 2.ii.1969, *W.B. Jacobsen* 8657 (PRE). W: Hwange Distr., Victoria Falls near the old Centre Car Park, 880 m, fl. 13.ii.1976, *L. Simpathu* 8 (K; SRGH). C: Zvimba Distr., Atlantica Ecological Research Station, Bulawayo road, c. 16 km from Harare (Salisbury), fr. 18.iv.1963, *Loveridge* 653 (BM; BR; SRGH). E: Mutare (Umtali), 1184 m, fl. 11.ii.1955, *Chase* 5462 (COI; K; LISC; S; SRGH). S: Mberengwa Distr., Buhwa (Bukwa) Mt., lower NW foot hills, c. 1100 m, fl. 30.iv.1973, *G.V. Pope* 1025 (K; SRGH). **Malawi**. N: Mzimba Distr., Mzimba Airport, 1370 m, fl. & fr. 28.ii.1959, *Robson* 1729 (BM; BR; K; LISC; SRGH). C: Dedza Distr., Chongoni Forest Reserve, fr. juv. 26.iii.1961, *Chapman* 1184 (FHO; SRGH). S: Mangochi Distr., Cape Maclear, Lake Malawi, on road to old Livingstone Mission, fr. 25.iii.1974, *I.H. Patel* 91 (SRGH). **Mozambique**. N: Lalaua Distr., 10 km from Lalaua to Ribáuè, 400 m, fl. 22.i.1964, *Torre & Paiva* 10113 (LISC). T: Cahora Bassa Distr., Songo, fl. 10.ii.1972, *Macêdo & Esteves* 4824 (LISC; LMA; LMU; SRGH).

Also in Kenya, Tanzania and Zanzibar. Usually on anthills in plateau woodland, miombo, floodplains and dambos, and in thickets and woodlands on rocky hillsides and outcrops, and in gully forest; 400–1700 m.

This taxon is very similar to *C. glabrum* E. Mey. and is perhaps only a subspecies of it. However, the two form a difficult complex and we prefer to treat them as separate species for the time being, as does Verdcourt loc. cit.

They may be distinguished by small morphological differences, usually not all simultaneously present in a specimen, and these are: indumentum of spreading longer usually stronger hairs in *C. eriophyllum* than in *C. glabrum* where the branchlets, petioles and leaf laminas are glabrous or have sparse, appressed hairs (except in forma *pubescens* R. Fern., where the hairs can be denser and subspreading); calyx with a slightly longer and broader tube and longer narrower nearly awl-like lobes in *C. eriophyllum*; corolla tube a little longer, wider and hairy in *C. eriophyllum*. Besides these differences, at least in the Flora Zambesiaca area, the two taxa appear to favour different habitats: *C. glabrum* is usually a plant of coastal areas, frequent in thickets of dunes at low altitudes, but also extends inland, while *C. eriophyllum* is found inland, often on anthills. *C. glabrum* is usually a subshrub or a shrub, while *C. eriophyllum* is described as a tree or shrub in the specimens seen from the Flora Zambesiaca area.

In dry specimens the inflorescences of *C. eriophyllum* are rather characteristic in the contrast in colour between the lower part (the pedicels, bracts and calyces) which have a ± dense whitish indumentum and the outer or upper part (the corollas) which is dark brownish-red.

18. **Clerodendrum tricholobum** Gürke in Bot. Jahrb. Syst. **18**: 178 (1893); in Engler, Pflanzenw. Ost-Afrikas **C**: 341 (1895). —J.G. Baker in F.T.A. **5**: 303 (1900). —Thomas in Bot. Jahrb. Syst. **68**: 56 (1936). —Brenan, Check-list For. Trees Shrubs Tang. Terr.: 634 (1949). — Verdcourt in F.T.E.A., Verbenaceae: 120 (1992). Type from Tanzania.

    Clerodendrum acerbianum sensu Thomas in Bot. Jahrb. Syst. **68**: 89 (1936), pro parte, quoad *Schlieben* 5866, non (Vis.) Benth.

    Clerodendrum lindiense Moldenke in Phytologia **5**: 83 (1954); op. cit. **61**: 492 (1987). Type from Tanzania.

    Clerodendrum eriophyllum sensu Vollensen in Opera Bot. **59**: 82 (1980), quoad specim. *Ludango* 1198 et *Rodgers* 862, fide *Verdcourt*, (op. cit.: 121) non Gürke.

A shrub or a scrambler, 1–3 m tall, sometimes with long trailing branches; older branches subglabrous with a slightly fissured, brownish bark; young shoots and flowering branchlets subcylindrical, ± densely villous with soft, whitish-yellow, somewhat long hairs. Leaves opposite or 3-whorled, shortly petiolate; lamina 3–8(14) × 2–5(7.5) cm, elliptic, acute or shortly acuminate and apiculate at the apex, cuneate at the base, entire, chartaceous, markedly discolorous, pubescent above with fine ± bulbous-based hairs making the upper surface ± densely papillose mainly after the fall of the hairs, ± densely pale yellowish-grey soft-woolly beneath with the indumentum sometimes hiding the nervation; petiole up to 9 mm long, slender, hairy as the branchlets. Flowers in condensed or capitate cymes up to 4 cm in diameter, axillary in the uppermost 1–2 whorls of leaves and sometimes appearing terminal; peduncles up to 7 cm long, ± equalling the subtending leaf, long-spreading hairy with indumentum similar to that on the bracts and pedicels; bracts ovate to lanceolate-linear, narrowly attenuate into a long point. Calyx white-hairy; tube 2–2.5 mm long, cylindrical; lobes longer than the tube, 3–5 mm long, attenuate into a filiform point, longer than the triangular base, pilose. Corolla white, scented; tube 9–11.5(18?) mm long, narrowly cylindrical, widening slightly upwards, ± densely covered with stipitate and minute sessile glands; lobes elliptic to round, 2.2–2.5 × 1.2–1.8 mm. Stamens and style exserted 3–8 mm. Fruit dull brown, 1–1.3 × 1–1.5 cm, globose, each lobe longitudinally grooved; fruiting calyx lobes c. 5 mm long, ovate, acuminate, ridged.

Mozambique. N: Ilha Distr., Ilha de Moçambique, Festland, 2.i.1889, *Stuhlmann* 444 (HBG). Also in Kenya and Tanzania. Riverine bushland, light woodland, scrub on coral rocks above beach; seasonal flooded forest, woodland and grassland mosaic; 0–300 m.

The above description is based on the types of *C. tricholobum* (*Stuhlmann* 587, HBG, isolectotype; *Stuhlmann* 444, HBG, syntype) and the type of *C. lindiense* (*Schlieben* 5866, HBG, P, isotypes), and also on the descriptions in Verdcourt (op. cit.: 120–121) and Moldenke (loc. cit., 1954). The habitat information is taken from Verdcourt.

Stuhlmann 444 (HBG) is the only specimen cited for Mozambique by Gürke, Baker, Thomas and Verdcourt.

19. **Clerodendrum ternatum** Schinz in Verh. Bot. Vereins Prov. Brandenburg **31**: 205 (1890). — Gürke in Engler, Pflanzenw. Ost-Afrikas **C**: 341 (1895); in Warburg, Kunene-Samb.-Exped. Baum: 351 (1903). —J.G. Baker in F.T.A. **5**: 312 (1900). —Thomas in Bot. Jahrb. Syst. **68**: 79 (1936). —Hutchinson, Botanist South. Africa: 672 (1946). —Brenan, Check-list For. Trees Shrubs Tang. Terr.: 637 (1949). —Friedrich-Holzhammer et al. in Merxmüller, Prodr. Fl. SW. Afrika, fam. 122: 5 (1967). —Verdcourt in F.T.E.A., Verbenaceae: 124, fig. 18/9–12 (1992). Type from Namibia.

    Clerodendrum wilmsii Gürke in Bot. Jahrb. Syst. **28**: 304 (1900). —H. Pearson in F.C. **5**, 1: 224 (1901). —Binns, First Check List Herb. Fl. Malawi: 103 (1968). Syntypes from South Africa (Mpumalanga).

    Clerodendrum simile H. Pearson in F.C. **5**, 1: 224 (1901). —Hutchinson, Botanist South. Africa: 672 (1946). Type from South Africa (Transvaal).

    Clerodendrum transvaalense B. Thomas in Bot. Jahrb. Syst. **68**: 78 (1936) *nom. nov.* Type as for *C. simile*.

    Clerodendrum ternatum var. *vinosum* B. Thomas in Bot. Jahrb. Syst. **68**: 79 (1936) *nom. inval.* Type from Angola.

    Clerodendrum lanceolatum sensu White, For. Fl. N. Rhod.: 367 (1962).

Rhizomatous suffrutex from an extensive woody rootstock, forming small colonies of scattered groups of 1–many stems. Stems woody-based, virgate, 10–47(60) cm high, simple or sometimes few-branched, pubescent with short ± appressed or spreading hairs; axillary leafy fascicles sometimes present. Leaves 3-whorled or opposite, sessile or very shortly petiolate, with an unpleasant smell when crushed; lamina 1.2–10 × 0.4–2.2 cm, narrowly lanceolate or lanceolate to oblanceolate, acute at the apex, and gradually narrowed to the base, serrate with 2–5 spaced sharp teeth in the upper half on each side or entire, ± pubescent, gland-dotted on the lower-face. Cymes mostly 3-flowered, axillary in the upper 4–6 axils, with one peduncle per axil, forming leafy, terminal racemes up to 13(20) cm long; peduncles of the lower cymes up to 2.2 cm long; pedicels 0–1.5 mm long. Buds with a very excentric globose head (corolla limb). Calyx 3–5 mm long, cupular; lobes 0.5–0.8 mm long, triangular. Corolla with a pungent odour usually described as unpleasant-smelling, usually white, sometimes creamy or greenish-white; tube 1.5–4.5 cm long, abruptly funnel-shaped at the top, pilose and glandular outside, the glands stalked or nearly sessile. Filaments dark red, violet, purple or mauve; anthers 3 mm long, brown or violet;

anthers and style exserted 3–3.5 cm. Fruit black, c. 7 × 8 mm, depressed subglobose, deeply 4-lobed.

Hairs of the indumentum short, curved-antrorse, subappressed, somewhat rigid; ovary usually glabrous ·· forma *ternatum*
Hairs of the indumentum longer, ± crisped, ± spreading, soft, distinctly many-celled; ovary ± hairy ··· forma *villosum*

Forma **ternatum**

Clerodendrum ternatum var. *lanceolatum* sensu Jacobsen in Kirkia **9**: 172 (1973) non *C. lanceolatum* Gürke.

Young shoots, branchlets, petioles, leaves (mainly beneath) and calyx with an indumentum of short, curved, subappressed, somewhat rigid, sparse to ± dense whitish hairs.

Botswana. N: Ngamiland Distr., Maun, c. 274 m from river, fl. ii.1967, *Lambrecht* 24 (SRGH). SW: Kgoutsa (Khoutsa) c. 960 m, veg. 19.iv.1972, *Doose* 21 (LISC; SRGH). SE: North East Distr., c. 19 km south of Francistown, c. 1120 m, fl. juv. 21.i.1960, *Leach & Noel* 290 (K; SRGH). **Zambia.** B: Sesheke Distr., near Sichinga Forest, fl. 28.xii.1952, *Angus* 1052 (FHO). C: South Luangwa National Park, near Katete R., c. 608 m, 8.i.1966, *Astle* 4317 (K; SRGH); Luangwa Distr., Katondwe, fl. 4.ii.1964, *Fanshawe* 8304 (FHO; K). S: Mazabuka Distr., fl. 16.i.1952, *White* 1929 (FHO; K). **Zimbabwe.** N: Mhangura Distr., Farm Gudubu, Mhangura (Mangula), 1216 m, fl. 3.i.1965, *W. Jacobsen* 2659 (PRE). W: Bulawayo town, Hillside, c. 1440 m, fl. xii.1956, *O.B. Miller* 4018 (K; LISC; SRGH). C: Makoni Distr., Rusape, Mona, fl. 29.xii.1952, *Munch* 405 (K; SRGH). E: Chipinge Distr., 5 km south of Rusongo, 400 m, 1.ii.1975, *Gibbs Russell* 2742 (SRGH). S: Zaka Distr., close to Fort Victoria–Birchenough Bridge road to Ndanga turnoff, fl. 19.i.1960, *Goodier* 830 (SRGH). **Malawi.** C: Kasungu Distr., Chimaliro Forest, Phaso road, fl. 10.i.1975, *Pawek* 8885 (K). S: Chikwawa Distr., Lengwe National Park, c. 128 m, fl. juv. 14.xii.1970, *Hall-Martin* 1131 (K; SRGH). **Mozambique.** N: Monapo, at km 11 in the recent "picada" from Namialo to Meserepane, c. 150 m, fl. juv. 25.xi.1963, *Torre & Paiva* 9279 (LISC). MS: Manica Distr., at km 2 from Manica Village on the road to Chimoio (Vila Pery), c. 900 m, fl. 25.xi.1965, *Torre & Correia* 13247 (LISC). GI: Massingir, left bank of Elefantes R., at km 36, west from Albufeira de Massingir (Lagoa Nova), fl. 13.ii.1973, *Lousā & Rosa* 303 (LMA). M: Namaacha Distr., between Goba railway station and Goba Fronteira, near the road, fr. 16.ii.1949, *Myre & Balsinhas* 377 (LMA; SRGH).

Also in Kenya, Tanzania, Angola, Namibia and South Africa. Dry deciduous woodland with mopane and *Commiphora spp.*, wooded grassland, *Brachystegia* woodland, Kalahari Sand woodland; occasionally on anthills; 150–1620 m.

Forma *glabricalyx* Moldenke (in Phytologia **22**: 46, 1975), with a glabrous calyx, represents the glabrous extreme in the intergrading variation in indumentum.

Forma **villosum** R. Fern. in Mem. Soc. Brot. **30**: 26 (1998). Type: Malawi, *Buchanan* 468 (B†, holotype; BM; K).

Clerodendrum lanceolatum Gürke in Bot. Jahrb. Syst. **18**: 181 (1893); in Engler, Pflanzenw. Ost-Afrikas **C**: 341 (1895). —J.G. Baker in F.T.A. **5**: 312 (1900). —Sim, For. Fl. Port. E. Africa: 93 (1909). —Thomas in Bot. Jahrb. Syst. **68**: 79 (1936). —Hutchinson, Botanist South. Africa: 301 (1946). —Brenan, Check-list For. Trees Shrubs Tang. Terr.: 636 (1949). —Martineau, Rhodesia Wild Fl.: 66 (1953). —Wild in Clark, Victoria Falls Handb.: 158 (1952). — Garcia in Estud. Ensaios Doc. Junta Invest. Ci. Ultramar [in Mendonça, Contrib. Conhec. Fl. Moçamb., II] **12**: 168 (1954). —White, For. Fl. N. Rhod.: 367 (1962). —Binns, First Check List Herb. Fl. Malawi: 103 (1968). Type: Malawi, without locality, *Buchanan* 468 (B†, holotype; BM; K) *nom. illegit.* non F. Müll. (1863).
Clerodendrum myricoides var. *discolor* sensu J.G. Baker in F.T.A. **5**: 310 (1900) quoad specim. *Whyte* s.n. (K) in Mt. Milange lectum p.p.
Clerodendrum ternatum var. *lanceolatum* (Gürke) Moldenke in Phytologia **5**: 98 (1954). Type as for *C. lanceolatum* Gürke.
Clerodendrum ternatum sensu Verdcourt in F.T.E.A., Verbenaceae: 124 (1992) pro parte.

Young shoots, branchlets, petioles, leaves (mainly beneath), calyx and sometimes the corolla lobes with a ± dense indumentum of somewhat long, spreading, multicellular hairs; ovary covered with white, straight, ± dense hairs, mainly at the top, rarely completely glabrous.

Botswana. N: Ngamiland Distr., Mogobewandehodi R., Okavango, 930 m, fl. 15.iii.1961, *Richards* 14722 (K). SW: Ghanzi, Farm 56, c. 960 m, fl. 10.i.1970, *R.C. Brown* 7575 (K). **Zambia.** N: Kasama Distr., c. 13 km north of Kasama on Mbala (Abercorn) road, c. 1376 m, fr.

22.xii.1961, *Astle* 1145 (K; SRGH). C: South Luangwa National Park, near Kapamba R., c. 640 m, fl. 12.i.1966, *Astle* 4347 (K; SRGH). E: Petauke Distr., Nyimba–Luembe road, 600 m, fl. 13.xii.1958, *Robson* 934 (BM; BR; K; LISC; SRGH). S: Kalomo Distr., 16 km north of Kalomo, 1400 m, fl. 31.xii.1971, *M.J.A. Weriger* 1568 (PRE). **Zimbabwe.** N: Guruve Distr., foothills of Chiruwa Mt., c. 600 m, fl. 30.i.1966, *Müller* 281 (K; SRGH). W: Hwange Distr., Victoria Falls, c. 960 m, fl. 12.ii.1912, *Rogers* 5530 (K; PRE; SRGH). C: Harare (Salisbury town), Hatfield, fl. 1.i.1957, *Whellan* 1156 (K; SRGH). E: Mutare (Umtali), 27.xii.1946, *Fischer* 1087 (PRE). S: Chiredzi Distr., Hippo Valley Estate, Chitsanga Hill, fl. 2.i.1971, *P. Taylor* 51 (K; SRGH). **Malawi.** N: Rumphi Distr., c. 19 km north of Rumphi (Rumpi) on Great North Road, c. 1216 m, fl. 26.xii.1970, *Pawek* 4149 (K). C: Kasungu Distr., near Kasungu Hill, 1100 m, fl. 14.i.1959, *Robson & Jackson* 1149 (BM; BR; K; LISC; SRGH). S: Mangochi Distr., hill 4 km NE of Mangochi, opposite Malindi turn-off, 515 m, fl. & fr. 24.xii.1979, *Brummitt & Patel* 15471 (K). **Mozambique.** N: Mecúfi, waterfalls of Lúrio R., c. 160 m, fr. juv. 17.xii.1963, *Correia* 60 (LISC). T: Cahora Bassa Distr., Mucangádeze R., 5 km from the dam, near Posto Policial nº 3, on the new road to Meroeira (?), c. 470 m, fl. 30.i.1973, *Torre, Carvalho & Ladeira* 18944 (LISC). Z: Morrumbala Distr., Campo Experimental do CICA, fl. 26.x.1945, *J.G.Pedro* 476 (LMA). MS: Sussundenga Distr., Mavita, Rotanda, at km 4 on the road Messambuzi–Chimoio (Vila Pery), fl. 25.i.1966, *Pereira, Sarmento & Marques* 1294 (BR; LMU).

Probably also in all countries where forma *ternatum* is found. Dry deciduous woodland with mopane and *Commiphora spp.*, wooded grassland, *Brachystegia* woodland, Kalahari Sand woodland, on rocky hillsides, riverside alluvium and dambo margins; occasionally on anthills and in old cultivation; 45–1500 m.

Forma *ternatum* and forma *villosum* are found in all countries of the Flora Zambesiaca area, sometimes in the same places and even in same gatherings (but represented by separate individuals). However, forma *ternatum* appears to be twice as common as forma *villosum* in Zambia, Zimbabwe and Botswana, while forma *villosum* is more common than forma *ternatum* in Malawi and Mozambique.

20. **Clerodendrum pusillum** Gürke in Bot. Jahrb. Syst. **30**: 390 (1901); in Warburg, Kunene-Samb.-Exped. Baum: 351 (1903). —Thomas in Bot. Jahrb. Syst. **68**: 78 (1936). —R. Good in J. Bot. **68**, Suppl. 2 (Gamopet.): 141 (1930). —Brenan, Check-list For. Trees Shrubs Tang. Terr.: 636 (1949). —White, For. Fl. N. Rhod.: 367 (1962). —Jacobsen in Kirkia **9**: 172 (1973). —Verdcourt, in F.T.E.A., Verbenaceae: 124, fig. 18/5–8 (1992). Type from Tanzania.

Dwarf rhizomatous pyrophytic suffrutex 1 to 25 cm tall, from a woody rootstock; rootstock with spreading, slender branches. Stems herbaceous, 1–several, erect, simple or few-branched, leafy, sparsely to densely pilose-pubescent with short white spreading hairs. Leaves opposite or sometimes 3-whorled; lamina 1.5–7.5(9) × 0.6–3.5 cm, lanceolate, elliptic, oblong or oblong-obovate, acute to obtuse or rounded at the apex, ± narrowed at the base, entire, thinly-chartaceous to somewhat fleshy, dark green on drying, concolorous, shortly pubescent with hairs more dense on the nervation and margins, sometimes subglabrous; petiole up to 3 mm long. Inflorescence capitate, 3–12-flowered, with cymes axillary in the usually crowded uppermost leaves, sometimes with 2 solitary axillary flowers below the capitate inflorescence; cymes 1–2-flowered, subsessile, the flowers sessile; bracteoles 5–7 × 1 mm, linear, hairy. Calyx (6)7–9.5 mm long, narrowly infudibuliform-tubular, with acute triangular lobes (1.5)2.5–3 mm long, sparsely pubescent, ciliate. Buds with the terminal part (the corolla limb) c. 12 mm in diameter, globose. Corolla usually white or sometimes pale cream-coloured, densely and softly white villose outside, the hairs rather long; tube (6)8.5–11(11.5) cm long; lobes c. 10 × 4 mm, elliptic; filaments, anthers and style apex purple, blackish-purple, violet or mauve; stamens and style exserted c. 4 cm. Fruit not seen.

Zambia. N: Kaputa Distr., Msanka-Lufubu divide, fl. 26.ix.1956, *Richards* 6300 (K). W: Mwinilunga Distr., south of Matonchi Farm, fl. 30.x.1937, *Milne-Redhead* 3023 (K). C: Lusaka Distr., c. 8 km from Lusaka along the main road to Chilanga, 1300 m, fl. 9.ix.1972, *Strid* 2074 (K). S: Mumbwa Distr., c. 8 km from Mumbwa on road to Kafue Hoek, fl. 7.xi.1959, *Drummond & Cookson* 6176 (K; SRGH). **Zimbabwe.** N: Hurungwe Distr., Zwipani Camp, Urungwe Native Reserve, fl. 1.xii.1957, *Goodier* 431 (K; SRGH). **Malawi.** C: Kasungu Distr., Chamama, near Chipata (Chipala) Hill, 1000 m, fl. 16.i.1959, *Robson & Jackson* 1221 (K).

Also in Angola, Dem. Rep. Congo and Tanzania. In recently burnt plateau and escarpment miombo, wooded grassland and dambo margins, on red sandy soil or hard stony ground; 1000–1300 m.

21. **Clerodendrum incisum** Klotzsch in Peters, Naturw. Reise Mossambique 6, part 1: 257 (1861), —Vatke in Linnaea **43**: 537 (1882). —Gürke in Engler, Pflanzenw. Ost-Afrikas **C**: 341 (1895). —J.G. Baker in F.T.A. **5**: 307 (1900). —Sim, For. Fl. Port. E. Africa: 93 (1909). —Thomas in Bot. Jahrb. Syst. **68**: 78 (1936) pro max. parte. —Brenan, Check-list For. Trees Shrubs Tang. Terr.: 636 (1949) pro parte. —Garcia in Estud. Ensaios Doc. Junta Invest. Ci. Ultramar [in Mendonça, Contrib. Conhec. Fl. Moçamb., II] **12**: 168 (1954). —Huber in F.W.T.A., ed. 2, **2**: 443 (1963). —Drummond in Kirkia **10**: 272 (1975). —Moldenke in Phytologia **60**: 271 (1986). —Verdcourt in F.T.E.A., Verbenaceae: 122, fig. 18/1–4 (1992). Syntypes: Mozambique, Rios de Sena, Quirimba (Querimba) and Boror, *Peters* (B†, syntypes).

    Clerodendrum macrosiphon Hook.f. in Bot. Mag. **109**: t. 6695 (1883). —Gürke in Engler, Pflanzenw. Ost-Afrikas **C**: 340 (1895). Type from Tanzania.

    Clerodendrum incisum var. *macrosiphon* (Hook.f.) Baker in F.T.A. **5**: 308 (1900). — Chiovenda, Fl. Somala **2**: 364 (1932). —Moldenke in Phytologia **60**: 277 (1986), pro parte excl. specim. *Schlieben* 5260. Type as for *C. macrosiphon.*

    Clerodendrum incisum var. *vinosum* Chiov., Fl. Somala **2**: 364 (1932). —Thomas in Bot. Jahrb. Syst. **68**: 78 (1936). —Brenan, Check-list For. Trees Shrubs Tang. Terr.: 636 (1949). —Moldenke in Phytologia **60**: 281 (1986). Type from Somalia.

    Clerodendrum incisum var. *longipedunculatum* B. Thomas in Bot. Jahrb. Syst. **68**: 78 (1936). —Moldenke in Phytologia **60**: 276 (1986). Type from Kenya.

    ?*Clerodendrum dalei* Moldenke in Phytologia **4**: 287 (1953); op. cit. **59**: 240 (1936). Type from Kenya.

    Clerodendrum amplifolium sensu Moldenke in Phytologia **57**: 476 (1985), quoad specim. *Whellan* 349, non S. Moore.

    Rotheca incisa (Klotzsch) Steane & Mabb. in Novon **8**: 205 (1998)*.

Suffrutex or a 2–5-stemmed shrub up to 1.6(2.5) m high; stems erect, straight, 2–3.5 mm thick at the base, terete, simple to few-branched, rarely much branched, shortly hairy, more sparsely so towards the base, with hairs patent, whitish to ferruginous. Leaves opposite or 3-whorled; lamina 1.7–15.5(20) × 1.2–7.5(6.8) cm, sometimes unequal at the same node, longer than the internodes, elliptic to lanceolate, rarely obovate, acute to acuminate or tapering attenuate at the apex, sometimes nearly caudate, long attenuate-cuneate towards the base, tapering into a subpetiole, entire or usually coarsely and irregularly serrate-dentate to pinnatilobed with 1–7 teeth or lobes on each side, the teeth large, broadly triangular and acute to subobtuse, or small, short and acute in smaller leaves, the largest teeth or segments sometimes 1–2-dentate, the blade membranous, usually slightly discolorous, drying darker mainly above, shortly pubescent with scattered thin hairs above, more densely so beneath mainly on the nerves; petiole usually not distinct from the narrowly attenuate leaf base; short branchlets sometimes present with 1 to few pairs of small, entire or few-dentate leaves. Cymes in condensed 5–18-flowered inflorescences, terminal on stem and main lateral branches, sometimes also axillary below to the terminal inflorescence forming a racemose leafy inflorescence; peduncles up to 2 cm long; apparent pedicels 2.5–8 mm long; bracts and bracteoles 2–3 mm long, linear to spathulate. Calyx 4.5–5 mm long, cupular, pubescent; lobes 1.5–2 mm long, narrowly triangular, acute. Buds with globose limb. Corolla white; tube 5–11.25 cm long, with short gland-tipped hairs, corolla tubes borne erect and ± parallel in the same inflorescence; limb 2.5 × 2 cm, lobes 8–15 × 4.5–7 mm, oblong, glabrous or with sparse to dense gland-tipped hairs mainly when in bud. Style purple and white, anthers purple, filaments crimson. Style and filaments exserted for 3–5 cm. Fruit depressed-globose, 7 × 9 mm, 3–4-lobed, glabrous.

Zimbabwe. E: Chimanimani Distr., Haroni–Rusitu (Lusitu) River confluence, on edge of Haroni R., fl. juv. 21.iv.1973, *Ngoni* 204 (K; SRGH). **Mozambique**. N: Murrupula, at km 12 on road to Alto Ligonha, 400 m, fl. 31.i.1968, *Torre & Correia* 17492 (LISC). Z: Milange Distr., at 4 km from Sabelua to the crossing to Mongoé, at the foot of serra Chiperone, north side, 11.ii.1972, *Correia & Marques* 2611 (LMU). MS: Mossurize Distr., on road to Búzi R. and Gogói, c. 800 m, fl. juv. i.1962, *Goldsmith* 7/62 (K; LISC; SRGH). GI: Bilene Distr., Macia, São Martinho do Bilene, 20.v.1968, *Balsinhas* 1089 (COI; LMA).

---

* This species, together with *C. ternatum, C. pusillum* and *C. lutambense*, belongs to *Clerodendrum* sect. *Konocalyx* Verdc. Although Steane & Mabberley (1998) transferred this section to *Rotheca* it does not have the corolla structure characteristic for that genus, and is here retained in *Clerodendrum*, see: Fernandes & Verdcourt, in Kew Bull. **55**: 147 (2000).

Also in Somalia, Kenya and Tanzania. Understorey of well developed *Brachystegia* woodland and evergreen forest, dense coastal dune-thickets, on sandy and compact red soils. Sometimes grown as a garden ornamental; 5–1000 m.

Salbany 105 (COI; LISC; LMU; SRGH), from Chimoio (Mozambique) is described on the label as a shrub or a tree, 3 m tall. This may be a mistake because the label data on the other sheets seen from the Flora Zambesiaca area refer to the plant as a perennial herb not exceeding 1.5 m.

22. **Clerodendrum lutambense** Verdc. in F.T.E.A., Verbenaceae: 124 (1992). —R. Fernandes in Mem. Soc. Brot. **30**: 14 (1998). Type from Tanzania.

  Clerodendrum incisum sensu Thomas in Bot. Jahrb. Syst. **68**: 78 (1936) pro parte, quoad *Schlieben* 5260. —sensu Brenan, Check-list For. Trees Shrubs Tang. Terr.: 636 (1949) pro parte, quoad *Schlieben* 5260, non Klotsch.

  Clerodendrum incisum var. *macrosiphon* sensu Moldenke in Phytologia **60**: 275 (1986) in adnot. et pag. 280 quoad specimen *Schlieben* 5260, non (Hook.f.) Baker.

Suffrutex up to 80 cm tall from a woody rootstock; stems several, herbaceous becoming thinly woody, erect, simple to somewhat branched, ± leafy, ± densely pilose-pubescent with spreading soft whitish to reddish hairs, often intermixed with gland-tipped hairs toward the apex, glabrescent below. Leaves opposite, shortly petiolate; lamina 3–14 × 1–6 cm, oblanceolate to obovate, acuminate to tapering-acute at the apex, long attenuate at the base nearly to the insertion, with 1–6 coarse teeth usually in the upper one third, or entire, membranous or somewhat stiffly textured, hardly discolorous on drying, ± densely pubescent above and on the venation beneath; petiole 1–2 mm long. Cymes in terminal spicate or racemose inflorescences 5.5–8 cm long, the inflorescence axis and peduncles densely pilose-pubescent with spreading hairs intermixed with gland-tipped hairs; flowers 1–7(9), usually sessile; peduncles of the cymes up to 1.8 cm long, longest in the lowermost cymes; bracts subtending the lowermost pedunculate cymes sometimes foliaceous, up to 4.5 × 2.2 cm, ovate, abruptly cuspidate; upper bracts smaller, c. 9 mm long, elongate-triangular; bracts and calyx with short gland-tipped hairs and sparse pilose spreading hairs. Calyx 9–10 mm long (including lobes); tube cylindric; lobes 3–4 mm long, narrowly triangular, subcuspidate, sometimes spreading after anthesis. Corolla white; tube narrow, c. 10 cm long, with sparse gland-tipped hairs outside; lobes c. 12 × 7 mm. Style and filaments exserted for 3.5 cm or more. Mature fruit not seen.

Mozambique. N: Ancuabe Distr., Pemba, Metoro, c. 5 km from Namatuca, fl. 30.i.1984, *de Koning & Maite* (J).

Also in Tanzania. In abandoned cultivation and at roadsides; 0–350 m.

This species closely approaches *C. incisum* Klotzsch but may be distinguished from it by: the presence of gland-tipped hairs in the indumentum of the inflorescence and calyx (in *C. lutambense*); by the leaves progressively attenuate towards the base (not cuneate-attenuate as in *C. incisum*), and acuminate to tapering-acute at the apex with shorter teeth on the margin; by the terminal spicate or narrowly racemose inflorescence; and by the longer calyx, with a cylindrical tube and longer, spreading lobes.

In K (ex B), there is a Peter specimen from Mozambique, s. loc., s.d., s.n., which is possibly an isosyntype.

DOUBTFUL TAXON

Clerodendrum stenanthum Klotzsch in Peters, Naturw. Reise Mossambique **6**, part 1: 258 (1861). Types: Mozambique, Zambezi R. (Rios de Sena), Boror and Querimba Island, *Peters* (B†, syntypes).

In the absence of type material this species is treated as a doubtful taxon.

From the description this taxon would appear to be a variety or form of *C. cephalanthum* (see: R. Fernandes in Mem. Soc. Brot. **30**: 24–26 (1998)). However, Thomas who had the type material available did not unite *C. stenanthum* and *C. cephalanthum*, instead he sank *C. stenanthum* into *C. mossambicense* (= *C. robustum*). This interpretation was borne out by Verdcourt who suggested, in F.T.E.A., Verbenaceae: 101 (1992), that *C. stenanthum* may be a form of *C. robustum* Klotzsch, a species described from Querimba Insel und Festland von Ilha de Moçambique (Mossambique), *Peters* s.n. (B†).

It is pointed out that if the taxon described as *C. stenanthum* Klotzsch were to be united with *C. cephalanthum* then the name *C. stenanthum* Klotzsch would take precedence over *C. cephalanthum* Oliv.

4. ROTHECA Raf.

Rotheca Raf., Fl. Tellur. **4**: 69 (1838). —Steane & Mabberley in Novon **8**: 204–206 (1998). —R. Fernandes & Verdcourt in Kew Bull. **55**: 147–154 (2000). —Harley et al. in Kadereit (ed.), Fam. Gen. Vasc. Pl. (Kubitzki, ed. in chief) **VII**: 198 (2004).
Spironema Hochst. in Schaed. Schimperi Iter Abyss., Sectio prima., Unio. Iter: 330 (?1841).
Cyclonema Hochst. in Flora **25**: 225 (1842).
Clerodendrum L. sect. *Cyclonema* (Hochst.) Gürke in Engler, Pflanzenw. Ost-Afrikas **C**: 341 (1895).
Clerodendrum L. subgen. *Cyclonema* (Hochst.) B. Thomas in Bot. Jahrb. Syst. **68**: 22 (1936).

Perennial herbs or suffrutices with woody rootstocks, sometimes geoxylic, shrubs, sometimes scandent, or small trees. Leaves simple, opposite-decussate or 3–4-whorled, usually petiolate; lamina entire, undulate, dentate or crenate, distinctly acrid when crushed. Inflorescences terminal or axillary, with flowers in lax or ± dense cymes, the cymes racemosely, corymbosely or paniculately arranged. Flowers non-resupinate. Calyx usually red or purple, campanulate to subspherical, ± deeply (4)5(6)-lobed, usually persistent and slightly accrescent in fruit; the lobes acute to rounded. Corolla predominantly blue to whitish, asymmetrical in bud, distinctly zygomorphic when open; tube relatively short and broad, anteriorly gibbose; limb 5-lobed with 2 posterior lobes subequalling 2 lateral lobes and 1-anterior lobe larger than the others, concave or spoon-like and often coloured differently. Stamens 4(5?) inserted at the base of corolla-tube, parallel, long exserted, arched; anthers usually basifixed, thecae parallel, not confluent. Ovary glabrous or hairy, 2-carpellate, each carpel incompletely divided into 2 locules; placentas not furcate; ovules 1 per locule. Style terminal, slender, ± equalling the filaments; stigma 2-fid, the lobes usually unequal. Fruit a 3–4-lobed drupe or drupaceous schizocarp, obovoid or globose; mesocarp ± fleshy; endocarp bony or crustaceous; each lobe containing a 1-seeded pyrene; seed without albumen.

A genus of 26–30 species, widespread throughout sub-Saharan Africa, with 1 species in tropical Asia eastwards to the Moluccas, and one species introduced and naturalized in Madagascar and the Indian Ocean islands.

1. Inflorescence raceme-like or narrowly paniculiform, terminal, consisting of sessile or shortly pedunculate cymes; peduncles 0–15 mm long, 1–3 per axil of opposite bracts; corolla limb when in bud ± densely covered with small sessile or shortly stalked spherical reddish-purple or orange-brown glands; leaves opposite · 2
   – Inflorescence not as above, the cymes with longer peduncles; corolla limb when in bud sometimes ± pilose but not glandular outside; leaves opposite or 3–4-whorled · · · · · · · 3
2. Leaves absent, or very young, at anthesis; petiole distinct, up to 5 cm long; leaf lamina elliptic or broadly elliptic, up to 13,5 × 7.2 cm, acuminate; a soft woody shrub or small tree · 11. *wildii*
   – Leaves present at anthesis; petiole apparently winged by the decurrent lamina; leaf lamina oblong-obovate to oblanceolate, 5–15(20) × 1.5–3.5(7) cm, rounded or subacute at the apex; a perennial herb or suffrutex · 12. *aurantiaca*
3. Perennial herbs or suffrutices with annual simple or few-branched stems, usually less than 1 m tall, from a woody rootstock; or sometimes subshrubs; petiole indistinct or short · · 4
   – Shrubs, subshrubs, lianes or trees · 6
4. Cymes solitary in the cauline leaf axils in the upper ¼–½ of the stem, 1–2(3)-flowered; leaves up to 5 × 1.3 cm, usually linear-elliptic to oblong, sessile or subsessile · · · 1. *hirsuta*
   – Cymes in terminal panicles; leaves larger ·5
5. Leaves up to 15 × 12 cm, with lamina decurrent into a winged petiole, markedly discolorous; petiole wings undulate, auriculate; calyx lobes triangular or subtriangular, c. 1.25 mm long · 3. *teaguei*
   – Leaves narrower, not winged-petiolate, ± concolorous; calyx lobes semicircular or oblong, larger · 2. *luembensis*
6. Calyx lobes usually oblong and reflexed at anthesis; leaf lamina ± attenuate at the base and decurrent into the petiole - · 7

– Calyx lobes semicircular, triangular or sometimes oblong, erect at anthesis, or if reflexed then leaf lamina not decurrent into the petiole $\cdots\cdots$ 9

7. Corolla lobes (or corolla limb when in bud) densely tomentose; leaf lamina up to 12.5 × 9.5 cm, entire or ± shallowly crenate, densely hairy to subtomentose beneath $\cdots\cdots$ $\cdots\cdots$ 4. *sansibarensis* var. *eratensis*

– Corolla lobes (or corolla limb when in bud) glabrous; leaf lamina relatively narrower, dentate or distinctly crenate, subglabrous or ± hairy beneath, not tomentose $\cdots\cdots$ 8

8. Leaf lamina usually rhombic, up to 13.8 × 5 cm, deeply dentate; petiole not distinctly winged $\cdots\cdots$ 5. *reflexa*

– Leaf lamina broadly ovate, up to 15 × 8 cm, regularly crenate; petiole undulate-winged towards the lamina $\cdots\cdots$ 6. *amplifolia*

9. Calyx lobes triangular or oblong-triangular, acute or subacute $\cdots\cdots$ 10

– Calyx lobes usually semicircular or oblong, rarely somewhat lanceolate or obsolete, or if ovate-triangular then the leaf lamina decurrent into a winged petiole $\cdots\cdots$ 11

10. Petiole winged, the wings undulate and articulate; leaf lamina up to 15 × 12 cm, whitish-green or ochraceous-tomentose beneath, not conspicuously reticulate; panicles somewhat dense $\cdots\cdots$ 3. *teaguei*

– Petiole not winged, slender; leaf lamina up to 14 × 9 cm, relatively narrower, sparsely hairy on the nerves beneath and conspicuously finely reticulate beneath; panicles lax $\cdots\cdots$ $\cdots\cdots$ 7. *verdcourtii*

11. Leaves sessile, subsessile or indistinctly petiolate $\cdots\cdots$ 12

– Leaves distinctly petiolate $\cdots\cdots$ 16

12. Plants glabrous or nearly so, particularly on the leaves (± sparsely pilose particularly on the nerves beneath in *quadrangulata*) $\cdots\cdots$ 13

– Plants ± hairy to tomentose $\cdots\cdots$ 15

13. Leaf lamina usually large, up to 22.5 × 10 cm, mostly oblanceolate or oblong-oblanceolate to subspathulate and sessile or the lower ⅓ attenuate or contacted into a narrow petiole-like base; panicle up to 30 × 20 cm, open and laxly-spreading; peduncles slender, 6–12 cm long in the lower cymes $\cdots\cdots$ 8. *quadrangulata*

– Leaf lamina mostly 6.5 × 2.8 cm; inflorescences much smaller, the cymes few-flowered and not forming distinct panicles; peduncles shorter $\cdots\cdots$ 14

14. Leaves sessile or subsessile, clustered at the apex of the short flowering branches; leaf lamina obovate or spathulate, entire or obscurely dentate in the upper ⅓, rounded or sub-obtuse at the apex, thinly coriaceous when dry, punctate beneath $\cdots\cdots$ $\cdots\cdots$ 10. *myricoides* subsp. *myricoides* var. *capiriensis*

– Leaves obscurely petiolate with lamina narrowly decurrent almost to the point of attachment, not clustered towards the apices of the flowering branchlets; leaf lamina obovate-rhombic or oblong-obovate, usually broadly crenate-dentate, shortly acuminate at the apex, membranous or thinly coriaceous, not punctate beneath $\cdots$ 10. *myricoides* subsp. *myricoides* var. *moldenkei*

15. Branches and leaf lower surface tomentose, at least when young; leaves discolorous, dark above and yellowish or greenish-yellow beneath; a shrub up to 2 m high $\cdots\cdots$ $\cdots\cdots$ 10. *myricoides* var. *discolor* forma *alatipetiolata*

– Stems, branches and petioles hispid; leaves not tomentose beneath, ± concolorous; a subshrub $\cdots\cdots$ 2. *luembensis* subsp. *luembensis* var. *luembensis*

16. Plants ± persistently hairy, sometimes tomentose on the upper part of the stem and leaf lower surface $\cdots\cdots$ 17

– Plants glabrous, the inflorescence branches sometimes pilose, and the nerves on the leaf lower surface slightly appressed hairy, the young leaves sometimes with an evanescent araneose indumentum $\cdots\cdots$ 18

17. Leaf lamina elliptic, broadest about the middle and tapering equally to both ends, hairy on both surfaces, concolorous; short lateral flowering branches absent $\cdots\cdots$ $\cdots\cdots$ 10. *myricoides* subsp. *myricoides* var. *eleanorae*

– Leaf lamina usually broadest above the middle, usually discolorous and whitish or yellowish tomentose beneath; branches often densely hairy or tomentose towards the apices; short lateral flowering branches often present $\cdots\cdots$ 10. *myricoides* subsp. *myricoides* var. *discolor*

18. Leaves ± closely spaced on the upper part of the flowering branches; leaf lamina mostly elliptic, usually entire, pale green, ± concolorous, thinly coriaceous when dry, densely punctate and with an inconspicuous reticulation beneath; inflorescence ample, up to 18 cm long; peduncles of the lower cymes 5–8 cm long, spreading or arched; calyx 6.5–10 mm in diameter, calyx lobes 1.5–3 × 3–4 mm, semicircular, pale green below, magenta or purple towards the margin, not drying dark $\cdots\cdots$ 9. *cyanea*

– Characters not combined as above ·· 19
19. Leaf lower surface conspicuously finely reticulate, not punctate; leaf lamina 4.5–14 × 2.9 cm, broadly elliptic or ovate to subcircular; panicle ample, up to 25 cm long; peduncles of the lower cymes 4.5–8 cm long; calyx lobes short or nearly absent ········ 7. *verdcourtii*
– Leaf lower surface coarsely reticulate (if nervation is visible), often punctate; leaf lamina relatively narrower, usually elliptic, ± attenuate and acute at both ends; panicle usually not ample, up to 15.5 cm long; calyx lobes distinct ···························
···························· 10. *myricoides* subsp. *myricoides* var. *myricoides*

1. **Rotheca hirsuta** (Hochst.) R. Fern. in Kew Bull. **55**: 148 (2000). Type from South Africa (KwaZulu-Natal).
　　Cyclonema hirsutum Hochst. in Flora **25**: 228 (1842). —Walpers, Repert. Bot. Syst. **4**: 101 (1845). —Schauer in A. de Candolle, Prodr. **11**: 676 (1847).

Perennial herb or suffrutex 4–77 cm tall, from a woody rootstock. Stems 1–several (up to 8 or more), erect, usually simple, 4-angled, ± densely leafy. Leaves usually in whorls of 3, rarely opposite or in whorls of 4, sessile or nearly so; lamina shorter to longer than the internodes, smallest and relatively widest in the lowermost leaves, up to 5 cm long and (1.5)2–13 mm wide, elliptic, linear-elliptic to oblong or sometimes obovate, acute to obtuse or rounded at the apex, attenuate towards the base, entire, coriaceous to membranous, profusely gland-dotted on lower surface; midrib rather prominent beneath and the lateral nerves inconspicuous. Cymes 1–2(3)-flowered, solitary, axillary, usually in the upper $^{1}/_{4}$–$^{1}/_{2}$ of the stems (rarely also in the lower part), subequalling the subtending leaf, or the uppermost somewhat exceeding the subtending leaf, the three leaves at each node ± equal in length; peduncles up to 3 cm long, slender, ascending or the lowermost sometimes spreading; pedicels 2–3 mm long, 2-bracteate at the base. Calyx campanulate, divided to about the middle; lobes up to 3 mm long, oblong or ovate to triangular, obtuse or acute, sometimes appearing acute or nearly subulate due to the margins becoming involute on drying. Corolla deep or vivid blue with a bifurcate violet stripe on the median lobe, or lilac-blue to violet-blue with a yellow stripe on each side of the median lobe, or livid-blue with broad, white stripes on the median lobe; tube 4–5 mm long; lateral lobes obliquely obovate or elliptic, obtuse; lower lobe obovate or oblong, exceeding the others. Fruit black, c. 8 mm long, deeply lobed, each lobe cylindrical, smooth, rounded at the apex, c. 3 mm in diameter.

Plant hirsute on stems, leaves and calyx, or at least the leaves and calyx ± hairy ··· forma *hirsuta*
Plant glabrous, or sometimes the calyx puberulous ····················· forma *triphylla*

Forma **hirsuta**
　　Clerodendrum hirsutum (Hochst.) H. Pearson in F.C. **5**, 1: 221 (1901). —Thomas in Bot. Jahrb. Syst. **68**: 80 (1936). —Hutchinson, Botanist South. Africa: 339 et 672 (1946). — Garcia in Estud. Ensaios Doc. Junta Invest. Ci. Ultramar [in Mendonça, Contrib. Conhec. Fl. Moçamb., II] **12**: 169 (1954). —Ross, Fl. Natal: 300 (1972) *nom. illegit.*, non G. Don (1824).
　　Clerodendrum hirsutum var. *ciliatum* H. Pearson in F.C. **5**, 1: 221 (1901). Type from South Africa (KwaZulu-Natal).
　　Clerodendrum pearsonii Moldenke in Rev. Sudamer. Bot. **5**: 1 (1937). Type as for *Cyclonema hirsutum* Hochst.
　　Clerodendrum triphyllum var. *ciliatum* (H. Pearson) Moldenke in Phytologia **5**: 8 (1954).
　　Clerodendrum triphyllum forma *hirsutum* (Hochst.) R. Fern. in Mem. Soc. Brot. **30**: 104 (1998). (Incorrect name)

Zimbabwe. E. Chimanimani Distr., Tarka Forest Reserve, c. 1088 m, fl. x.1968, *Goldsmith* 133/68 (K; LISC; SRGH). **Mozambique.** MS: Barué Distr., Serra de Chôa, at 29 km from Catandica (Vila Gouveia), on the road to the frontier, c. 1000 m, fl. 9.xii.1965, *Torre & Correia* 13449 (LISC). M: Namaacha Distr., M'Ponduine (Ponduine) Mts., fl. 20.xi.1966, *Moura* 115 and 126 (COI; LMU).
　　Also in Swaziland and South Africa (Transvaal and KwaZulu-Natal). Gully forest and margins of evergreen forest on mountain sides, and in firebreaks in montane grassland; 700–1100 m.

Forma **triphylla** (Harv.) R. Fern. in Kew Bull. **55**: 149 (2000). Type from South Africa (Gauteng).
　　Cyclonema triphyllum Harv., Thes. Cap. **1**: 17, t. 27 (1859).

Clerodendrum natalense Gürke in Bot. Jahrb. Syst. **18**: 183 (1893). Type from South Africa (KwaZulu-Natal).

Clerodendrum triphyllum (Harv.) H. Pearson in F.C. **5**, 1: 220 (1901). —Thomas in Bot. Jahrb. Syst. **68**: 80 (1936). —Hutchinson, Botanist South. Africa: 339 et 400 (1946). — Garcia in Estud. Ensaios Doc. Junta Invest. Ci. Ultramar [in Mendonça, Contrib. Conhec. Fl. Moçamb., II] **12**: 169 (1954). —Letty, Wild Fl. Transvaal: 280, t. 140 fig. 2 (1962). — Watt & Breyer-Brandwijk, Medic. & Pois. Pl. S. & E. Africa: 1048 (1962). —Ross, Fl. Natal: 300 (1972). —Herman in Bothalia **25**: 100 (1995). —R. Fernandes in Mem. Soc. Brot. **30**: 104 (1998).

Siphonanthus triphyllus Hiern ex S. Moore in J. Bot. **41**: 405 (1903), as "*triphylla*". Type as for *Clerodendrum triphyllum.*

Clerodendrum triphyllum var. *vernum* B. Thomas in Bot. Jahrb. Syst. **68**: 80 (1936) *nom. inval.* Type from South Africa (Gauteng).

Clerodendrum triphyllum forma *angustissimum* Moldenke in Phytologia **32**: 46 (1975). Type from South Africa (Transvaal).

Zimbabwe. E. Chimanimani Distr., Nyahode R. bridge, fl. 18.xi.1965, *Plowes* 2728 (K; SRGH).
Mozambique. M: Namaacha Distr., Goba Fronteira, fl. 12.xii.1947, *Barbosa* 697 (BR; COI; LISC).

Also in South Africa (Free State, Transvaal, KwaZulu-Natal and Cape Prov.). In wooded grassland, open xerophytic bush and in burnt open grassland; 350–1150 m.

Both forms occur together in the eastern border region of Zimbabwe, mainly in the Chimanimani area, and in Mozambique along the Swaziland border. The two forms are sometimes found intermixed within the same gathering, e.g. *Plowes* 2547 from Zimbabwe where the LISC duplicate is forma *hirsuta* and the SRGH duplicate is forma *triphylla*. Similar examples are seen in the duplicates of *Gomes e Sousa* 127 and 127A (from Goba in Mozambique) where the LISC duplicate is forma *hirsuta* and the LMU and another LISC duplicate are forma *triphylla*; and in *Torre & Correia* 13541 from Baruè, Chôa Mts., both in LISC.

2. **Rotheca luembensis** (De Wild.) R. Fern. in Kew Bull. **55**: 149 (2000). Type from Dem. Rep. Congo.

Clerodendrum luembense De Wild. in Bull. Jard. Bot. État **3**: 267 (1911). —Thomas in Bot. Jahrb. Syst. **68**: 84 (1936). —Moldenke in Phytologia **62**: 127 (1987) pro parte. —R. Fernandes in Mem. Soc. Brot. **30**: 39 (1998).

Perennial herb or suffrutex 25–80(100) cm tall, from a long creeping woody rootstock. Stems few to many, erect, unbranched or occasionally somewhat branched, usually with a ± dense indumentum of spreading, ± rigid, short hairs. Leaves in whorls of 3 or opposite, rarely in whorls of 4, sessile to shortly petiolate; internodes up to 10 cm long; lamina 4–18.5 × 1.2–6 cm, oblong-obovate, oblong-lanceolate, elliptic or obovate-cuneate, up to 5-times longer than wide but more usually 2.5-times longer than wide, widest above the middle, ± long-attenuate towards the base, obtuse to acute or shortly acuminate at the apex, entire to coarsely serrate-dentate in the upper $^1/_3$, ± concolorous, usually not becoming dark on drying, pubescent on both surfaces with hairs similar to those of stems and branches but shorter, with longer hairs on the midrib beneath, the hairs becoming rigid at the base making the lamina ± scabridulous; petiole 1–2 mm long or absent. Inflorescence a terminal, pubescent panicle, 7–23 cm long, formed by opposite 2–5-flowered cymes, sometimes with 2–3 additional cymes in the axils of the uppermost leaves. The uppermost cymes subtended by foliaceous sessile bracts; the lowermost bracts up to 6 × 2 cm. Peduncles of the lowermost cymes up to 5.5 cm long. Calyx pale green, red in the upper part when fresh, red-purple on drying, (4)4.5–6 mm long and c. 7 mm in diameter, obconical-campanulate, subglabrous to ± appressed-hairy; lobes 3–4 mm wide at the base, rounded or oblong and obtuse, ciliate. Corolla variable in colour, green to greenish-white except for the median bright blue lobe, or pale blue, or pale blue and yellow-green, or white with the lateral lobes light green and the median lobe purple, or 4 lobes pale muddy blue and the median lobe royal-blue, or 4 lobes greenish-mauve and the median lobe violet, green without and brownish within; corolla tube 5–6 mm long, corolla lobes ciliate; limb of corolla bud 5–6 mm in diameter, subspherical, glabrous or subglabrous to ± densely hairy. Anthers c. 2 mm long, yellow turning brown; filaments, style and stigma mauve to purple.

1. Inflorescence c. 10 cm long, ± condensed; peduncles of the lower cymes c. 1.5 cm long; fruit 5 mm long and 8–9 mm wide · subsp. *niassensis*
- Inflorescence up to 23 cm long, loose; peduncles of the lower cymes up to 5.5 cm long; fruit c. 8 mm long and c. 11 mm wide (subsp. *luembensis*) · 2

2. Leaves relatively wider, usually 1.7–2.1 times longer than wide, sometimes the broadest part of the lamina ± circular, deeper dentate, subtomentose beneath; calyx lobes oblong, obtuse
· subsp. *luembensis* var. *malawiensis*
 – Leaves 2.4–5 times as long as wide, hairy on both surfaces; calyx lobes semi-circular (subsp. *luembensis* var. *luembensis*) · 3
3. Leaves mostly 2.5 times longer than wide, but can range from 2.4–3.8 times longer than wide, widest in the upper $^1/_3$, oblong-obovate, obovate-cuneate, long-attenuate towards the base, shortly acuminate or obtuse at the apex, entire or coarsely dentate in the upper $^1/_3$
· var. *luembensis* forma *luembensis*
 – Leaves relatively narrower, widest about the middle, elliptic or lanceolate, attenuate towards the base and acute at the apex, entire or with some small acute teeth · · · · · · · · · · · · · ·
· var. *luembensis* forma *herbacea*

Subsp. **luembensis**

Inflorescence up to 23 cm long, loose; peduncles of the lower cymes up to 5.5 cm long; calyx 4–6 mm long; corolla limb when in bud 5–6 mm in diameter; fruit c. 8 mm long and 11 mm wide.

Var. **luembensis** forma **luembensis**

Clerodendrum prittwitzii B. Thomas in Bot. Jahrb. Syst. **68**: 105 et 84 (1936). —Brenan, Check-list For. Trees Shrubs Tang. Terr.: 653 (1949). —Verdcourt in F.T.E.A., Verbenaceae: 129 (1992). Type from Tanzania.
Clerodendrum milne-redheadii Moldenke in Phytologia **3**: 264 (1950). —White, For. Fl. N. Rhod.: 366 (1962). Type: Zambia, east of R. Matonchi, *Milne-Redhead* 3526 (K, holotype).
Clerodendrum discolor var. *kilimandscharense* sensu Moldenke in Phytologia **59**: 264 (1986) quoad specim. *Borle* s.n., non B. Thomas.
Clerodendrum luembense subsp. *luembense* var. *luembense* forma *luembense* R. Fern. in Mem. Soc. Brot. **30**: 41 (1998).
Rotheca prittwitzii (B. Thomas) Verdc. in Kew Bull. **55**: 152 (2000).

Zambia. B: Sesheke, 1925, *Borle* s.n. (PRE); Sefula Forest, s.d., *Kieser* (P). N: Kasama Distr., 30 km south of Kasama on road to Mpika, 1250 m, immat. fl. 13.i.1975, *Brummitt & Polhill* 13762 (K). W: Kitwe, fl. 19.i.1955, *Fanshawe* 1812 (BR; K; SRGH). C: Lusaka Distr., c. 10 km east of Lusaka, c. 1344 m, fl. 16.i.1956, *King* 275 (K). **Malawi.** C: Kasungu National Park, c. 1088 m, fl. 10.i.1971, *Hall-Martin* 1608 (PRE).
Also in Dem. Rep. Congo and Tanzania. In mixed deciduous woodland, miombo woodland and riverine forest, sometimes on termite mounds; 1120–1350 m.
The Zambian specimens of *Macaulay* 879 pro parte (K) from Mumbwa, *Cruse* 461 (BR; K) from Mufulira, *Fanshawe* 422 (BR; K) from Solwezi, *Holmes* 1417 (K) from Solwezi, and *Strid* 2515 (K) from Solwezi, differ in that the limb of the floral bud is densely hairy, not glabrous or sparsely hairy as in the other specimens.
Clerodendrum erectum De Wild. is considered to be synonymous with *C. myricoides* = *Rotheca myricoides* (Hochst.) Steane & Mabberley by E. Persson (in her Ph.D. thesis, 1990) and as a "variant of *C. myricoides*" = *R. myricoides* by Verdcourt (op. cit.: 133). Examination of the type (*Hock* s.n., BR) has shown that it is probably the same as *R. luembensis* (De Wild.) R. Fern., also from Katanga. Specimens of *R. luembensis* with leaves narrower than usual, and completely glabrous as in *C. erectum*, are also found in Zambia. In typical *R. myricoides* the leaves are petiolate, and not decurrent into a long base, and the calyx is usually smaller (not 5.5–7 mm long as in *Hock* s.n.), with relatively shorter lobes. Thus *C. erectum* is probably a synonym of *R. luembensis* forma *luembensis*, corresponding to plants with narrower leaves.

Var. **luembensis** forma **herbacea** (Hiern) R. Fern. in Kew Bull. **55**: 149 (2000). Type from Angola.

Siphonanthus myricoides var. *herbaceus* Hiern, Cat. Afr. Pl. Welw. **1**, 4: 845 (1900), as "*herbacea*".
Clerodendrum myricoides var. *camporum* Gürke in Bot. Jahrb. Syst. **28**: 299 (1900) pro parte quoad *Newton* 186.
Clerodendrum myricoides var. *savanarum* sensu B. Thomas in Bot. Jahrb. Syst. **68**: 87 (1936), as "*savanorum*" pro parte quoad *Antunes* 338, *Welwitsch* 5768 (et specim. Angol.), et syn. var. *herbacea* Hiern, non *C. savanarum*, as "*savanorum*" De Wild. nom. illegit.
Clerodendrum luembense subsp. *luembense* var. *luembense* forma *herbaceum* (Hiern) R. Fern. in Anales Jard. Bot. Madrid **54**, 1: 293 (1996); in Mem. Soc. Brot. **30**: 42 (1998).

Zambia. W: Solwezi Distr., c. 10 km from Solwezi along the road to Mwinilunga, fl. 20.xi.1972, *Strid* 2482 (K).
Also in Angola. Dense miombo woodland on deep brownish-red lateritic soils; c. 1350 m.
Strid 2482 is very similar to *Welwitsch* 5768, the type of forma *herbacea*. It has been incorrectly

determined as *C. alatum* Gürke, a species from which it differs by the lack of wings on its stem, and by the hairy indumentum of its stem, leaves and inflorescence, its leaves more attenuate and acute at the apex and not rounded at the base, its inflorescence shorter, smaller and not so dense, and by its flowers having larger calyces.

Var. **malawiensis** (R. Fern.) R. Fern. in Kew Bull. **55**: 149 (2000). Type: Malawi, Chongoni Forest Reserve, *Salubeni* 862 (SRGH, holotype; K).
   Clerodendrum luembense subsp. *luembense* var. *malawiense* R. Fern. in Mem. Soc. Brot. **30**: 45 (1998).

Malawi. C: Dedza, Danya village, fl. 10.xii.1969, *Salubeni* 1432 (K; SRGH).
Unknown elsewhere. Habitat unknown.

Subsp. **niassensis** (R. Fern.) R. Fern. in Kew Bull. **55**: 149 (2000). Type: Mozambique, Niassa Prov., Mandimba, fl. 3.xi.1942, *A.J.W. Hornby* 3720 (PRE, holotype).
   Clerodendrum luembense sensu Moldenke in Phytologia 62: 128 (1987) quoad specim. *A.J.W. Hornby* 3720, non De Wild. sens. str.
   Clerodendrum luembense subsp. *niassense* R. Fern. in Mem. Soc. Brot. **30**: 45 (1998).

Inflorescence c. 10 cm long, ± condensed; peduncles of the lower cymes c. 1.5 cm long; calyx 3.5–4 mm long (4–6 mm in the typical subsp.); corolla limb when in bud 4(5) mm in diameter (5–6 mm in the typical subsp.); fruit 5 mm long and 8–9 mm wide.

Mozambique. N: Mandimba, fr. 14.v.1948, *Pedro & Pedrógão* 3455 (LMA).
Unknown elsewhere. Habitat unknown.
Milne-Redhead 625, determined by Moldenke (in sched., 1949) as *C. discolor* var. *kilimandscharense* B. Thomas and cited by him (in Phytologia **62**: 128, 1987) as *C. luembense* De Wild., is very different from the latter species, and possibly belongs to the complex of *Rotheca myricoides* var. *discolor*.

3. **Rotheca teaguei** (Hutch.) R. Fern. in Kew Bull. **55**: 153 (2000). Type Zimbabwe, Mutasa Distr., Odzani R. Valley, 1915, *Teague* 552 (K, holotype; BOL).
   Clerodendrum teaguei Hutch. in Ann. Bolus Herb. **3**: 9 (1920).
   Clerodendrum discolor sensu Garcia in Estud. Ensaios Doc. Junta Invest. Ci. Ultramar [in Mendonça, Contrib. Conhec. Fl. Moçamb., II] **12**: 170 (1954). —sensu Moldenke in Phytologia **59**: 261 (1986) pro parte quoad specim. *Mendonça* 3874, non (Klotzsch) Vatke.
   Clerodendrum myricoides sensu Drummond in Kirkia **10**: 272 (1975) pro parte quoad specim. *Teague* 552.

A many-stemmed perennial herb, or shrub(?). Stems simple, somewhat thick, cylindric or c. 4-angular, densely hairy towards the apex, glabrescent and striate below. Leaves opposite, petiolate; lamina 6–15 × 3.8–12 cm, broadly ovate, attenuate and decurrent at the base into a winged petiole, acuminate acute at the apex, regularly dentate in the upper $^1/_2$–$^2/_3$ with broad deep teeth up to 12 mm wide, or subentire, strongly discolorous, drying nearly black and sparsely hairy-scabrid with short hairs on the upper surface, ± densely whitish-green or brownish-yellow tomentose beneath; midrib rather thick. Petiole up to 2 cm long, winged; the wings up to 5 mm wide, undulate, similar to the lamina in texture, colour and indumentum, each wing expanded at the base into a broad, rounded auricle. Inflorescence a terminal panicle up to 10 cm long, formed by opposite cymes; rhachis densely hairy; peduncles hairy, up to 5 cm long; pedicels up to 2.5 mm long; cymes few-flowered, the lowermost in the axils of the upper leaves, the rest subtended by foliaceous bracts, becoming progressively smaller towards the apex. Calyx 4–5 mm long, blackish on drying, appressed hairy mainly on the tube; lobes c. 1.75 mm long, triangular, acute or subacute, separated by wide sinuses, only slightly enlarged after anthesis. Corolla limb when in bud glabrous; corolla lobes not ciliate. Anthers 1.25–1.5 mm long. Fruit 7–10 mm long, black.

Zimbabwe. E: Mutasa Distr., Odzani River Valley, fl. & fr. 1915, *Teague* 552 (BOL; K).
Mozambique. MS: on the road between Macequece and Mutare (Umtali), near the frontier Customs Post, fl. 4.iv.1948, *Mendonça* 3874 (LISC; LMA).
Not known elsewhere. In sparsely wooded grassland; 800–1400 m.
It approaches *R. myricoides* subsp. *austromonticola* (Verdc.) Verdc. in the shape of its leaves, but differs in having a denser indumentum on the leaf lower surface, broader undulate wings to the

petiole, triangular and acute or subacute calyx lobes (oblong or oblong-semicircular and rounded at the apex in subsp. *austromonticola*), and the corolla limb glabrous when in bud (pubescent in subsp. *austromonticola*). It is also somewhat similar to *R. luembensis* var. *malawiensis* (R. Fern.) R. Fern. with respect to the leaf shape and the strongly dentate leaf margin, but in var. *malawiensis* the leaves are smaller, not distinctly winged-petiolate and the calyx lobes are neither triangular nor acute.

4. **Rotheca sansibarensis** (Gürke) Steane & Mabb. in Novon **8**: 205 (1998). —R. Fernandes in Kew Bull. **55**: 153 (2000). Type from Tanzania.
    Clerodendrum sansibarense Gürke in Bot. Jahrb. Syst. **18**: 181 (1893). —J.G. Baker in F.T.A. **5**: 312 (1900). —Thomas in Bot. Jahrb. Syst. **68**: 81 (1936). —Verdcourt in F.T.E.A., Verbenaceae: 126 (1992). —R. Fernandes in Mem. Soc. Brot. **30**: 101 (1998).

Subsp. **sansibarensis**
    Clerodendrum sansibarense subsp. *sansibarense* —Verdcourt in F.T.E.A., Verbenaceae: 127 (1992).

Var. **eratensis** (R. Fern.) R. Fern. in Kew Bull. **55**: 153 (2000). Type Mozambique, Nampula, Mte. Cheovi (Geovi), 8.i.1964, *Torre & Paiva* 9884 (LISC, holotype; COI; SRGII).
    Clerodendrum reflexum sensu Moldenke in Phytologia **59**: 261 (1986), quoad specim. *Torre & Paiva* 9884, non H. Pearson.
    Clerodendrum sansibarense subsp. *sansibarense* var. *eratense* R. Fern. in Mem. Soc. Brot. **30**: 103 (1998).

Shrub 2–3 m high. Leaves large, up to 12.5 × 9.5 cm, broadly ovate, subtruncate at the base with the lamina produced into a narrowly cuneate portion about the midrib base, becoming somewhat decurrent on the upper part of the petiole, attenuate acute at the apex in the younger leaves, shortly acuminate in the oldest leaves, shallowly crenate, submembranous, hairy on both surfaces (more densely so to nearly tomentose beneath), ± concolorous; petiole up to 2.7 cm long, hairy. Inflorescence, pedicels and calyx subtomentose, with very fine soft spreading whitish hairs. Calyx c. 6 mm long; lobes oblong, obtuse, reflexed before and at anthesis, up to 5–6 × 3 mm in the fruiting-calyx. Corolla limb when in bud densely tomentose.

Mozambique. N: Namapa Distr., 12 km from Namapa to Alua, Cheovi (Geovi) Mte., c. 500 m, fl. & fr. 8.i.1964, *Torre & Paiva* 9884 (COI; EA; LISC; LMU; LUA; SRGH; Z).
    Not known elsewhere. Amongst rocks on granite outcrops; c. 500 m.
    Var. *sansibarensis* from Kenya and Tanzania, with leaves entire or more distinctly toothed on the margins, and with the inflorescence and calyx lacking fine soft spreading whitish hairs, has not been recorded in the Flora Zambesiaca area.
    R. sansibarensis subsp. *caesia* (Gürke) Steane & Mabb., from Tanzania, differs in its entirely glabrous corollas with lobes up to 9 mm wide.

5. **Rotheca reflexa** (H. Pearson) R. Fern. in Kew Bull. **55**: 152 (2000). —M. Coates Palgrave, ed. 3 of K. Coates Palgrave, Trees South. Africa: 991 (2002). Type: Zimbabwe, Bulawayo, *Eyles* 1006 (BOL, holotype; SRGH).
    Clerodendrum reflexum H. Pearson in Trans. S. African Philos. Soc. **15**: 182 (1905); in Repert. Spec. Nov. Regni Veg. **4**: 27 (1907). —Thomas in Bot. Jahrb. Syst. **68**: 91 (1936). —Wild in Clark, Victoria Falls Handb.: 158 (1952). —R. Fernandes in Mem. Soc. Brot. **30**: 99 (1998).
    Clerodendrum myricoides sensu White, For. Fl. N. Rhod.: 366 (1962) pro parte quoad *Angus* 1010. —sensu Drummond in Kirkia **10**: 272 (1975) pro parte quoad specim. *Eyles* 1006, non (Hochst.) Vatke.
    Clerodendrum discolor var. *oppositifolium* sensu Wild in Clark, Victoria Falls Handb.: 158 (1952). —sensu Moldenke in Phytologia **59**: 266 (1986) pro parte quoad specim. *C.E.F. Allen* 102, non Thomas.
    Clerodendrum myricoides var. *grosseserratum* sensu Moldenke in Phytologia **62**: 468 (1987) pro parte quoad specim. *Bainbridge* 635, non Gürke.

A weak-stemmed shrub up to 3 m tall, or a slender tree 3–4 m tall, sometimes an under-shrub 0.65–1 m tall, profusely branched. Branches slender, 4-angled or the old ones terete, smooth or striate, lenticellate, usually not swollen at the nodes, subglabrous or sparsely to ± densely hairy, brownish to deep brown, usually persistently leafy, sometimes with axillary fascicles of leaves. Leaves opposite or subopposite; lamina 5–13.8 × 2–5 cm (smaller in the leaves subtending the lower cymes and in those of the axillary fascicles, sometimes not more than 1.3 × 0.7 cm), usually rhombic or sometimes lanceolate or subobovate-oblong, attenuate or cuneate at the base and decurrent into the petiole, acuminate at the apex, deeply dentate on the margins with broad subapiculate teeth, the cuneate base entire, or some leaves completely entire, membranous, concolorous or somewhat discolorous, deep green above, paler beneath, often very sparsely appressed shortly pubescent on both surfaces, the hairs a little longer on the nerves beneath; midrib and lateral nerves rather slender; petiole up to 1.2 cm long, slender. Cymes 1–few-flowered, arranged in loose terminal panicles up to 9 cm long, the panicle not usually distinct, the uppermost pair of normal leaves sometimes subtending the 2 basal cymes; peduncles 4–5 cm long, not densely appressed or spreading hairy. Calyx 5–7 mm long, campanulate, usually sparsely appressed hairy, rose-purple; lobes 2–3 mm long, oblong or oblong-semicircular, at least some lobes reflexed or spreading, pale pink. Corolla usually with the four lateral lobes white or pale blue or sometimes green, and the median lobe blue, mauve-purple or purple; corolla tube c. 10 mm long. Anthers c. 1.5 mm long.

Botswana. N: near the Kwando River main stream, at 18°03'S, 23°19.5'E, fl. 29.i.1978, *P.A. Smith* 2305 (K; PRE; SRGH). **Zambia.** B: Sesheke Distr., Simungoma (Simongoma) Forest Reserve, near Masese, st. 26.i.1952, *White* 1967 (FHO; K). S: Namwala Distr., Kafue National Park, c. 72 km north of Ngoma, st. 16.i.1963, *B.L. Mitchell* 17/23 (SRGH). **Zimbabwe.** N: Mount Darwin Distr., Kandeya C.L. (Native Reserve), slopes of Mavuradonha Mts., 1024 m, 17.i.1960, *Phipps* 2276 (K, fl.; PRE; SRGH, st.). W: Hwange Distr., Victoria Falls, fl. 10.xii.1971, *van den Berghen* 30 (BR). C: Makoni Distr., Chiduku, fl. xii.1960, *Davies* 2864 (SRGH).

Known only from the Flora Zambesiaca area. In riverine forest, dense mixed woodland, understorey thicket and the shrub layer of dry deciduous Kalahari Sand forest not subject to heavy fires ("mutemwa"); 960–1220 m.

Rotheca reflexa usually approaches *R. myricoides* var. *myricoides* more closely than it does *R. myricoides* var. *discolor* (in indumentum and leaf shape). However, some specimens, mainly from Zimbabwe, with leaf bases ± attenuate, leaf margins ± deeply and acutely dentate, and calyces (4)5.5–6 mm long, approach *R. myricoides* var. *discolor* in their denser indumentum on the leaves and calyces and in their distinct, more floriferous panicle. See also the notes under *C. myricoides* var. *discolor* forma *reflexilobatum* in Mem. Soc. Brot. **30**: 75 (1998).

The specimens from Mozambique: MS., *Barbosa* 1018 (LISC) and 1582 (LISC; Z) and *Garcia* 323 (LISC), referred to *Clerodendrum reflexum* by Moldenke (in Phytologia **87**: 261, 1986), are a form of *R. myricoides* var. *discolor* with a very dense indumentum and large leaves (see note: R. Fernandes in Mem. Soc. Brot. **30**: 76 (1998).

Pole Evans 2598 (PRE) from Botswana (between Tutume and Sebina?), with leaves smaller than usual for this species, probably belongs to *R. reflexa*.

6. **Rotheca amplifolia** (S. Moore) R. Fern. in Kew Bull. **55**: 148 (2000). —M. Coates Palgrave, ed. 3 of K. Coates Palgrave, Trees South. Africa: 990 (2002). Type: Zimbabwe, Chirinda Forest, 1184–1280 m, *Swynnerton* 335 (BM, holotype; K, err. no. 524*).

    Clerodendrum amplifolium S. Moore in J. Linn. Soc., Bot. **40**: 167 (1911). —B. Thomas in Bot. Jahrb. Syst. **68**: 90 (1936). —Moldenke in Phytologia **57**: 475–476 (1985) pro parte, excl. specim. *Whellan* 349. —R. Fernandes in Mem. Soc. Brot. **30**: 28 (1998).

    Clerodendrum myricoides sensu Drummond in Kirkia **10**: 272 (1975) pro parte quoad specim. *Swynnerton* 335, non (Hochst.) Vatke.

A tall shrub; branches leafy, somewhat flattened and ± pubescent towards the apex, subterete and glabrous in the lower part with a pale brown bark. Leaves 9–15 × 6–8 cm, broadly ovate, rounded at the base with the lamina produced into a narrowly cuneate portion about the midrib base to become somewhat decurrent on the upper part of the petiole, acuminate at the apex, regularly crenate on the margins from near the base to the apex with crenations 4–11 mm wide and 1–1.5 mm high; lamina thickly membranous, appressed sparsely hairy and olivaceous on the upper face, hairy to nearly glabrous and greyish beneath; petiole c. 10 mm long, with the decurrent lamina forming undulate wings in its apical part, pubescent. Inflorescence a terminal panicle up to 8 cm long, with the rhachis and branches rather hairy; cymes loosely many-flowered; peduncles up to c. 4.5 cm long, pedicels 5–25 mm long, pubescent. Calyx tube 4 mm long, campanulate; lobes c. 3.5 mm

---

*See Mem. Soc. Brot. **30**: 30, 31 (1998).

long and c. 3 mm wide, oblong, obtuse at the apex, reflexed, minutely lepidote on the inner face. Corolla tube 11 mm long; corolla lobes 12–14 mm long and 5–8 mm wide, obovate-oblong, obtuse, glabrous on the outside. Stamen filaments exceeding the corolla tube by c. 3 cm; anthers c. 2 mm long.

Zimbabwe. E: Chirinda, 1184–1280 m, iv.1905, *Swynnerton* 335 (BM; K, err. nº 524). Not known elsewhere. Afromontane evergreen rainforest; 1100–1200 m.

This taxon approaches *R. sansibarensis* subsp. *sansibarensis* in having large leaves and reflexed calyx lobes. It differs from it in leaf shape (broadly-ovate, not elliptic) with the decurrent base forming two undulate wings on the upper part of the petiole, and with the margin regularly crenate (not dentate-mucronate), and the lamina appressed hairy above (not glabrous), and hairy beneath (not completely glabrous or only sparsely hairy on the nervation); and in the hairy inflorescence (not glabrous) and the corolla lobes glabrous outside (not pubescent or puberulous as in subsp. *sansibarensis*).

7. **Rotheca verdcourtii** (R. Fern.) R. Fern. in Kew Bull. **55**: 153 (2000). —M. Coates Palgrave, ed. 3 of K. Coates Palgrave, Trees South. Africa: 991 (2002). Type: Zimbabwe, Mutare Distr., Engwa, 1700 m, *Exell, Mendonça & Wild* 44 (SRGH, holotype; BM; LISC).
   Clerodendrum myricoides sensu S. Moore in J. Linn. Soc., Bot. **40**: 167 (1911) pro parte quoad specim. *Swynnerton* 6089 non (Hochst.) Vatke.
   Clerodendrum myricoides var. *camporum* sensu Moldenke in Phytologia **62**: 465 (1987) pro parte quoad specim. *Zimb.*, non Gürke.
   Clerodendrum verdcourtii R. Fern. in Mem. Soc. Brot. **30**: 105 (1998).

A spreading or bushy shrub 2–6 m tall, or sometimes a slender tree up to c. 9.6 m tall; old branches with a brown or grey rough bark, glabrous; young branches appressed hairy. Leaves opposite, distinctly petiolate; lamina 4.5–14 × 2–9 cm, usually less than twice as long as wide or nearly as long as wide, broadly ovate, broadly elliptic to subcircular or subrhombic, cuneate or rounded at the base, short to long-acuminate at the apex, distinctly dentate-mucronate to subentire or entire, membranous, somewhat discolorous, drying nearly black on the upper surface, usually with sparse slender ± appressed hairs on the nerves beneath, the finer reticulation darker and very conspicuous beneath; petiole (0.5)1–2(2.7) cm long, slender; the leaves subtending the lowermost cymes (the lowermost foliaceous bracts) up to 5 × 4 cm, rounded, obtuse or acuminate at the apex, with petioles short or nearly absent. Cymes 1–5-flowered, arranged in (4)5–8 opposite pairs in a loose terminal panicle up to 25 cm long, indumentum similar to that of the leaves; peduncles divaricate or erect-patent, very slender, the lowermost 4.5–8.5 cm long; pedicels of the dichotomies up to 8 mm long. Calyx 3–5(6) mm long and 3–5 mm in diameter at the apex of the lobes, subturbinate, usually glabrous, rarely with some hairs at the base; calyx lobes 1–1.5(2.5) mm long, shorter than the tube or nearly absent, triangular or shortly triangular-oblong, subacute or rarely rounded. Corolla bright blue throughout or with the median lobe dark blue or blue-purple; or corolla pale mauve with the median lobe violet or dark blue, or white and blue; tube 6.5–8.5 mm long and c. 4 mm in diameter; lateral lobes 9 × 4–5 mm, median lobe c. 13 mm long, all ciliate; corolla limb spherical when in bud, glabrous. Anthers c. 1.25–1.75 mm long. Fruit 8–10 mm long, coppery-red (immature?).

Zimbabwe. E: Mutasa Distr., Stapleford Forest Reserve, fl. 12.vi.1934, *Gilliland* 287 (BM; FHO; K). **Mozambique.** MS: Báruè Distr., Catandica (Vila Gouveia), Chôa Mt., 1300 m, fl. 26.ii.1968, *Torre & Correia* 17763 (LISC).

Known only from Zimbabwe and Mozambique. Afromontane evergreen forest, *Podocarpus*, mist and gulley forests, and forest margins; 1300–2080 m.

This species, from the eastern region of Zimbabwe, mainly in the Mutare area, is similar to *R. myricoides* var. *myricoides*. It may be distinguished from that taxon by the taller habit, by the leaves relatively broader and not so attenuate towards the base and not attenuate towards the apex, by the more distinct and finer reticulation beneath, and by the cymes with longer peduncles forming long ample panicles. It has also been confused with *R. sansibarensis* and *R. violacea* by various collectors.

8. **Rotheca quadrangulata** (B. Thomas) R. Fern. in Kew Bull. **55**: 152 (2000). Type from Dem. Rep. Congo.
   Clerodendrum myricoides var. *myricoides* sensu J.G. Baker in F.T.A. **5**: 310 (1900) pro parte quoad specim. *Whyte* s.n. (K) Misuku Hills (Masuku Plateau) Malawi non (Hochst.) Vatke.

Clerodendrum myricoides var. *attenuatum* De Wild. in Repert. Spec. Nov. Regni Veg. **13**: 143 (1914). Syntypes from Dem. Rep. Congo.

Clerodendrum attenuatum (De Wild.) De Wild. in Bull. Jard. Bot. État **7**: 182 (1920) *nom. illegit.* non R. Br. (1810).

Clerodendrum quadrangulatum B. Thomas in Bot. Jahrb. Syst. **68**: 88 (1936). —Brenan, Check-list For. Trees Shrubs Tang. Terr.: 633 (1949). —White, For. Fl. N. Rhod.: 366 (1962). —Richards & Morony, Check List Fl. Mbala: 237 (1969). —Gilli in Ann. Naturhist. Mus. Wien **77**: 30 (1973). —R. Fernandes in Mem. Soc. Brot. **30**: 96 (1998). Syntypes as for *Clerodendrum myricoides* var. *attenuatum* De Wild.

Clerodendrum quadrangulatum var. *reclinans* B. Thomas in Bot. Jahrb. Syst. **68**: 88 (1936). —Brenan, Check-list For. Trees Shrubs Tang. Terr.: 633 (1949) *nom. inval.* Type from Tanzania.

Clerodendrum scheffleri sensu Richards & Morony, Check List Fl. Mbala: 237 (1969) non Gürke.

Clerodendrum sp. *aff. sansibarense* Gürke, Richards & Morony, Check List Fl. Mbala: 237 (1969).

Clerodendrum myricoides var. *camporum* sensu Moldenke in Phytologia **62**: 465 (1987) quoad specim. *Richards* 18778, non Gürke.

Clerodendrum sansibarense subsp. *sansibarense* sensu Verdcourt in F.T.E.A., Verbenaceae: 127 (1992) pro parte quoad synonym *C. myricoides* var. *attenuatum*, *C. attenuatum* et *C. quadrangulatum* et distr. geogr. pro parte, non *C. sansibarense* Gürke.

Shrub 1–4 m tall, erect or sometimes scrambling up to 5(10?) m, weak-stemmed. Branches 4-angled, glabrous or with a caducous floccose whitish or fulvous indumentum towards the apices, flowering branches slender, sometimes pendulous; bark brown with prominent lenticels. Leaves opposite decussate, ± closely spaced on the upper part of the flowering branches, sessile or very shortly petiolate, spreading; lamina mostly sessile, (3)8–22.5 × (2)2.4–10 cm, (including the narrowly tapering base), broadly elliptic to orbicular in the upper ²/₃, oblanceolate or oblong-oblanceolate, sometimes subspathulate, ± long attenuate to broadly cuneate in the c. lower ¹/₃ or ± abruptly contracted and tapering into a ± broadly winged pseudopetiole, rounded to subcordate or semi-amplexicaul at the base, rounded to obtuse or ± acuminate at the apex, entire or serrate-dentate in the upper broader part of the leaf, teeth 12(14) on each side, up to 7 × 2.2 mm, sometimes crenulate; the blade thinly membranous, drying very dark to nearly black on the upper surface, less so beneath, usually glabrous or very sparsely pubescent above and more hairy beneath with short whitish hairs, the young terminal leaves sometimes with a floccose yellowish-brown very caducous indumentum at the base of the midrib beneath, and with scattered hairs along the midrib and the margins; leaves subtending the cymes, sessile, up to 10 × 5.5 cm, ovate-oblong, or obovate to broadly elliptic, rounded or cordate at the base, rounded or shortly acuminate at the apex, entire or serrate-dentate. Cymes up to 8-paired, forming a terminal leafy laxly-spreading slender-peduncled open panicle, up to c. 30 × 20 cm; lowermost floral internode up to 7.5 cm long; peduncles slender to filiform, the lowermost 6.5–12 cm long, spreading or sometimes arched and reflexed, purpurascent, glabrous or sparsely shortly hairy; bracts linear; pedicels c. 6 mm long, glabrous or sparsely hairy. Calyx 3–4.5(5) mm long and 4–5 mm in diameter about the tips of the lobes, reddish-purple, usually glabrous, sometimes minutely sparsely pubescent, or rarely densely so; lobes shorter than the tube, ovate-triangular to oblong or ± semicircular, obtuse or rounded at the apex, remaining erect even in the fruiting calyx. Corolla usually with the four lateral lobes green, bluish-green, creamy-green or greenish-violet, and the median lobe deep violet, bright blue to purple or deep purple-blue; tube 7–8 mm long, glabrous; lateral lobes 7.5–9.5 × 4 mm, oblong, obtuse, the median lobe up to 12.5 × 5.5 mm; corolla limb spherical when in bud, glabrous or appressed-hairy, the hairs very short and thin, usually sparse. Anthers (1)1.25–1.5(1.75) mm long. Fruit 7–9 mm broad and 5–6 mm long, ovoid, glossy, red, reddish-brown or bronze-brown.

Zambia. N: Mpika Distr., 30 km SW of Mpika by road to Lusaka, 1450 m, fl. 14.i.1975, *Brummitt, Chisumpa & Polhill* 13783 (K; SRGH). W: Solwezi Distr., Kifubwa (Chifulwa) Gorge, fl. 10.xii.1961, *Holmes* 1401 (FHO). C: Serenje Distr., Kundalila Falls, 1360 m, fl. 4.ii.1973, *Kornaś* 3160 (FHO). E: Chama Distr., Nyika Plateau, upper slopes of Kangampande Mt., fl. 3.v.1952, *White* 2566 (BR; FHO). S: Choma Distr., c. 2 km below confluence of Siamambo and Bunchele streams, fl. & fr. 6.ii.1960, *White* 6887 (FHO; K). **Malawi.** N: Mzimba Distr., Mzuzu, Marymount, towards Tung Estate, c. 1440 m, fr. 15.vi.1972, *Pawek* 5464 (K; SRGH). C: Dedza Mt., slopes, fr. 4.v.1968, *Salubeni* 1089 (SRGH). S: Ntcheu Distr., Chirobwe, fl. & immat. fr. 10.ii.1967, *Salubeni* 555 (K; SRGH).

Also in Tanzania and Dem. Rep. Congo. Afromontane evergreen forest, *Juniper* rainforest, gulley forests and forest margins, swamp forest (mushitu) and riverine forest, and in high rainfall woodland, often on termitaria; 570–2135 m.

Robson 1370 (BM; BR; K; LISC; SRGH) and *Adlard* 238 (FHO) from Malawi have the calyces more pilose and the corolla limb (in bud) more densely puberulous than is usual for *R. quadrangulata*, while in *Mutimushi* 236 (K), from Kitwe (Zambia), the teeth on the leaf margins are more acute and spreading, and the calyx lobes relatively longer than wide. However, these specimens are closer to *R. quadrangulata* in all other characters and are therefore retained in this species.

Wright 284 (K; SRGH) approaches *R. quadrangulata* in its broad leaves, nearly circular in the lowermost, and in its ample inflorescence, but it differs in that the leaves are not broadly winged pseudopetiolate at the base, and in the larger calyx lobes and the indumentum of the branches, leaves and calyx.

R. quadrangulata, treated by Verdcourt (loc. cit.: 127 (1993)) as *C. sansibarense* Gürke subsp. *sansibarense*, may be distinguished from that species by the shape of its leaves. In *Rotheca sansibarensis* the leaves are subtruncate at the base with only a short narrowly cuneate portion about the midrib base and are not long attenuate at the base with the lamina tapering into a ± broadly winged pseudopetiole. Moreover, the corolla limb when in bud is spherical in *R. quadrangulata*, whereas it is ovoid or oblong-ovoid and sometimes mamillate at the top in *R. sansibarensis*, the corolla lobes are indistinctly nerved in *R. quadrangulata* but those of typical *R. sansibarensis* have the nerves rather raised beneath, etc.

R. quadrangulata is closer to *R. bukobensis*, which has very similar leaves. However, in the former the calyx is usually glabrous, smaller and not so deeply divided, and its lobes are narrower, usually narrowing only slightly towards the apex (often subtriangular), whereas in *R. bukobensis* the calyx is pubescent, particularly at the base, with lobes broader semicircular and rounded at the top, often reflexed.

9. **Rotheca cyanea** (R. Fern.) R. Fern. in Kew Bull. **55**: 148 (2000). TAB. **17**. Type: Zambia, Mbala (Abercorn) to Sumbawanga road, c. 16 km from Kawimbe, 1620 m, *Richards* 6132 (K, holotype).

  Clerodendrum myricoides sensu R.E. Fries, Wiss. Ergebn. Schwed. Rhod.-Kongo-Exped.: 275 (1916) non (Hochst.) Vatke.

  Clerodendrum violaceum sensu White, For. Fl. N. Rhod.: 366 (1962). —sensu Richards & Morony, Check List Fl. Mbala: 237 (1969).

  Clerodendrum near *C. caesium* of Richards & Morony, Check List Fl. Mbala: 236 (1969).

  Clerodendrum sp. aff. C. violaceum of Richards & Morony, Check List Fl. Mbala: 237 (1969).

  Clerodendrum cabrae sensu Moldenke in Phytologia **58**: 355 (1985) quoad specim. *Burtt* 6380 non De Wild.

  Clerodendrum caesium sensu Moldenke in Phytologia **58**: 357 (1985) quoad specim. *Burtt* 6124 et *Milne-Redhead* 1047 non Gürke.

  Clerodendrum myricoides var. niansanum sensu Moldenke in Phytologia **62**: 469 (1987), quoad specim. *Robinson* 3848 non Thomas.

  Clerodendrum cyaneum R. Fern. in Mem. Soc. Brot. **30**: 36 (1998).

An erect few- to much branched shrub 1.2–4 m tall, or sometimes a tree to c. 6.5 m tall, often also flowering on herbaceous sucker or coppice shoots. Branches 4-angled, somewhat thick, with a well developed central pith, glabrous or yellow-brownish puberulous towards the apices soon glabrescent, often with prominent scattered lenticels; bark on young branches, dark brown, soft, becoming longitudinally wrinkled when dry. Leaves opposite, distinctly petiolate, ± closely spaced on the upper part of the flowering branches; lamina 6–17 × 2–7.7 cm, narrowly to broadly elliptic or lanceolate, rarely ovate-elliptic, acute to acuminate or sometimes obtuse at the apex, cuneate or rarely rounded at the base, usually entire to obscurely ± sparsely crenulate or nearly so, ± concolorous, ± coriaceous when dry, usually densely minutely punctate beneath, glabrous; petiole 0.8–3 cm long, somewhat thick, reddish-purple, glabrous or puberulous and soon glabrescent in young leaves. Inflorescence a loose terminal glabrous panicle, 8–18 cm long, consisting of opposite cymes subtended by foliaceous bracts, the lowermost pair of cymes in the axils of the two uppermost leaves; sometimes with additional smaller panicles terminal on lateral branchlets; the lower floral internode 2–7.5 cm long, those above successively shorter; peduncles of the lowermost pair of cymes 5–8 cm long, spreading or subreflexed, the upper successively shorter; bracts up to 4 × 2.8 cm, subpetiolate; pedicels 2–7 mm long, glabrous, reddish-purple. Calyx 6–8 mm long and 6.5–10 mm in diameter about the tips of the lobes, obliquely obconical, pale green below, magenta or purple towards the apex, glabrous or rarely sparsely hairy; lobes 1.5–3 mm long, 3–4 mm wide at the

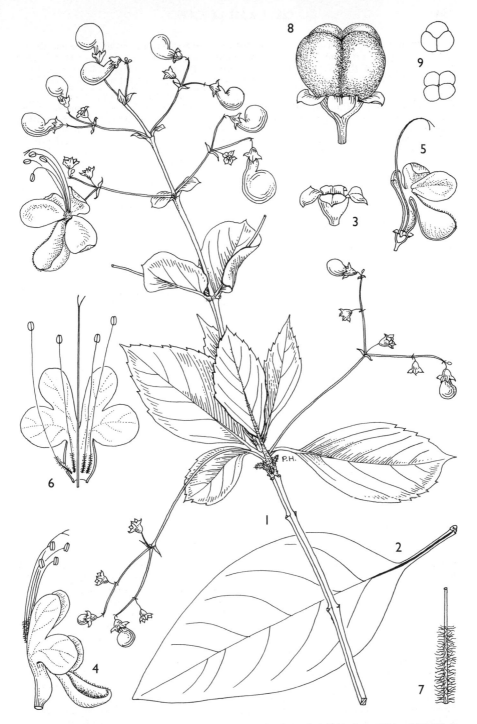

Tab. 17. ROTHECA CYANEA. 1, portion of flowering branch (× 2/3), from *Richards* 13262; 2, mature leaf (× 2/3), from *Fanshawe* 3662; 3, flowering calyx (× 2); 4, profile of flower (× 1); 5, longitudinal section of flower, stamens removed (× 1); 6, corolla, opened out, median lobe removed (× 1); 7, base of filament, enlarged, 3–7 from *Simon & Williamson* 1035; 8, fruit (× 2); 9, apical views of fruit, diagrammatic, 8 & 9 from *Fanshawe* 3662. Drawn by Pat Halliday.

base, equalling or shorter than the tube, ± semicircular, erect or partially to completely reflexed (the 3 stages often present on the same specimen), ciliolate. Corolla pale or bright to deep blue, or mauve to lilac, or the 4 lateral lobes pale blue and the median lobe dark blue, or the 4 lateral lobes pale violet and the median lobe deeper violet; tube 7–11.5 mm long and 4–4.5 mm in diameter; lateral lobes 9.5–13(17) × 6–8.5 mm, oblong-obovate, the median one 14–20 × 8.5–10 mm, deeply concave, all ciliolate at the margins; corolla limb when in bud 10–12 mm in diameter, spherical, glabrous. Anthers 1.75–2.5 mm long. Style c. 5 cm long, whitish; stigma reddish-purple. Fruit c. 12 mm in diameter, subglobose, glossy, at first reddish or reddish-brown, black or blue-black when ripe; fruiting calyx saucer-shaped, c. 13 mm in diameter, papery, pale green in the centre, red-purple towards the margin.

Zambia. C: dambo c. 112 km south of Mpika, c. 1536 m, fl. 16.x.1967, *Simon & Williamson* 1035 (K; LISC; SRGH). W: c. 80 km SW of Mwinilunga Boma, fl. 11.vii.1955, *Holmes* 1155 (K); Chingola Distr., Luano Forest Reserve, fl. 21.viii.1969, *Mutimushi* 3532 (SRGH). **Zimbabwe**. E: Mutare, fl. 27.xii.1974, *Cannell* 632* (SRGH).

Also in Angola. Swamp-forest (mushitu) and dambo margins, and in riverine forest; 1160–1760 m.

Some specimens, mainly of Dem. Rep. Congo, cited by authors as *Clerodendrum violaceum* Gürke, *C. caesium* Gürke or *C. cabrae* De Wild., belong to *R. cyanea*.

10. **Rotheca myricoides** (Hochst.) Steane & Mabb. in Novon **8**: 205 (1998). —R. Fernandes & Verdcourt in Kew Bull. **55**: 150 (2000). —M. Coates Palgrave, ed. 3 of K. Coates Palgrave, Trees South. Africa: 990 (2002). Type from Ethiopia.

Spironema myricoides Hochst. in Schaed. Schimperi Iter Abyss., Sectio prima., Unio. Iter: 330 (?1841).

Cyclonema myricoides (Hochst.) Hochst. in Flora **25**: 226 (1842).

Clerodendrum myricoides (Hochst.) Vatke in Linnaea **43**: 535 (1882). —Engler, Hochgebirgsfl. Afrika: 356 (1892). —Gürke in Engler, Pflanzenw. Ost-Afrikas **C**: 341 (1895); in Engler & Prantl, Nat. Pflanzenfam. IV, 3a: 176 (1895); in Bot. Jahrb. Syst. **28**: 298 (1900). —E.G. Baker in F.T.A. **5**: 310 (1900) pro parte. —B. Thomas in Bot. Jahrb. Syst. **68**: 86 (1936). —Brenan, Check-list For. Trees Shrubs Tang. Terr.: 632 (1949). —Drummond in Kirkia **10**: 272 (1975) pro parte quoad *Goldsmith* 101/61. —Malaisse in Troupin, Fl. Rwanda **3**: 276 (1985) pro parte. —Moldenke in Phytologia **62**: 328, 452 (1987). —Verdcourt in F.T.E.A., Verbenaceae: 130 (1992). —R. Fernandes in Mem. Soc. Brot. **30**: 47 (1998)**.

Siphonanthus myricoides (Hochst.) Hiern, Cat. Afr. Pl. Welw. **1**, 4: 844 (1900).

Clerodendrum myricoides subsp. *myricoides* var. *myricoides* Verdc. in F.T.E.A., Verbenaceae: 132 (1992). —R. Fernandes in Mem. Soc. Brot. **30**: 48 (1998).

Shrub up to 3 m tall, irregularly branched, or a small tree up to 10 m tall; older branches with rough deeply fissured bark; twigs pale brown, ridged or 4-angular, pithy in the centre, glabrous to velvety hairy. Leaves opposite or in whorls of 3–4; lamina 2–16(19.5) × 0.4–6(10) cm, narrowly to broadly elliptic, ovate-elliptic or oblanceolate, oblong or obovate, usually small but in cultivation can attain large dimensions, acute to acuminate at the apex, cuneate to ± attenuate at the base, entire to coarsely serrate, glabrous or pubescent to densely velvety hairy, glandular-punctate beneath, ± sessile or petiole up to 15 mm long, with unpleasant smell when crushed. Flowers in few–several-flowered dichasial cymes arranged in unelaborated and lax to quite extensive and elongate panicles 6.5–15(30) cm long; peduncles 0–7 cm long, secondary peduncles up to 4 cm long; apparent stalks 1–2.5 cm long but true pedicels 3–5 mm long. Calyx often entirely purplish or crimson-margined, glabrous to hairy; tube cupular, c. 2.5 mm long; lobes semicircular to ovate or triangular, 1.2–5 mm long, quite rounded, obtuse or ± acute. Corolla asymmetrical in bud expanding abruptly on anterior side, usually greenish with a white to pale blue to lilac limb, the median lobe dark blue; tube 5–7 mm long, pubescent at the throat; lobes unequal, (0.6)1–1.9(2) cm × (1.5)3.5–7.5 mm, the upper obovate, the lower one spathulate and much larger than the other four. Stamens and style long-exserted and curving upwards. Fruit ± black, 5–6 × 8–10 mm, subglobose, depressed, mostly deeply 4-lobed, glabrous.

---

* *Cannell* 632, the only specimen of this taxon collected in Zimbabwe, is probably cultivated. It is unlikely to occur naturally so far south of its normal geographical range.
** see Moldenke, loc. cit. for a more complete bibliography.

Subsp. **myricoides**

1. Leaves glabrous or sparsely pubescent on the nerves beneath, sometimes punctate beneath, reticulation on lower surface conspicuous with large meshes; corolla lobes glabrous · · · 2
- Leaves sparsely to ± densely pubescent on both surfaces, or tomentose beneath, not punctate beneath, reticulation on lower surface inconspicuous; corolla lobes sometimes hairy · 4
2. Leaves obovate or spathulate, sessile or subsessile, clustered at the apex of the short flowering branches; calyx lobes c. 1 mm long · · · · · · · · · · · · · · · · · · iii) var. *capiriensis*
- Leaves not usually obovate or spathulate, petiolate or with the lamina ± decurrent on the petiole, not clustered at the apex of the short flowering branches; calyx lobes usually longer · 3
3. Leaves ovate-rhombic or oblong-obovate, obtuse or subobtuse at the apex, ± coarsely crenate-dentate, lamina narrowly long decurrent almost to the point of attachment; inflorescence not paniculiform, cymes laxly arranged in the axils of the uppermost leaves · iv) var. *moldenkei*
- Leaves elliptic or lanceolate, usually tapering towards the base and apex, acute at the apex, not as wide, entire or with very small acute mucronate teeth, distinctly petiolate; cymes usually grouped in distinct terminal panicles · · · · · · · · · · · · · · · · · · · i) var. *myricoides*
4. Leaves usually ovate, sometimes elliptic or oblanceolate, acute or obtuse at the apex, usually up to 7.5(9) × 2(4) cm, up to 18 × 8 cm on sucker or coppice growth, pubescent on upper surface and ± tomentose beneath, ± discolorous (± dark to blackish above and greyish to ochraceous beneath); petiole tomentose · v) var. *discolor*
- Leaves elliptic, rather acute at both extremities, up to 10 × 2.4 cm, hairy on both faces, concolorous, light green, not becoming dark on drying; petiole hairy · · ii) var. *eleanorae*

Subsp. *namibiensis*, from Namibia, differs mainly in having the flowers arranged in axillary cymes not terminal panicles, and in the corolla being white or with 4 lobes whitish-green and the median lobe blue, not with all lobes violet. Subspecies *mafiensis* Verdc., *ussukumae* Verdc., *austromonticola* Verdc., *napperae* Verdc. and *muenzneri* Verdc. occur in Tanzania only.

i) Var. **myricoides**

Shrub 1.5–4 m tall, glabrous or sparsely finely appressed pubescent on the young branches, calyx and nerves of the leaf lower surface. Leaves opposite, or sometimes 3-whorled, petiolate; lamina in mature leaves (5)6.5–13.5(15–17) × (2.5)3–6.5 cm, usually more than twice as long as wide, elliptic, or sometimes ovate or lanceolate, acute at the apex, entire or ± remotely denticulate on the margin, conspicuously coarsely reticulate and sometimes minutely punctate beneath, membranous and ± concolorous, sometimes drying nearly black; petiole up to 2 cm long, slender. Cymes arranged in loose terminal panicles up to 15.5 cm long, the two lowermost cymes sometimes in the axils of the upper pair of leaves; peduncles and pedicels glabrous or sometimes pubescent. Calyx (2.75)3.25–5 mm long, campanulate, usually glabrous; lobes less than half as long as the tube, usually semicircular or sometimes oblong or lanceolate. Corolla usually with the 5 lobes all of the same colour, either lilac, blue or bluish-purple, rarely with 4 lobes greenish and the fifth (median lobe) coloured. Fruit black when ripe.

Calyx lobes semicircular or oblong-circular · forma *myricoides*
Calyx lobes lanceolate, subacute · forma *lanceolatilobata*

Forma **myricoides**
Clerodendrum myricoides var. *microphyllum* Gürke in Bot. Jahrb. Syst. **28**: 299 (1900). —R. Fernandes in Mem. Soc. Brot. **30**: 49 (1998) in adnot.
Clerodendrum myricoides var. *camporum* Gürke in Bot. Jahrb. Syst. **28**: 299 (1900). —R. Fernandes in Mem. Soc. Brot. **30**: 50 (1998) in adnot.
Clerodendrum myricoides var. *laxum* Gürke in Bot. Jahrb. Syst. **28**: 299 (1900). —R. Fernandes in Mem. Soc. Brot. **30**: 50 (1998) in adnot.
Clerodendrum schlechteri Gürke in Bot. Jahrb. Syst. **28**: 302 (1900). —R. Fernandes in Mem. Soc. Brot. **30**: 50 (1998) in adnot.
Clerodendrum ugandense Prain in Bot. Mag. **135**: t. 8235 (1909). —R. Fernandes in Mem. Soc. Brot. **30**: 53 (1998) in adnot.

Clerodendrum savanarum De Wild. in Bull. Jard. Bot. État **7**: 183 (1920), as "*savanorum*".
—R. Fernandes in Mem. Soc. Brot. **30**: 53 (1998) in adnot.
Clerodendrum myricoides var. *savanarum* (De Wild.) B. Thomas in Bot. Jahrb. Syst. **68**: 87 (1936), as "*savanorum*". —R. Fernandes in Mem. Soc. Brot. **30**: 53 (1998) in adnot.
Clerodendrum myricoides var. *niansanum* B. Thomas in Bot. Jahrb. Syst. **68**: 87 (1936) *nom. inval.* —R. Fernandes in Mem. Soc. Brot. **30**: 54 (1998) in adnot.
Clerodendrum myricoides var. *involutum* B. Thomas in Bot. Jahrb. Syst. **68**: 88 (1936) *nom. inval.*

Zimbabwe. C: Harare, c. 1524 m, fr. 5.xi.1921, *Eyles* 3187 (K), ? cultivated. E: Nyanga Distr., Sawunyama C.L., Gairesi R. camp, x.1958, *Davies* 2540 (SRGH). **Mozambique.** N: Ribáuè Distr., slopes of serra de Mepáluè, c. 700 m, fl. & fr. 12.xii.1967, *Torre & Correia* 16451 (LISC). Z: Maganja da Costa Distr., Floresta de Gobene, prox. Praia de Raraga, 43 km de Vila de Maganja, c. 20 m, fl. & fr. 20.i.1968, *Torre & Correia* 17067 (LISC). MS: Nhamatanda (Vila Machado), banks of Mucúzi R., fl. & fr. 23.iv.1948, *Mendonça* 4033 (LISC). GI: Inharrime Distr., near Chacane, c. 96 m, 26.x.1935, *Lea* 108 (PRE). M: Matutuíne Distr., entre Salamanga e Zitundo, fl. & fr. 9.xii.1961, *Lemos & Balsinhas* 273 (BM; COI; K; LISC; PRE).

Also in Ethiopia, Sudan, Somalia, Uganda, Kenya, Tanzania, Rwanda, Dem. Rep. Congo and Angola. Coastal plains and dunes on sandy soil, usually in dense thickets; also in wooded grassland, deciduous woodland, riverine forest and low altitude mixed evergreen forest; 5–1000 m.

Var. *myricoides* appears to be more coastal in its distribution, and although it extends somewhat inland at lower altitudes it has not been recorded growing naturally in those areas where var. *discolor* (Klotzsch) Baker is more frequent. The specimens of var. *myricoides* seen from Harare are probably from cultivated plants.

Davies 2540 (SRGH), cited above, differs somewhat in its smaller leaves, small floral buds and small calyx 2.75–3 mm long, with lobes c. 1.5 mm long, but cannot otherwise be kept separate from var. *myricoides*.

In a few specimens from Mozambique the leaves are very large, up to c. 17 cm long (cf. R. Fernandes, op. cit.: note on *C. schlechteri*). These may come from sucker or coppice shoots or plants from shady places.

Forma **lanceolatilobata** (R. Fern.) R. Fern. in Kew Bull. **55**: 150 (2000). Type: Zimbabwe, Mutare (Umtali), Dora Ranch, *Chase* 3705 (SRGH, holotype).
Clerodendrum myricoides subsp. *myricoides* var. *myricoides* forma *lanceolatilobatum* R. Fern. in Mem. Soc. Brot. **30**: 55 (1998).

Differs from typical *R. myricoides* in its calyx lobes which are c. 2.75 mm long, slightly exceeding the calyx tube in length, and which are lanceolate narrowing slightly towards the acute apex.

Zimbabwe. E: Mutare Distr., Dora Ranch, west slope of Aukuchaie Mt., c. 1024 m, 1951, *Chase* 3705 (SRGH).
Not known elsewhere. In shelter of granite boulder; c. 1024 m.

ii) Var. **eleanorae** (R. Fern.) R. Fern. in Kew Bull. **55**: 151 (2000). Type: Malawi, c. 80 km (50 miles) south of Nkhata Bay, *E. Phillips* 2249 (MAL, holotype).
Clerodendrum milne-redheadii sensu Moldenke in Phytologia **62**: 460 (1987) pro parte quoad specim. *E. Phillips* 2249, non Moldenke (1950).
Clerodendrum myricoides subsp. *myricoides* var. *eleanorae* R. Fern. in Mem. Soc. Brot. **30**: 88 (1998).

Shrub 0.9–1.3 m tall. Branches straight, 4-angled and ribbed on the angles, longitudinally striate between, branches somewhat densely hispid with short spreading hairs, younger shoots and flowering branches with longer denser hairs. Leaves opposite, subsessile to shortly petiolate; lamina up to 10 × 2.4 cm, elliptic, acute, entire or sparsely dentate, sparsely shortly pilose with hairs appressed on upper surface and spreading and more densely arranged on the nerves beneath, thinly coriaceous, concolorous, light green and paler beneath, not becoming dark on drying, not punctate with reticulation inconspicuous; midrib rather prominent and whitish beneath. Cymes few, axillary, solitary, 3-flowered in cymes borne in the axils of the uppermost leaves, 1-flowered in the sessile foliaceous bracts above; lowermost peduncles 3.3–4 cm long, curving downwards, the upper successively shorter, peduncles and pedicels spreading pilose. Calyx 2.5–3 mm long, campanulate, ± pilose below; lobes c. 1 mm long and 1.75 mm wide at the base, purple edged. Corolla deep sky-blue, glabrous outside.

Malawi. N: Nkhata Bay Distr., c. 80 km south of Nkhata Bay, fl. 8.v.1977, *E. Phillips* 2249 (Z). Not known elsewhere. Sand forest patch; c. 480 m.

This variety approaches *C. luembense* De Wild. in the nature of its indumentum, but differs in the shape of the leaves.

Some specimens of var. *discolor* from Malawi, [*Richards* 10575 (K; SRGH) and *Pawek* 6278 (K; SRGH) from Karonga Distr., and *Brummitt, Polhill & Banda* 16109 (K) from Kasungu Distr.], approach var. *eleanorae* in the shape of their leaves. However, the smaller distinctly discolorous leaves, and the somewhat crisped hairs of the indumentum place these specimens clearly in var. *discolor*.

iii) Var. **capiriensis** (R. Fern.) R. Fern. in Kew Bull. **55**: 150 (2000). Type: Zambia, Kapiri Mposhi, *Mutimushi* 3920 (K, holotype).
  Clerodendrum myricoides subsp. *myricoides* var. *capiriense* R. Fern. in Mem. Soc. Brot. **30**: 55 (1998).

Shrub 1–1.8 m tall. Branches subterete to sulcate and greyish-black towards the base, 4-angled and brownish-red towards the apex; lateral branches short, leafy and flowering, sparsely pilose towards the base. Leaves opposite, ± sessile, ± clustered towards the top of the lateral branches, small, up to 3.8 × 2.7 cm, obovate or spathulate, rounded or sub-obtuse at the apex, ± attenuate at the base, entire or obscurely dentate in the upper $1/3$, thinly coriaceous when dry, concolorous, drying dark, glabrous or sparsely pubescent on the nerves beneath, minutely punctate beneath; midrib and nerves impressed above, prominent beneath, reticulation conspicuous on the lower surface. Cymes 2–3, few-flowered, axillary from the leaf cluster on the short lateral branches, not forming a paniculate inflorescence; peduncles up to 2.5 cm long, spreading, slender, glabrous, black; pedicels 2.5–3 mm long. Calyx c. 3 mm long, glabrous, drying black; lobes c. 1 mm long, semicircular. Corolla deep sky-blue; limb when in bud c. 5.5 mm in diameter, subspherical, glabrous, almost black. Fruit shiny black when ripe.

Zambia. C: Kabwe Distr., Kapiri Mposhi, fl. 8.xii.1969, *Mutimushi* 3920 (K). Not seen elsewhere. On rocky hill top, amongst granite rocks; c. 1250 m.
Fanshawe 1842 (FHO; K), from Serenje (Zambia, C:), referred to *Clerodendrum myricoides* by White (For. Fl. N. Rhod.: 366), approaches var. *capiriensis*, but has longer branches ending in panicles, and has a pubescent indumentum particularly on the branches but also on the leaves and inflorescences. This is considered to be intermediate between var. *capiriense* and var. *discolor*. Other Zambian material, e.g., *Robinson* 2613 (K; SRGH) from Mkushi and *Richards* 1975, from Mbala, also approaches var. *capiriense* in possessing obovate leaves and short calyx lobes (c. 1 mm long). However, the ± dense indumentum, the larger discolorous leaves and the terminal inflorescences place these in var. *discolor*.

iv) Var. **moldenkei** (R. Fern.) R. Fern. in Kew Bull. **55**: 151 (2000). Type: Zimbabwe, Matobo Distr., south of Matopos Mission, *Norris-Rogers* 565 (SRGH, holotype).
  Clerodendrum dekindtii var. *dinteri* sensu Moldenke in Phytologia **59**: 249 (1986) quoad specim. *Olsson* s.n. non B. Thomas.
  Clerodendrum discolor var. *crenatum* sensu Moldenke in Phytologia **59**: 262 (1986) quoad specim. *Davies* 255 and *McGregor* 15/51 (in *GHS* 32448) non B. Thomas.
  Clerodendrum discolor var. *oppositifolium* sensu Moldenke in Phytologia **59**: 266 (1986) quoad specim. *Davies* 255 non B. Thomas.
  Clerodendrum myricoides subsp. *myricoides* var. *moldenkei* R. Fern. in Mem. Soc. Brot. **30**: 89 (1998).

Shrub or a small tree up to 3.8 m tall; branches dark brown, glabrous. Leaves opposite, not clustered toward the apex of the branches, up to 6.5 × 2.8 cm, obovate-rhombic to oblong-obovate, obtuse or subobtuse at the apex, broadly cuneate in the basal half and narrowly long decurrent almost to the point of attachment; margins distinctly coarsely crenate-dentate, rarely obscurely so; blade ± discolorous, thinly coriaceous to membranous, usually glabrous on both surfaces or sometimes sparsely pilose beneath, with nerves fine, inconspicuously reticulate. Cymes few, lax, glabrous, borne in the axils of the uppermost leaves, not forming a terminal panicle. Calyx small, 2–3.5 mm long, campanulate, usually glabrous or rarely sparsely appressed pilose; lobes 0.5–1.5 mm long, semicircular. Fruit shiny black when ripe.

Botswana. SE: South East Distr., Kgale (Kgali) Mt., c. 8 km south of Gaborone, c. 1440 m, fl. 29.xii.1974, *Mott* 508 (SRGH). **Zambia**. B: Sesheke Distr., Masese Mission Station, fr. 29.xi.1947,

Olsson (S). **Zimbabwe.** W: Matobo Distr., near Matopos Dam., fr. xii.1952, *O.B. Miller* 1439 (K, pro parte; SRGH). C: Chirumhanzu Distr., near Shashe R., st. 23.iii.1951, *T.R. Clarke* 25/51 (SRGH).

Not known elsewhere. Sand veld, on rocky outcrops and amongst boulders; also in riverine forest; 1000–1400 m.

v) Var. **discolor** (Klotzsch) Verdc. in Kew Bull. **55**: 150 (2000). Type: Mozambique, Rios de Sena, *Peters* s.n. (B†, holotype).

Cyclonema discolor Klotzsch in Peters, Naturw. Reise Mossambique **6**, part 1: 262 (1861).

Shrub 1–5 m tall, sometimes scandent to c. 7 m, or a tree up to 8 m tall. Penultimate branches ascending, 4-angled with flattish-sulcate faces, swollen at the nodes and with very marked leaf scars, usually leafy only towards the apex, often with ± short lateral leafy flowering branchlets (these sometimes so short that the cymes they bear appear to be inserted directly on the main branch), ± densely greyish-whitish pubescent, particularly on the young parts, with soft, slightly crispate to spreading hairs; internodes up to 8.5 cm long. Leaves opposite or 3-whorled, petiolate, clustered towards the branch apices, sometimes the flowering branches and branchlets leafy along the entire length; lamina usually twice as long as wide, 3–7.5(9) × 2.2–4.2 cm, sometimes up to 16.5(23.5) × 10(11.5) cm in sucker or coppice growth, smallest on lateral branchlets, oblanceolate to spathulate with the apical $^2/_3$ elliptical to obovate or ovate, acute to obtuse or somewhat acuminate at the apex, cuneate to ± attenuate into the petiole in the basal $^1/_3$, sometimes narrowly long decurrent almost to the point of attachment producing a winged petiole, entire to dentate or sometimes irregularly lobed or crenate-lobed on the margins, usually strongly discolorous, drying nearly black on upper surface and greyish-white or yellowish beneath, ± pubescent on both surfaces or densely tomentose beneath, particularly in young leaves. Cymes 2–5-flowered, usually forming terminal panicles with only the two lowermost cymes in the axils of the upper pair of leaves, or sometimes the cymes all subtended by leaves towards the ends of the branches and branchlets and then not forming distinct panicles; peduncles of the lower cymes 1.8–4.5 cm long; pedicels 2–4 mm long; inflorescence axis and branches spreading-hairy; bracts c. 12 × 3.5 mm, discolorous and with an indumentum similar to that of the leaves. Calyx (2–3)3.5–5.5(7) mm long, ± pubescent; lobes (1.5)2–2.5(3.5) mm long, usually semicircular, sometimes oblong-semicircular or lanceolate, usually erect or sometimes ± reflexed, ciliate on the margin. Corolla usually greenish with a white to pale blue to lilac or purplish limb, the median lobe dark blue; ± pubescent or tomentose, limb glabrous when in bud. Fruit 7 × 9–12 mm.

1. Calyx lobes lanceolate, acute at the apex · · · · · · · · · · · · · · · · vi) forma *angustilobata*
 – Calyx lobes semicircular or oblong-semicircular, rounded or obtuse at the apex · · · · · · 2
2. Calyx lobes c. 3 mm long, ± equalling the calyx tube in length, reflexed in all flowers · ii) forma *reflexilobata*
 – Calyx lobes usually slightly shorter, erect in all flowers or sometimes 1–few reflexed in some flowers · 3
3. Calyx tube subcylindrical; calyx lobes c. 1.25 mm long, length equalling c. $^1/_3$ of the calyx tube · iii) forma *brevilobata*
 – Calyx tube campanulate; calyx lobes usually more than 1.25 mm long · · · · · · · · · · · · 4
4. Petiole distinct, not winged · i) forma *discolor*
 – Petiole winged or the leaf lamina decurrent on the petiole · · · · · · · · · · · · · · · · · · · 5
5. Wings of the petiole entire, flat; leaf margin entire or dentate · · · iv) forma *alatipetiolata*
 – Wings of the petiole lobed, undulate; leaf margin irregularly lobed or crenate-lobed · v) forma *lobulata*

i) Forma **discolor**

Clerodendrum discolor (Klotzsch) Vatke in Linnaea **43**: 536 (1882). —Gürke in Engler, Pflanzenw. Ost-Afrikas **C**: 341 (1895); in Mildbraed, Wiss. Ergebn. Deutsch. Zentr.-Afrika-Exped., Bot. **2**: 281 (1914). —Fries, Wiss. Ergebn. Schwed. Rhod.-Kongo-Exped.: 275 (1916). —Thomas in Bot. Jahrb. Syst. **68**: 84 (1936). —Robyns, Fl. Sperm. Parc Nat. Alb. **2**: 147 (1947). —Brenan, Check-list For. Trees Shrubs Tang. Terr.: 632 (1949) pro parte. —Garcia in Estud. Ensaios Doc. Junta Invest. Ci. Ultramar [in Mendonça, Contrib. Conhec. Fl. Moçamb., II] **12**: 169 (1954) pro max. parte. —Moldenke in Phytologia **59**: 255 (1986).

Clerodendrum myricoides subsp. *myricoides* var. *discolor* (Klotzsch) Baker in F.T.A. **5**: 310 (1900). —Malaisse in Troupin, Fl. Rwanda **3**: 276 (1985) in adnot. —Verdcourt in F.T.E.A., Verbenaceae: 132 (1992). —R. Fernandes in Mem. Soc. Brot. **30**: 56 (1998).

Clerodendrum discolor var. *pluriflorum* Gürke in Bot. Jahrb. Syst. **30**: 301 (1901). —Thomas in Bot. Jahrb. Syst. **68**: 85 (1936). —Brenan, Check-list For. Trees Shrubs Tang. Terr.: 632 (1949). —Moldenke in Phytologia **59**: 267–268 (1986). —Verdcourt in F.T.E.A., Verbenaceae: 135 (1992) in adnot. —R. Fernandes in Mem. Soc. Brot. **30**: 58 (1998) in adnot. Type from Tanzania.

Clerodendrum phlebodes C.H. Wright in Bull. Misc. Inform., Kew **1907**: 54 (1907). — Thomas in Bot. Jahrb. Syst. **68**: 84 (1936). —Brenan, Check-list For. Trees Shrubs Tang. Terr.: 632 (1949). Type from Uganda.

Clerodendrum myricoides sensu S. Moore in J. Linn. Soc., Bot. **40**: 167 (1911) pro parte quoad specim. *Swynnerton* 173, 1295, 1296, 1297. —White, For. Fl. N. Rhod.: 366 (1962) quoad specim. *White* 1951 & 3677, non (Hochst) Vatke. —R. Fernandes in Mem. Soc. Brot. **30**: 60 (1998) in adnot.

Clerodendrum bequaertii De Wild. in Bull. Jard. Bot. État **7**: 185 (1920); Pl. Bequaert. **1**: 256 (1922). —R. Fernandes in Mem. Soc. Brot. **30**: 63 (1998) in adnot. *nom. illegit.*, non De Wild. (1914). Type from Dem. Rep. Congo.

Clerodendrum bequaertii var. *debeerstii* De Wild. in Bull. Jard. Bot. État **7**: 186 (1920). —R. Fernandes in Mem. Soc. Brot. **30**: 65 (1998) in adnot. Type from Dem. Rep. Congo.

Clerodendrum villosulum De Wild., Contrib. Fl. Katanga: 165 (1921). Type as for *C. bequaertii* De Wild.

Clerodendrum villosulum var. *debeerstii* (De Wild.) De Wild., Contrib. Fl. Katanga: 165 (1921). Type as for *C. bequaertii* var. *debeerstii* De Wild.

Clerodendrum phlebodes var. *pilosicalyx* B. Thomas in Bot. Jahrb. Syst. **68**: 84 (1936) *nom. inval.* Type from Tanzania.

Clerodendrum discolor var. *duemmeri* B. Thomas in Bot. Jahrb. Syst. **68**: 86 (1936). —Moldenke in Phytologia **59**: 262 (1986). —R. Fernandes in Mem. Soc. Brot. **30**: 68 (1998) in adnot. *nom. inval.* Type from Uganda.

Clerodendrum discolor var. *kilimandscharense* B. Thomas in Bot. Jahrb. Syst. **68**: 85 (1936). —Brenan, Check-list For. Trees Shrubs Tang. Terr.: 632 (1949). —Moldenke in Phytologia **59**: 263–264 (1986). —R. Fernandes in Mem. Soc. Brot. **30**: 68 (1998) in adnot. *nom. inval.* Type from Tanzania.

Clerodendrum discolor var. *oppositifolium* B. Thomas in Bot. Jahrb. Syst. **68**: 86 (1936). — Moldenke in Phytologia **59**: 264–267 (1986) pro parte. —R. Fernandes in Mem. Soc. Brot. **30**: 71 (1998) in adnot. *nom. inval.* Type: Malawi, Blantyre, *Buchanan* 6982.

Clerodendrum dekindtii sensu Moldenke in Phytologia **59**: 248 (1986) quoad specim. *Debeerst* 13 (err. Descamps) and *O.B. Miller* 1226 (K; PRE) non Gürke. —R. Fernandes in Mem. Soc. Brot. **30**: 67 (1998) in adnot.

Clerodendrum myricoides subsp. *myricoides* var. *kilimandscharense* Verdc. in F.T.E.A., Verbenaceae: 133 (1992). —R. Fernandes in Mem. Soc. Brot. **30**: 68–70 (1998) in adnot. Type from Tanzania.

Clerodendrum sansibarense subsp. *occidentale* Verdc. in F.T.E.A., Verbenaceae: 127 (1992). —R. Fernandes in Mem. Soc. Brot. **30**: 71 (1998) in adnot. Type from Tanzania.

Clerodendrum reflexum sensu Moldenke in Phytologia **59**: 216 (1986) quoad specim. *Barbosa* 1018, 1059, 1582 and *Garcia* 313, non H. Pearson.

Zambia. B: Sesheke Distr., Masese, fl. & fr. 12.iii.1960, *Fanshawe* 5469 (FHO; K). N: Mbala Distr., Mpulungu, shore of L. Tanganyika, fl. 16.xi.1952, *White* 3668 (FHO; K). W: Solwezi, fl. 16.x.1953, *Fanshawe* 422 (SRGH). C: Lusaka Distr., Chalimbana, 2 km south of Liempe along road joining Great East Road and Leopards Hill Road, c. 20 km east of Lusaka, c. 1200 m, fl. 24.xii.1972, *Strid* 2683 (LD). E: Chadiza Distr., Mwangazi Valley, fl. 26.xi.1958, *Robson* 712 (BM; BR; LISC; K). **Zimbabwe**. N: Gokwe Distr., Sengwa Research Station, fl. & fr. 8.i.1976, *P.R. Guy* 2371 (BR; SRGH). W: Matobo Distr., Figtree, fl. 17.xii.1950, *Orpen* 107/50 (SRGH). C: Harare Distr., Epworth Mission, fl. 21.i.1967, *Biegel* 1792 (SRGH). E: Mutare Distr., railway cutting at Darlington, c. 1030 m, fl. 14.x.1958, *Chase* 7002 (BR; K; LISC; SRGH). S: Masvingo Distr., Umshandige, c. 960 m, fl. 6.x.1949, *Wild* 2988 (SRGH). **Malawi**. N: Nkhata Bay Distr., 8 km east of Mzuzu, Rose Falls, Roseveare's, c. 1344 m, fr. 26.i.1974, *Pawek* 8015 (SRGH). C: Kasungu Distr., near Chamama, c. 1000 m, fl. 16.i.1959, *Robson & Jackson* 1232 (BM; K; LISC; SRGH). S: Blantyre Distr., side of Soche Mt., c. 1280 m, fl. 2.x.1937, *Lawrence* 480 (BR; K). **Mozambique**. N: Monapo Distr., andados 17 km de Itoculo para Netia, c. 200 m, fl. & fr. 5.xii.1963, *Torre & Paiva* 9426 (LISC). T: Angónia Distr., Domué Mt., c. 1450 m, fl. 9.iii.1964, *Torre & Paiva* 11082 (LISC; LMU). Z: Gurué Distr., ao km 3 de Lioma, Serra Namuli (Gurué Mt.), c. 600 m, fl. 10.xi.1967, *Torre & Correia* 16053 (LISC). MS: Gondola Distr., near Chimoio (Vila Pery), fl. 22.vii.1941, *Torre* 3206 (LISC; LMA).

Also in Ethiopia, Uganda, Kenya, Tanzania, Rwanda, Burundi, Dem. Rep. Congo and Angola. Miombo and open mixed deciduous woodlands, on sandy soils and rocky slopes, on lake shores and beside seasonal rivers often among boulders, in riverine forest, and on margins of swamp forest and evergreen rainforest, in thickets on termitaria and mutemwa thickets on Kalahari Sand; also in old cultivation; 70–1600 m.

Swynnerton (1296, K) records that: i) the chiNdao name is "Bukusa", and ii) the wood is much used for "fire sticks", for lighting fires by friction.

Specimens with leaves very attenuate towards the base and a looser indumentum than is usual are taken to be intermediate between var. *discolor* forma *discolor* and var. *myricoides*. Some specimens with unusually large leaves, e.g. *Torre & Correia* 14969 (LISC) from Mozambique [Zambesia Prov., 3 km from Ile Mt.] with leaves up to 24.5 × 11.6 cm, and *Faulkner* 104 (K) from Mozambique [Quelimane Distr., Tacuane] with leaves up to 17.8 × 8 cm, probably come from sucker or coppice growth.

Macêdo 5130, from Mozambique [Tete Prov., between Marueira and Songo (LISC; LMA; LMU; SRGH)], appears to be different because it has internodes longer towards the end of the branches with leaves persistent on the flowering branches and all ± similar in shape size and colour. However, such characters often in combination also occur in material that is clearly forma *discolor*. *Macêdo* 5130 is therefore included in forma *discolor*.

In the border area between Manica, in Mozambique, and Mutare to Chimanimani, in Zimbabwe, plants of forma *discolor* tend to be more robust and have larger leaves, inflorescence and flowers, but they cannot be treated as a separate variety since they intergrade in these and other features with typical var. *discolor*. Features which, mostly in combination, characterize these border area plants include: shrub or tree size 5–7 m tall; leaves densely whitish pubescent particularly on the lower surfaces, or lanate on both surfaces; laminas up to 17.5 × 9 cm, broadly elliptic or ovate-elliptic, acuminate and ⊥ acute at the apex, not to somewhat attenuate at the base, regularly dentate along the upper ²/₃–³/₄ with up to 23 teeth per side, sometimes crenulate to entire; inflorescences large, up to 18.5 × 16 cm, with the lowermost peduncles up to 6.5 cm long and spreading or reflexed; calyx 4.5–6(8.5) mm long, and larger than is usual in forma *discolor*. Specimens exhibiting this variation include *Andrada* 1243 (LISC); *Barbosa* 1018*, 1059* (LISC; M), 1313 (BR; COI; LISC), 1582* (LISC; Z), *Garcia* 313* (LISC), *Müller* 512 (K; LISC; SRGH), *Simão* 15/48 and 373/48 (LISC) and from Zimbabwe [Mutare (Umtali), *Armitage* 50/54 (K; SRGH), *Chase* 8103 (K; SRGH)] and Melsetter, *McGregor* 17/45 and 59/46 (FHO). The specimens marked with * in the above list were cited by Moldenke (in Phytologia **59**: 261, 1986) as *C. reflexum* H. Pearson. They differ from this by many characters, including the very different indumentum.

The specimens *McClounie* 27 and 107, both from Nyika Plateau (Malawi), collected in Nymkowa Hills and Mwanemba respectively, have narrow lanceolate leaves acute at the base and apex, and these are considered to represent one extreme in the range of leaf variation in var. *discolor*.

ii) Forma **reflexilobata** (R. Fern.) R. Fern. in Kew Bull. **55**: 151 (2000). Type: Zambia, Lusaka, *Fanshawe* 10490 (K, holotype).

    Clerodendrum myricoides subsp. *myricoides* var. *discolor* forma *reflexilobatum* R. Fern. in Mem. Soc. Brot. **30**: 75 (1998).

Zambia. C: Lusaka, fl. 13.xii.1968, *Fanshawe* 10490 (K).

Not known elsewhere. In thicket understorey of *Acacia* woodland "munga" on a limestone scarp; c. 1280 m.

The reflexed calyx lobes are found in all flowers of this forma, whether opened or still in bud. Sometimes the calyx lobes of forma *discolor* can be reflexed, but in these cases it appears that this has been the result of placing the specimens in a plant press. Specimens of forma *discolor* in which reflexed calyx lobes have been observed include: *Gilges* 507 (K; PRE; SRGH) and *Robinson* 2935 (K; M) from Zambia; *Goldsmith* 101/61 (K; LISC; SRGH); *Gosnell* 115 (SRGH); *Leach* 11286 (BR; COI; K; LISC; M; SRGH); *O.B. Miller* 1140 (K; PRE; SRGH); *Noel* 2374 (BM; COI; LISC; SRGH); *Norrgrann* 399 (K; SRGH); *Monro* s.n. and 2284 (BM) from Zimbabwe and *Torre & Correia* 13983 (LISC) from Mozambique.

iii) Forma **brevilobata** (R. Fern.) R. Fern. in Kew Bull. **55**: 150 (2000). Type: Mozambique, Cahora Bassa Distr., Songo, *Macêdo* 5413 (LISC, holotype).

    Clerodendrum myricoides subsp. *myricoides* var. *discolor* forma *brevilobatum* R. Fern. in Mem. Soc. Brot. **30**: 74 (1998).

Distinguished from forma *discolor* by the subcylindrical calyx tube, and distinguished from forma *reflexilobata* and from specimens of forma *discolor* with some reflexed calyx lobes, by the relatively shorter calyx lobes.

Mozambique. T: Songo, highlands on the road to the old airport, fl. 5.xii.1973, *Macêdo* 5413 (LISC).

Not known elsewhere. Ecology not known.

iv) Forma **alatipetiolata** (R. Fern.) R. Fern. in Kew Bull. **55**: 150 (2000). Type: Mozambique, MS: Chimoio, Vanduzi, *Mendonça* 3612 (LISC, holotype; BR).

    Clerodendrum discolor sensu Garcia in Estud. Ensaios Doc. Junta Invest. Ci. Ultramar [in Mendonça, Contrib. Conhec. Fl. Moçamb., II] **12**: 170 (1954) pro parte quoad specim. *Mendonça* 3612 non (Klotzsch) Vatke.

Clerodendrum discolor var. *oppositifolium* sensu Moldenke in Phytologia **59**: 261 (1986) pro parte quoad specim. *Mendonça* 3612 non Thomas.
Clerodendrum myricoides subsp. *myricoides* var. *discolor* forma *alatipetiolatum* R. Fern. in Mem. Soc. Brot. **30**: 72 (1998).

Distinguished from forma *discolor* by the leaf lamina narrowly long decurrent almost to the point of attachment, producing a winged petiole with entire margins.

Mozambique. MS: Manica Distr., on road, 2 km from Manica (Vila de Manica) to Chimoio (Vila Pery), c. 900 m, fl. 25.xi.1965, *Torre & Correia* 13258 (LISC).
Not known elsewhere. Ecology unknown.

v) Forma **lobulata** (R. Fern.) R. Fern. in Kew Bull. **55**: 150 (2000). Type: Mozambique, Milange Distr., serra de Tumbine, *Mendonça* 1418 (LISC, holotype; M).
   Clerodendrum discolor sensu Garcia in Estud. Ensaios Doc. Junta Invest. Ci. Ultramar [in Mendonça, Contrib. Conhec. Fl. Moçamb., II] **12**: 170 (1954) pro parte quoad specim. *Mendonça* 1418 non (Klotzsch) Vatke.
   Clerodendrum discolor var. *oppositifolium* sensu Moldenke in Phytologia **59**: 261 (1986) pro parte quoad specim. *Mendonça* 1418 non Thomas.
   Clerodendrum myricoides subsp. *myricoides* var. *discolor* forma *lobulatum* R. Fern. in Mem. Soc. Brot. **30**: 74 (1998).

Distinguished from forma *alatipetiolata* by the larger broader leaves, up to 12.5 × 8.7 cm, the leaf margin irregularly lobed or crenate-lobed, and by the lobed usually undulate petiole wings.

Mozambique. Z: Milange Distr., Tumbine Mt., fl. 12.xi.1942, *Mendonça* 1418 (LISC; M).
Not known elsewhere. In deciduous forest; 700–1500 m.

vi) Forma **angustilobata** (R. Fern.) R. Fern. in Kew Bull. **55**: 150 (2000). Type from Angola.
   Clerodendrum cuneiforme sensu Moldenke in Phytologia **59**: 121 (1986) quoad specim. *Dehn* 558 non Moldenke (= *C. cuneatum* Gürke).
   Clerodendrum myricoides subsp. *myricoides* var. *discolor* forma *angustilobatum* R. Fern. in Mem. Soc. Brot. **30**: 73 (1998).

Zimbabwe. E: Mutare Distr., Maranke C.L., near Bazeley Bridge, c. 960 m, fl. 4.xii.1952, *Chase* 4735 (SRGH).
Also in Angola. Granite outcrops, amongst rocks; c. 800 m.

11. **Rotheca wildii** (Moldenke) R. Fern. in Kew Bull. **55**: 154 (2000). —M. Coates Palgrave, ed. 3 of K. Coates Palgrave, Trees South. Africa: 992 (2002). Type: Zimbabwe, Hurungwe Distr., Mwami (Miami), K34, Experimental Farm, *Wild* 1321 (NY, holotype; SRGH).
   Clerodendrum wildii Moldenke in Phytologia **3**: 110 (1949). —White, For. Fl. N. Rhod.: 366 (1962). —Drummond in Kirkia **10**: 272 (1975). —R. Fernandes in Mem. Soc. Brot. **30**: 107 (1998).

A soft, woody, much- to sparsely-branched shrub, (0.64)1–3(4) m tall, from a woody rootstock, sometimes scandent or a small tree, usually leafless when in flower; branches thick and ± fleshy, stiff and somewhat brittle, obtusely 4-angled toward the apex, pithy in the centre, with internodes shorter towards the end of the branches; bark smooth, purplish-grey when young, glabrescent, becoming pale to dark greyish-brown and corky, longitudinally wrinkled when dry, with whitish prominent lenticels and concave cordate-elliptic to semi-circular leaf scars surrounded by raised corky margins. Leaves opposite, petiolate, slightly fleshy; lamina 3–13.5 × 1.5–7.2 cm, ovate or elliptic to broadly elliptic, acuminate at the apex, attenuate at the base and ± decurrent into the petiole, entire to subentire often ± coarsely serrate in the apical half, membranous to chartaceous, concolorous, pale green, lighter beneath, glabrous or sometimes sparsely pubescent on the upper surface, more densely so beneath with hairs particularly on the nerves; petiole up to c. 5 cm long, ± winged by the ± decurrent lamina. Inflorescences (2.2.)4–15.5 cm long, raceme-like or narrowly paniculiform resembling a 'labiate' with dense whorls of flowers, axillary in the apical leaf cluster (but appearing terminal) or on short lateral branches which are sometimes so reduced that the inflorescences appear cauline; inflorescence shoot (rhachis) nodes 3–9, the lowermost internode up to 30(35) mm long, successively shorter towards the apex, pubescent or glabrous, each node with 1–2(3) cymes in each axil of opposite bracts, the uppermost nodes sometimes sterile; cymes subsessile or with peduncles up to c. 10 mm long, 1–7-

flowered; peduncles longest and more branched at the rhachis base; pedicels 3–8 mm long, slender; bracts lanceolate-ovate, up to 15 mm long and leaf-like at the lower nodes. Calyx (4)4.5–6(7–8) mm long, campanulate, green or purple-tinged at the base, becoming deep violet or purplish-red towards the upper margin; lobes (2)2.5–3.5 mm long, usually oblong and ± rounded at the apex, or sometimes semicircular (subtriangular and subacute on drying), erect, ciliate. Corolla tube (5)6–10(11) mm long and 3–4.25 mm in diameter, curved; lower lobe 13–15 × 7 mm, concave spathulate, violet, purple, mauve, lilac or blue, the other lobes 8–12 × 5.5–7 mm, oblong, obtuse, cream-coloured, greenish-white or green inside, all blackish or dark grey-green and ± covered with sessile or subsessile glands on the outside; corolla limb when in bud 6.5–9 mm in diameter, ± densely covered with sessile to subsessile spherical reddish-purple glands. Stamens erect and arcuate, exserted 12–13 mm; anthers (3)3.5–4 mm long, mauve, turning very dark on drying. Fruit 7–8 × 8–12 mm, green, turning black when mature; calyx persistent.

Inflorescence axis, peduncles and pedicels ± pubescent to tomentulose with crisped sometimes
   reddish-purple hairs; calyx ± pubescent; corolla limb when in bud ± pubescent with short
   whitish hairs · forma *wildii*
Inflorescence axis, peduncles pedicels and corolla limb when in bud glabrous; calyx glabrous
   or very sparsely and shortly hairy · forma *glabra*

Forma **wildii**

Zambia. C: Mkushi Distr., Chiwefwe, fl. 16.ix.1964, *Mutimushi* 1027 (K; SRGH). S: Mazabuka Distr., edge of Kafue R. downstream from the road bridge, fl. 10.ix.1960, *Angus* 2426 (FHO; K; SRGH). **Zimbabwe.** N: Makonde Distr., Mhangura (Mangula), Molly South Hill, c. 1322 m, fl. 30.viii.1963, *W.B.G. Jacobsen* 2201 (PRE). C: Chegutu Distr., Poole Farm, fl. 30.ix.1954, *R.M. Hornby* 3347 (SRGH). S: Kariba Distr., Chitudzi R. area, c. 640 m, fr. ix.1955, *Davies* 1554 (SRGH). **Malawi.** S: Mulanje Distr., Njobvu (Njobru), fl. 16.xi.1955, *G. Jackson* 1764 (BM; FHO; K). **Mozambique.** T: Cahora Bassa Distr., Mecangádzi R. Valley, on the Songo–Barrage road, at the Maroeira cross-roads, fl. 17.xi.1973, *M. Correia, Marques & B. Correia* 3843 (LMU).
   Known only from the Flora Zambesiaca area. A semi-succulent shrub flowering leafless before the rains, on dry rocky hillsides and sandveld, in miombo, *Combretum* and *Colophospermum mopane* woodlands, in thickets in grassland and on termite-mounds; 640–1536 m.

Forma **glabra** (R. Fern.) R. Fern. in Kew Bull. **55**: 154 (2000). Type: Mozambique, Nhandué, *Le Testu* 834 (P, holotype; BM).

   Clerodendrum makanjanum sensu Meeuse in Dyer, Fl. Pl. S. Africa **32**, 4: t. 1724 (1958), non H. Winkler.
   Clerodendrum wildii sensu Gillett in Kew Bull. **14**: 343–344 (1960) quoad specim. *Chase* 1349; *Eyles* 3187; *McClounie* 167; *Pole Evans* 4597; *Wild* 1255, non Moldenke.
   Clerodendrum wildii forma *glabrum* R. Fern. in Mem. Soc. Brot. **30**: 107 (1998).

Botswana. N: Ngamiland, south of Nunga vlei, fl. & fr. 20.ix.1949, *Pole Evans* 4597 (K; PRE). **Zambia.** S: Kalomo Distr., Chidi, fl. 8.x.1955, *Gilges* 461 (K; PRE). **Zimbabwe.** N: Mutoko Distr., near Nyaderi, between Wutu and Nyakaranga R., Fungwi Reserve, fr. 18.x.1955, *Lovemore* 452 (K; LISC; SRGH). W: Hwange Distr., Victoria Falls, 4th Gorge, 880 m, st. 15.ii.1977, *Moyo* 18 (SRGH). C: Harare Distr., edge of Seke Dam (Prince Edward Dam), immat. fr. 12.x.1960, *Rutherford-Smith* 234 (LISC; SRGH). E: Nyanga Distr., on pathside from Nyahokwe Museum to ancient village, Ziwa Farm, c. 1500 m, fl. & fr. 23.x.1965, *Chase* 8324 (K; SRGH). S: Birchenough Bridge, fl. 10.ix.1961, *Lord Methuen* 198 pro parte (K). **Malawi.** N: Karonga Distr., c. 40 km south of Karonga, c. 512 m, fl. & fr. ix.1977, *Pawek* 13103; & st. 6.i.1978, *Pawek* 13549 (K; SRGH). S: Chikwawa Distr., Lengwe National Park, c. 160 m, fl. 7.ix.1970, *Pawek* 3732 (K). **Mozambique.** N: Montepuez, in a plain near a Montepuez kitchen-garden, fl. & fr. 17.x.1942, *Mendonça* 902 (LISC; LMU; SRGH). Z: Gurué Distr., ao km 3 de Lioma, Serra Namuli (Gurué Mt.), c. 600 m, fl. & immat. fl. 10.xi.1967, *Torre & Correia* 16054 (LISC). MS: Cheringoma Distr., Inhamitanga, fl. 10.ix.1944, *Mendonça* 2019 (LISC).
   Also in South Africa (Northern Province). A semi-succulent shrub flowering leafless before the rains, in sandy soils around granite outcrops and on dry rocky hillsides, in dry wooded-grassland, *Acacia* and baobab woodlands, and in thicket vegetation with *Acacia, Grewia* and *Commiphora* spp.; 150–1500 m.
   Some Mozambique specimens determined by   Moldenke as *Clerodendrum makanjanum* [*Barbosa* 2473, 2505; *Mendonça* 902, 1242 (err. 1212) & 2019, and *Torre & Paiva* 9356, 9448 & 9574] and Phillips 2919 from Malawi, belong to *R. wildii* forma *glabra* (R. Fern.) R. Fern.
   Forma *wildii* and forma *glabra* occur with equal frequency in Zimbabwe, however, forma *wildii* appears to be more common in Zambia and forma *glabra* more common in Malawi and Mozambique.

12. **Rotheca aurantiaca** (Baker) R. Fern. in Kew Bull. **55**: 148 (2000). Type: Malawi, Manganja Hills, immat. fl. Oct. 1861, *Meller* s.n. (K, holotype).

Clerodendrum aurantiacum Baker in F.T.A. **5**: 313 (1900). —Thomas in Bot. Jahrb. Syst. **68**: 89 (1936). —Moldenke in Phytologia **57**: 490 (1985). —R. Fernandes in Mem. Soc. Brot. **30**: 31 (1998).

Suffrutex up to 1 m tall from a woody tuberiform rootstock. Stems 1 to few, erect, cylindric, hollow, up to c. 10 mm in diameter near the base, striate, simple or few-branched, annual and herbaceous, or becoming woody with a smooth greyish-brown corky bark, longitudinally wrinkled when dry, sparsely puberulous or glabrous, with prominent leaf scars surrounded by raised corky margins; branches of annual stems herbaceous long slender, leafy and sterile; lateral branchlets on woody growth short, leafy and flowering, sometimes greatly reduced, arising from above the scars of the previous seasons leaves. Leaves opposite, subsessile, spaced along the upper part of the stem and main branches with internodes up to 16 cm long, or numerous and clustered towards the apex of the short lateral branchlets; lamina 5–15.5(40) × 1.8–10 cm, smaller and relatively narrower on the lateral branchlets, oblanceolate to oblong-obovate or elliptic, rounded or subacute at the apex, long attenuate at the base and decurrent into an apparent winged petiole, crenulate to crenate-serrate on the margins in the apical part, or entire, somewhat succulent becoming membranous when dry, concolorous, pale green, lighter beneath, drying darker, sparsely puberulous or glabrous. Inflorescences raceme-like or spike-like resembling a labiate with dense whorls of flowers, 7–26 cm long, terminal on the stem and lateral branchlets, sometimes with two smaller axillary inflorescences in the axils of the uppermost leaves; inflorescence shoot (rhachis) nodes 3–10, the lowermost internode up to 6 cm long, internodes successively shorter towards the apex, puberulous or glabrous, with 1–3 cymes in each axil of opposite bracts; cymes 1–6-flowered, pedunculate to subsessile; peduncles up to c. 15 mm long and more branched at the rhachis base, progressively shorter to ± absent above with the flowers pedicellate and fasciculately arranged; pedicels 3–8 mm long, slender; lowermost bracts up to 3.8 cm long and 5 mm wide, and leaf-like, progressively smaller and becoming linear toward the rhachis apex. Calyx 5–7 mm long at anthesis, campanulate; lobes not overlapping at the margins, 3 × 2–3(3.5) mm, oblong or semicircular, obtuse (acute when margins become inrolled on drying), puberulous or glabrous, ciliolate. Corolla purple and green, violet or blueish; tube c. 8 mm long and c. 3 mm in diameter, exceeding the calyx; corolla limb when in bud spherical, c. 7 mm in diameter, very densely to entirely covered with small orange-brown glands (dark red-brown on drying). Stamens erect and arcuate, exserted 12–13 mm; anthers 3–3.5 mm long, linear. Fruit c. 12 mm long, green, turning black when mature; calyx persistent, accrescent.

Plants puberulous (except sometimes on the lower part of stem) ········ forma *aurantiaca*
Plants glabrous ··· forma *faulknerae*

Forma aurantiaca

Clerodendrum macrostachyum Baker in Bull. Misc. Inform., Kew **1898**: 159 (1898); in F.T.A. **5**: 313 (1900) non Turcz. (1863). Type: Malawi, Zomba and vicinity, immat. fr. xii.1897; *Whyte* s.n. (K, holotype).
Clerodendrum aurantiacum forma *aurantiacum* R. Fern. in Mem. Soc. Brot. **30**: 34 (1998).

Malawi. S: Chiradzulu Distr., Magomera road, fl. delaps. 23.i.1981, *Salubeni* 2903 (COI; MAL). **Mozambique**. N: Pemba Distr., at 53 km from Pemba (Porto Amélia) to Ancuabe, c. 200 m, fl. & fr. 21.xii.1963, *Torre & Paiva* 9636 (LISC).
Not known elsewhere. *Brachystegia* and *Colophospermum mopane* woodlands, on clayey sandy soils; c. 200–900 m.

Forma faulknerae (Moldenke) R. Fern. in Kew Bull. **55**: 148 (2000). Type: Mozambique, Lugela Distr., Mocuba, Namagoa Estate, fl. xii.1948, drawing No. 536, *Faulkner* Kew No.115 (NY, holotype; K; SRGH).

Clerodendrum faulknerae Moldenke in Phytologia **4**: 47 (1952).
Clerodendrum aurantiacum forma *faulknerae* (Moldenke) R. Fern. in Mem. Soc. Brot. **30**: 34 (1998).

Mozambique. Z: 64 km from Maganja da Costa village, on the road to Mocuba, c. 100 m, fl. & fr. 9.i.1968, *Torre & Correia* 16984 (LISC).

Not known elsewhere. In *Brachystegia* and mixed deciduous woodlands; c. 150–200 m.

Rotheca aurantiaca is very similar to the Tanzanian species *Clerodendrum kissakense* Gürke of Verdcourt in F.T.E.A., Verbenaceae 141–142, fig. 18/17–18 (1992), based on *Goetze* 42 (B†, holotype). *R. kissakensis* (Gürke) Verdc. may be distinguished from *R. aurantiaca* by its longer calyx with imbricately overlapping broader than long calyx lobes, and by the corolla tube being less exserted. Gürke, in Bot. Jahrb. Syst. **28**: 304 (1900), refers to the type, the only specimen seen by him, as having opposite leaves, while Verdcourt (op. cit.: 141) describes the *R. kissakensis* leaves as 3-whorled. All *R. aurantiaca* specimens seen so far from Mozambique have opposite leaves. It is possible that when more material is available the distinction between these two taxa will become less apparent, suggesting that they should be recognized as being synonymous. In this case *R. kissakensis*, described one month earlier, would be the correct name.

The specimen *Stuhlmann* 713, referred to *R. kissakensis* by B. Thomas, was probably collected on Mozambique Island or in the Quelimane area in Mozambique (cf. Verdcourt, loc. cit., in adnot.). However, this specimen has not been seen by any botanist since Thomas in 1936, and it is not possible to know if it is true *R. kissakensis* or *R. aurantiaca*. It should be pointed out that *R. kissakensis* is glabrous as are the three specimens seen so far of *R. aurantiaca* forma *faulknerae*, and that *Stuhlmann* 713 was probably collected in the same area as they were.

Baker, in F.T.A. **5**: 313 (1900), mistakenly described the corolla colour as 'light orange', probably because specimen data accompanying the holotype included the note "Color. light orange" (possibly by the collector's hand). These collector's notes almost certainly referred to the apparent colour of the corolla buds, the outsides of which were densely covered in orange-brown stalked glands. The corollas in Meller's specimen were all in bud when he collected it, with only the outer surface of the corolla limb and lobes visible. The corollas in subgen. *Cyclonema* are never orange or yellow. Another interpretation is that the collector might have been referring to the colour suggested by the indumentum of the plant itself.

5. KALAHARIA Baillon

Kalaharia Baillon, Hist. Pl. **11**: 110 (1892)*.
Clerodendrum L. subgen. *Kalaharia* (Baillon) B. Thomas in Bot. Jahrb. Syst. **68**: 23 (1936).

Spinescent spreading subshrubs. Stems divaricately branched; stems and branches with recurved sharp spines (homologues of the peduncles) in axils of the lower leaves and straight divaricate spines in axils of the upper leaves (the latter spines derived from the peduncles after the flowers have fallen). Leaves simple, undivided, entire. Flowers solitary in numerous upper leaf axils along the stems and branches; pedicels short, terminal on the peduncles, 2-bracteolate at the base; peduncles persisting as straight horizontal stiff spines after the flowers have fallen. Calyx regular, 5-lobed or -fid, somewhat accrescent in fruit. Corolla large; tube subcylindric, straight; limb 2-lipped, the upper lip 2-lobed, the lower lip 3-lobed nearly to the base, reflexed; corolla limb ovoid in bud with the upper lip partially covering the lower lip. Stamens didynamous, inserted near the throat of the corolla tube, long exserted; filaments slender, subparallel or somewhat divergent, slightly curved in the opened flowers, anthers deeply lobed at the base. Ovary l-locular, with 4 placentas, each l-ovulate; style long, ± equalling and curved like the stamens; stigma short, 2-fid. Fruit a drupe with 3(4?) pyrenes.

A tropical African genus with only one species.

Kalaharia uncinata (Schinz) Moldenke in Phytologia **5**: 132 (1955); Résumé Verbenaceae: 458 (1959). TAB. **18**. Type from Namibia.
　　Cyclonema spinescens Oliv. in J. Linn. Soc., Bot. **15**: 96 (1876); in Hooker's Icon. Pl. **13**: 18, tab. 1221 (1877), non Klotzsch (1861) *nom. illegit.* Type from Tanzania.
　　Clerodendrum uncinatum Schinz in Verh. Bot. Vereins Prov. Brandenburg **31**: 206 (1890). —Brenan, Check-list For. Trees Shrubs Tang. Terr.: 633 (1949). —Brenan in Mem. New York Bot. Gard. **9**: 37 (1954). —White, For. Fl. N. Rhod.: 366 (1962). —Friedrich-Holzhammer et al. in Merxmüller, Prodr. Fl. SW. Afrika, fam. 122: 5 (1967). —Richards & Morony, Check List Fl. Mbala: 237 (1969). —Gibbs-Russell et al. in Mem. Bot. Surv. S. Africa **56**: 171 (1987). —Verdcourt in F.T.E.A., Verbenaceae: 142, fig. 20 (1992).

---

* Placed in synonymy with *Clerodendrum* L. by P.D. Cantino in Harley et al. in J.W. Kadereit (ed.), Fam. Gen. Vasc. Pl. (Kubitzki, ed. in chief) **VII**: 199 (2004).

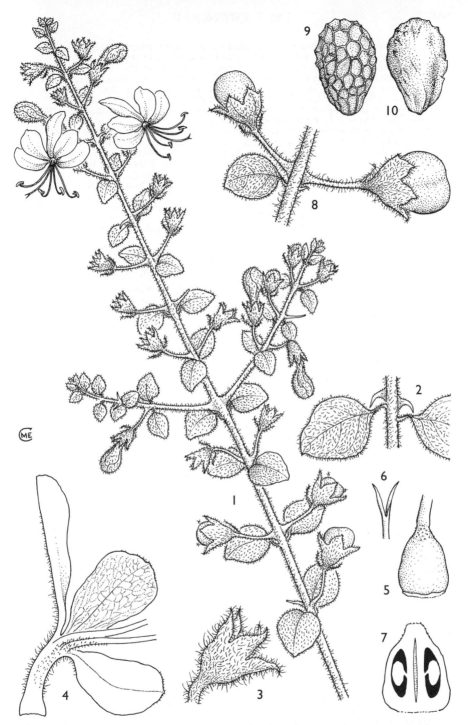

Tab. 18. KALAHARIA UNCINATA. 1, apical portion of flowering stem (× 2/3); 2, part of stem, showing paired spines (× 1), 1 & 2 from *Bullock* 3007; 3, calyx (× 2); 4, half corolla (× 2); 5, ovary (× 6); 6, stigma (× 6); 7, longitudinal section of ovary (× 8), 3–7 from *Sanane* 212; 8, fruits, showing stout peduncles (× 1); 9, 10, pyrene, 2 views, 8–10 from *B.D. Burtt* 3333. Drawn by M.E. Church. From F.T.E.A.

Kalaharia spinipes Baillon, Hist. Pl. **11**: 111, in adnot. (1892). —Briquet in Engler & Prantl, Nat. Pflanzenfam. **4**, 3a: 172 (1897). Type from South Africa.
Clerodendrum spinescens (Oliv.) Gürke in Bot. Jahrb. Syst. **18**: 180 (1893). —J.G. Baker in F.T.A. **5**: 313 (1900). —H. Pearson in F.C. **5**, 1: 221 (1901). —De Wildeman, Études Fl. Bas-Moyen-Congo **1**: 72 et 310 (1903); in Ann. Mus. Congo, Sér. V, Bot. **3**: 136 (1909). — Thomas in Bot. Jahrb. Syst. **68**: 89 (1936), *comb. illegit.* Type as above.
Clerodendrum spinescens var. *parviflorum* Schinz ex Gürke, op. cit.: 181 (1893), as *"parviflora"*. Type from Dem. Rep. Congo.
Kalaharia spinescens (Oliv.) Gürke in Engler, Pflanzenw. Ost-Afrikas **C**: 340 (1895); in Bot. Jahrb. Syst. **30**: 390 (1901); in Warburg, Kunene-Samb.-Exped. Baum: 350 (1903). — Henriques in Bol. Soc. Brot. **16**: 69 (1899). —R.E. Fries, Wiss. Ergebn. Schwed. Rhod.-Kongo-Exped.: 274 (1916). —Wild in Clark, Victoria Falls Handb.: 158 (1952). Type as above.
Kalaharia spinescens var. *parviflora* (Schinz ex Gürke) R.E. Fries, loc. cit. (1916).
Kalaharia spinescens var. *hirsuta* Moldenke in Phytologia **3**: 418 (1951). Type from Tanzania.
Kalaharia uncinata var. *hirsuta* (Moldenke) Moldenke in Phytologia **5**:132 (1955); loc. cit.: (1959). Type as above.
Rotheca uncinata (Schinz) P.P.J. Herman & Retief in Bothalia **32**(1): 81 (2002).

A low spinescent many-stemmed suffrutex from a woody rootstock, forming dense patches with stems erect-ascending and 0.2–1.2 m high or radiating procumbent and up to 1.2 m long. Stems herbaceous, ± angular, stiffly divaricately branched; branches straight, up to c. 60 cm long; stems and branches with recurved sharp orange-brown spines, 1.75–10(17) mm long, in axils of the lower sterile leaves and straight divaricate spines in axils of the upper leaves (the latter spines derived from the peduncles after the flowers have fallen); indumentum of stems and branches ± densely villous with soft patent hairs 0.5–1 mm long, some gland-tipped, becoming shortly puberulous or glabrescent. Leaves opposite, sometimes subopposite to alternate, shortly petiolate; lamina (0.7)1.5–6.3(9) × (0.5)1–3(6) cm, ovate or broadly-ovate to suborbicular or elliptic, sometimes broader than long, subacute to mostly rounded and shortly apiculate at the apex, rounded, broadly cuneate at the base, entire, ± densely hispid pubescent on both surfaces with soft erect hairs, some gland-tipped, or blades shortly crisped puberulous, ± concolorous, nerves inconspicuous above, prominent beneath; petiole 1.5–4.5(10–18) mm long. Flowers solitary in the axils of the upper leaves of stems and branches; true pedicels up to 3 mm long; bracts 2–9 mm long, linear to filiform; peduncles 0.7–1.6(2.5) cm long, spreading or ascending, terete, rigid, persisting as straight spines after the flowers have fallen. Calyx 9–15 mm long, campanulate, 5-lobed ± to the middle; lobes ovate and acuminate to lanceolate and then somewhat attenuate, acute. Corolla large, vermillion-scarlet shading into yellow at the base of the limb and throat or orange-red with cream-yellow throat; tube slightly curved in the bud but straight or nearly so at anthesis, 10–13 mm long; lobes hispidulous or glandular-pubescent on the outside, upper lip 14–17 × 7–8 mm, oblong, rounded at the apex, median lobe of the lower lip concave, 17–24 × 5.5–9 mm, oblong-obovate, rounded or subtruncate at the apex, the lateral lobes similar to the upper lip but slightly broader. Filaments 22–25 mm long, anthers c. 3 mm long, oblong. Style 20–30 mm long. Style and stamens exserted 1.5–2 cm. Fruit sitting in the persistent accrescent calyx, black, shining, smooth, 13–17 × 10–15 mm, ovoid, shortly hispidulous, slightly 2–4-lobed at maturity, with pulpy flesh; pyrenes 1–3(4?), bony, c. 11 × 8 mm.

Botswana. N: Chobe Distr., c. 15 km south of Hyaena camp on Linyanti R., 26.x.1972, *Biegel, Pope & Russell* 4075 (K; LISC; SRGH). SW: Ghanzi Distr., along the road between Ghanzi and Kalkfontein, c. 21°50'S, 21°20'E, 17.ix.1976, *Skarpe* 82 (K; S). SE: Kgatleng Distr., Mochudi, 24°10'S, 26°05'E, l.iv.1944, *C.C. Harbor* s.n. [Herb. *Rogers* 6445 (BM; PRE; SRGH; Z)]. **Zambia**. B: Mongu Distr., Kataba, xi.1959, *Armitage* 200/59 (LD; SRGH). N: Mbala (Abercorn), Lake Chila, 1550 m, 5.xi.1958, *Robson & Fanshawe* 492 (BM; BR; K; LISC; SRGH). W: Kawambwa Distr., roadsides in Luapula Valley, 6.vii.1957, *Savory* 245 (K; SRGH). C: Lusaka Distr., c. 8 km east of Lusaka on Great East Road, 18.ix.1935, *Galpin* 15108 (K; PRE). E: Chipata (Fort Jameson), 1088 m, 6.i.1936, *Winterbottom* 58 (K). S: Mazabuka, vi.1920, *Rogers* 26112 (PRE; S; Z). **Zimbabwe**. N: Binga Distr., Shambolo area on road to Lusulu camp, 12.viii.1956, *Lovemore* 462 (BR; K; LISC, S; SRGH). W: Hwange (Wankie) National Park, near Shumba, Picnic Site, 14.vii.1974, *R.F. Raymond* 271 (B; LD; P; SRGH). C: near Harare (Salisbury), 5.xii.1913, *A. Peter* 51070 (B). **Malawi**. C: Lilongwe Distr., Chitedze Agricultural Research Station, 5.viii.1965, *Salubeni* 337 (K; LISC; SRGH). S(?): main road west of Lake Nyasa from Zomba to the north of the lake, vii.1935, *J. Smuts* 2131 (COI; K; LISC).

Also in Tanzania, Dem. Rep. Congo, Angola, Namibia and South Africa (Transvaal?). Wooded grassland and open woodlands, on alluvial soils and in Kalahari Sand; often in overgrazed grassland, on roadsides, in disturbed ground and sometimes as a weed in old cultivation.

Flowers parasitized by an insect have an abnormal, much thickened subregular corolla, with a short tube and deeply 5-lobed limb; the lobes rounded, concave and connivent. The style and stamens in such flowers are very short and included, with the filaments inserted nearly at the base of corolla tube.

6. KAROMIA Dop

Karomia Dop in Bull. Mus. Hist. Nat. (Paris), sér. 2, **4**: 1052 (1932). —R. Fernandes in Garcia de Orta, Sér. Bot. **7**: 36 (1988). —Harley et al. in Kadereit (ed.), Fam. Gen. Vasc. Pl. (Kubitzki, ed. in chief) **VII**: 198 (2004).

Trees or shrubs, unarmed or with small axillary spines; stems sometimes subscandent. Leaves deciduous, petiolate, opposite, simple, entire dentate or sublobed. Inflorescences cymose, axillary or crowded at apices of shoots and appearing terminal. Calyx often coloured; tube shortly funnel-shaped; limb broadly expanded rotate or patelliform, shallowly to ± deeply 5-lobed, very accrescent and papyraceous with a ± raised reticulation in fruit. Corolla strongly zygomorphic; tube usually rather short, gibbous, contracted below the limb (mainly in bud), split dorsally to near base of tube giving a 1-lipped appearance; limb 5-lobed with the lobes all on anterior side arranged in a single lip comprising 2 separate posterior lobes, 2 lateral lobes and a larger median anterior labelliform lobe. Stamens 4, didynamous, inserted near the mouth of corolla tube; filaments rolled into a circle in the bud and distinctly exserted from split in the corolla tube at anthesis, parallel, ascending then curving downwards, with a tuft of scales at the base; anthers with parallel or ± divergent thecae. Ovary hairy, 2 carpellate, each carpel divided into 2 loculi; ovules one per locule; style long, ± equalling and curved like the stamens; stigma short, 2-fid. Fruit dry, hard, turbinate, obpyramidal or obconical, sometimes ± strongly 4-lobed above, rounded or ± flat at the apex, separating at maturity in 4 (?sometimes 2) trigonous mericarps.

A genus of nine species, one in Viet Nam the others in Madagascar and eastern and southern Africa.

The Madagascan and African species were formerly placed in *Holmskioldia*, however, they differ from this genus mainly in the strongly zygomorphic corolla with a rather short, gibbous, fissured tube (in *Holmskioldia* the corolla tube is rather long, subcylindrical and not gibbous or dorsally split). *Karomia* is further distinguished by its sub-1-lipped corolla limb; by its stamens inserted near the mouth of corolla tube (not in the middle and not prolonged below the insertion by raised pubescent lines, as in *Holmskioldia*), and rolled into a ring in the bud but not so in *Holmskioldia*; and by the hard dry fruit, not drupaceous. The affinities of *Karomia* are closest to *Rotheca* Raf.

Labellum of the corolla limb flabellate, laciniate, without a tuft of hairs at the base; scales on the filament-base short, whitish; horn-like appendages of the fruit 2–4-crested; leaf lamina up to 11 × 6.5 cm, broadest above the middle · 1. *tettensis*
Labellum 2-lobed, not flabellate, with a tuft of hairs at the base; scales on the filament-base longer (up to 1 mm), hyaline; horn-like appendages of the fruit not crested; leaf lamina up to 5 × 4 cm, broadest below the middle · 2. *speciosa*

1. **Karomia tettensis** (Klotzsch) R. Fern. in Garcia de Orta, Sér. Bot. **7**: 39 (1988). —M. Coates Palgrave, ed. 3 of K. Coates Palgrave, Trees South. Africa: 993, illust. 276 (2002). TAB. **19**. Type: Mozambique, Tete, near Carangache, near petrified forest, *Macêdo & Esteves* 4847 (LMU, neotype; LISC; LMA; SRGH).
    Cyclonema tettense Klotzsch in Peters, Naturw. Reise Mossambique **6**, part 1: 261 (1861). Type: Mozambique, Tete, Peters (B†).
    Cyclonema mucronatum Klotzsch in Peters, Naturw. Reise Mossambique **6**, part 1: 260 (1861). Type: Mozambique, Manica e Sofala, Vila de Sena, *Peters* (B†).
    Cyclonema spinescens Klotzsch in Peters, Naturw. Reise Mossambique **6**, part 1: 262 (1861). Type: Mozambique, Tete, *Peters* (B†).
    Holmskioldia tettensis (Klotzsch) Vatke in Linnaea **43**: 536 (1882). —Gürke in Engler, Pflanzenw. Ost-Afrikas C: 342 (1895). —J.G. Baker in F.T.A. **5**: 314 (1900). —Sim, For. Pl.

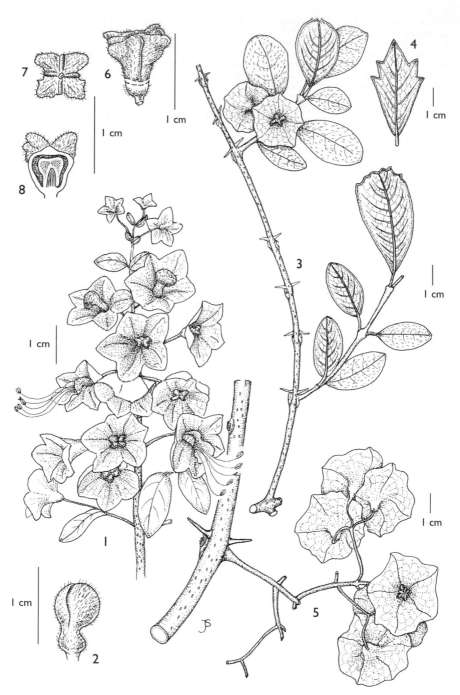

Tab. 19. KAROMIA TETTENSIS. 1, inflorescence; 2, bud, 1 & 2 from *Davies* 2509; 3, portion of fruiting branch, showing persistent peduncles, from *Gomes e Sousa* 4761; 4, leaf, coarsely serrately lobed, from *Simon* 701; 5, part of fruiting branch; 6, fruit, side view; 7, fruit, from above, showing horn like appendages; 8, fruit, longitudinal section, 5–8 from *Brummitt* 8882. Drawn by Judi Stone.

Port. E. Africa: 93 (1909). —White, For. Fl. N. Rhod.: 368 (1962). —Moldenke in Phytologia **48**: 354–356 (1981) pro min. parte.

Holmskioldia spinescens (Klotzsch) Vatke in Linnaea **43**: 536 (1882). —Gürke in Engler, Pflanzenw. Ost-Afrikas **C**: 342 (1895). —J.G. Baker in F.T.A. **5**: 314 (1900). —Sim, For. Pl. Port. E. Africa: 93 (1909). —Moldenke op. cit.: 352–353 (1981). Type as above.

Holmskioldia mucronata (Klotzsch) Vatke in Linnaea **43**: 536 (1882). —Gürke in Engler, Pflanzenw. Ost-Afrikas **C**: 342 (1895). —J.G. Baker in F.T.A. **5**: 314 (1900). —Sim, For. Pl. Port. E. Africa: 93 (1909). —Moldenke op. cit.: 331–332 (1981). Type as above.

Holmskioldia subintegra Moldenke in Bol. Soc. Brot., sér. 2, **40**: 122 (1966); op. cit.: 353–354 (1981). Type: Mozambique, Tete, near Furancungo, *Andrada* 1755 (LISC, holotype; COI; LMU).

Shrub with 2–several clustered erect stems becoming arched or scandent, 2–6 m high, or small bushy tree 4–7 m high, unarmed or with paired straight woody spines (persistent peduncle bases?) up to 2 cm long on stems and branches, inserted just above the nodes; older branches grey to reddish-brown, glabrescent, lenticellate; young branches and branchlets pubescent to puberulous with bristle-like or crisped hairs and with scattered sessile glands, branchlets and peduncles also with sparsely scattered thicker soft yellowish (glandular?) hairs; leaf scars raised and shallowly cup-shaped with a dense cushion of hairs in the axils. Leaves deciduous, lamina up to 11 × 6.5 cm, variable in shape sometimes on the same branch, elliptic-oblong, obovate to oblanceolate or lanceolate, acute at the apex, cuneate to narrowly rounded at the base, entire or coarsely serrate-dentate to lobed with 1–3(7) teeth or lobes on each in the upper $^1/_3$ to $^1/_2$, ± pubescent on both surfaces, usually sparsely so or hairs mainly on the nerves beneath, blades chartaceous, the nerves ± prominent beneath; petiole up to 8 mm long. Cymes axillary or crowded at apices of shoots, 1–3-flowered, up to 6.5 cm long; peduncles 2.5–3 cm long; bracts c. 4.5 mm long and 6–8 mm broad, broadly ovate; pedicels villous. Calyx green, grey-green to pinkish-green, c. 1.5 cm in diameter, rotate, shallowly 5-lobed, the lobes broadly ovate-triangular, densely pubescent and gland-dotted outside, fruiting calyx brownish, very accrescent, 2.5–4 cm in diameter, rotate-patelliform, with a shallowly to distinctly 5-lobed limb, chartaceous-coriaceous, venose with reticulation prominent. Corolla pale to dark blue, lilac, pinkish, purple or purplish-brown (?); tube 2.5–4 mm long, with short gland-tipped hairs distally; limb 16.5–18 mm long, narrow, with short gland-tipped hairs on the lower surface towards the base and densely villous towards the apex and on the lateral lobes; labellum 8.5–11 mm long, flabellate and irregularly narrowly laciniate; lateral lobes c. 4 × 3.5 mm, rounded at the top, entire. Filaments c. 4 cm long, densely covered with short gland-tipped hairs towards the base and provided with a tuft of small white obtuse scales near the insertion. Ovary densely villous at ± the upper half, glabrous below. Style and stamens long exserted. Fruit c. 6 mm long, turbinate-obconical or somewhat obpyramidal, ± flat at the top with 4 horn-like appendages arranged in the shape of a cross, the horns 2–3(4)-crested, densely villous to shortly hispidulous in the upper half, glabrous below.

Zambia. C: Luangwa Distr., Katondwe, fr. 12.xi.1963, *Fanshawe* 8114 (FHO; K). S: Gwembe Distr., near Kariba Hills, fl. 4.xii.1957, *R. Goodier* 442 (COI; K; LISC; SRGH). **Zimbabwe**. N: Guruve Distr., foothill of Nyadutu Mts., 25 km west of Kanyemba, near Tunsa-Zambezi R. junction, fr. 2.ii.1966, *Müller* 331 (B; FHO; SRGH). E: Nyanga Distr., Inyanga North Reserve, near banks of Gairezi R. c. 1120 m, fl. x.1958, *Davies* 2529 (SRGH). **Malawi**. S: Blantyre Distr., Matope, fl. & fr. 7.i.1956, *Jackson* 1780 (BM; FHO; K). **Mozambique**. T: Changara Distr., near Luenha R., 16°50'S, 33°15'E, 270 m, fr. 17.v.1962, *Gomes e Sousa* 4761 (COI; K; LMA; P). MS: Tambara Distr., 5 km on the road from Nhacolo (Tambara) to Mungári, c. 100 m, fl. 16.xii.1965, *Torre & Correia* 13705 (LISC).

Not known from elsewhere. Valley floor of the main low altitude rivers and at foot of bordering escarpments, in dry areas, mopane woodlands and mixed deciduous woodlands (baobab, *Sterculia*, *Acacia*) and deciduous thickets, sometimes in riverine vegetation; 90–500 m.

2. **Karomia speciosa** (Hutch. & Corbishley) R. Fern. in Garcia de Orta, Sér. Bot. **7**: 43 (1988). —M. Coates Palgrave, ed. 3 of K. Coates Palgrave, Trees South. Africa: 992 (2002). Type from South Africa (Mpumalanga, Komati Poort).

Holmskioldia speciosa Hutch. & Corbishley in Bull. Misc. Inform., Kew **1920**: 332, fig. 1 (1920). —Pole Evans, Fl. Pl. S. Africa **2**: t. 49 (1922). —Garcia in Estud. Ensaios Doc. Junta Invest. Ci. Ultramar [in Mendonça, Contrib. Conhec. Fl. Moçamb., II] **12**: 177 (1954).

Holmskioldia tettensis sensu Dyer, Verdoorn & Codd in Letty, Wild Fl. Transvaal: 280, t. 140 fig. 1 (1962). —sensu van der Schijff, Check List Vasc. Pl. Kruger Nat. Park: 81 (1969). — sensu Moldenke, Fifth Summ. Verbenaceae: 251 quoad distrib. in S Mozambique, 254 et 256 (1971); in Phytologia **48**: 354–356 et 384 (1981) pro max. parte. —sensu Palmer & Pitman, Trees South. Africa **3**: 1969, photo p. 1968, fig. p. 1970 (1972). —sensu J.H. Ross, Fl. Natal: 300 (1972), non (Klotzsch) Vatke (1882).

Slender shrub up to 4 m high with many upright stems, or a slender bushy tree up to 6 m high, unarmed or sometimes with paired straight woody spines (persistent peduncle bases?) inserted just above the nodes; indumentum of branchlets, petioles, peduncles and pedicels pubescent with bristle-like or antrorse-appressed or crisped hairs or ± densely tomentose, branchlets and peduncles also with sparsely scattered thicker soft yellowish (glandular?) hairs; older branches glabrescent, lenticellate; leaf axils with dense cushions of hairs, leaf scars raised and shallowly cup-shaped with a dense cushion of hairs in the axils. Leaves deciduous, dark green above paler beneath, lamina up to 5 × 4 cm, broadly ovate or subtriangular, acute at the apex, rounded at the base and shortly cuneate into the petiole, coarsely serrate-dentate to lobed with 1–3 obtuse teeth or lobes on each side or entire, softly puberulous on both surfaces, usually more densely so on the nerves beneath, with minute sessile glands beneath, reticulation ± conspicuous beneath; petiole up to 1.3 cm long. Cymes axillary, crowded toward the ends of the shoots, 1–3-flowered, up to 4 cm long; peduncles 1–1.8 cm long; bracts up to 7 mm long, obovate-spathulate; pedicels c. 12 mm long, bracteolate, the bracteoles c. 2 mm long, linear. Calyx rose-coloured, mauve, lilac or pale green suffused with purple, 8.5–15 mm long and 11–20 mm in diameter, rotate, the tube c. 2.5 × 2.5 mm, the limb shallowly 5-lobed, the lobes broadly ovate-triangular, softly ± densely puberulous and gland-dotted outside; fruiting calyx very accrescent, 20–30 mm in diameter, rotate-patelliform, with a shallowly to distinctly 5-lobed limb, papyraceous, venose with reticulation prominent. Corolla blue, violet, purple to mauve or lilac, or corolla white, 20–29 mm long; tube c. 10 mm long, concave at the back, convex below, somewhat contracted at the mouth, split along the back, indumentum of short gland-tipped hairs with scattered sessile spherical glands; corolla limb 12–18 mm long, with short gland-tipped hairs on the lower surface towards the base and densely villous towards the apex and on the lateral lobes; labellum 8–11 mm long, 2-lobed with lobes rounded and entire, ± densely villose-pilose at the base outside, glabrous above; lateral lobes shorter, oblong, obtuse, entire, with gland-tipped hairs and sessile glands outside becoming villose-pilose toward the apex. Filaments glandular towards the base and with a tuft of scales near the insertion, the scales c. 1 mm long, linear, hyaline; anthers c. 1.25 mm long, suborbicular. Ovary obovoid, densely glandular and pubescent in the upper half; style c. 3 cm long. Style and stamens long exserted. Fruit. c. 7 mm long, obconical, ± flat at the top with 4 horn-like appendages arranged in the shape of a cross, the horns not crested, ± densely puberulous in the upper half, glabrous below.

Forma **speciosa**

Corolla blue, violet, purple to mauve or lilac.

Mozambique. M: Namaacha, Swaziland frontier, 600 m, ii.1931, *Gomes e Sousa* 401 (K; LMA; SRGH).
Also in Swaziland and South Africa (Mpumalanga and KwaZulu-Natal). Hot dry rocky hillsides and along seasonal water courses, in mixed bushveld, sometimes cultivated as a garden ornamental; 70–600 m.

Forma **alba** (Moldenke) R. Fern. in Garcia de Orta, Sér. Bot. **7**: 44 (1988). Type: Mozambique, Namaacha Distr., near Goba, 15.xi.1940, *Torre* 2015 (LISC, holotype).
    Holmskioldia tettensis forma *alba* Moldenke in Bol. Soc. Brot., sér. 2, **40**: 123 (1966); in Phytologia **48**: 385 (1981).

Corolla white.

Mozambique. M: Namaacha Distr., near Goba, 15.xi.1940, *Torre* 2015 (LISC).
Not known from elsewhere; c. 140 m. Only the type specimen seen.

7. TEUCRIUM L.

Teucrium L., Sp. Pl.: 562 (1753); Gen. Pl., ed. 5: 247 (1754). —J.G. Baker in F.T.A. **5**: 500 (1900). —Skan in Fl. Cap. **5**, 1: 384 (1910). —Codd in Bothalia **12**: 177 (1977); in F.S.A. **28**, 4: 9 (1985). —Harley et al. in Kadereit (ed.), Fam. Gen. Vasc. Pl. (Kubitzki, ed. in chief) **VII**: 201 (2004).

Perennial herbs with few stems arising from a woody rhizome. Leaves opposite, decussate. Inflorescence terminal, thyrsoid with cymes opposite and arranged in a raceme-like manner. Flowers non-resupinate. Calyx green, campanulate, 5-lobed, persistent and slightly accrescent in fruit; lobes lanceolate. Corolla white; tube short with a pubescent annulus at throat, split dorsally to near base of tube giving a 1-lipped appearance; lobes 5, all on anterior side, with the anterior lobe cucullate, elliptic, longer than the others; lateral and posterior lobes rounded. Stamens 4, attached to corolla tube, parallel, ascending, arched, exserted from split corolla, pubescent at base; anthers with thecae divergent, confluent to such an extent that the separate thecae are hard to distinguish. Ovary hairy, 2 carpellate, each carpel divided into 2 loculi; ovules one per locule. Disk indistinct. Style subterminal, ± equalling filaments, bifid with equal lobes. Fruit 4-lobed eventually splitting into 4 dry reticulate mericarps each containing a seed; mericarps conspicuously scarred with a areole where previously fused, pubescent at apex. Seeds albuminous.

A genus of about 300 species, largely in the temperate areas of the world, concentrated in the Mediterranean region, but also found in the New World, Australia and South Africa.

Teucrium trifidum Retz., Observ. Bot. **1**: 21 (1779). —Codd in Bothalia **12**: 177 (1977); in F.S.A. **28**, 4: 10 (1985). Type from South Africa.
　Teucrium capense Thunb., Prodr. Pl. Cap. part 2: 95 (1800). —Skan in Fl. Cap. **5**, 1: 385 (1910). Type from South Africa.
　Ajuga capensis (Thunb.) Pers., Syn. Pl. **2**:109 (1806).
　Teucrium africanum sensu Barnes & Turton, List Fl. Pl. Botswana (1986), non Thunb.

Perennial herb 0.2–0.3(1) m tall with a woody rhizome; stems 1–few, simple or branched near base, woody at base, slender, 4-angular, pubescent with retrorse eglandular hairs and scattered pale sessile glands. Leaves sessile; lamina deeply divided with 3–5 ± linear lobes, 25–35 mm long and with lobes 1–3 mm broad, margin revolute, apex of lobes rounded or sometimes 3-fid, attenuate at base, pubescent with pale sessile glands. Inflorescence a leafy panicle in the upper one third of the stem; flowers usually in 3–7-flowered pedunculate cymes; peduncles opposite, 8–13(20) mm long; pedicels 3–5(12) mm long; bracteoles small, linear. Calyx 3–4 mm long, campanulate, equally lobed; teeth lanceolate, 2 mm long. Corolla white, 4–5 (6) mm long. Stamens exserted up to 5(8) mm, curved or coiled. Mericarps brown, 2 mm long, reticulate, pubescent at apex; areole paler, covering three quarters of the length of the mericarp.

Botswana. SE: South East Distr., c. 4.8 km north of Lobatse, fl & fr., 16.i.1960, *Leach & Noel* 126 (K; SRGH).
　Also in South Africa, only just entering Botswana. In dry woodland, often in overgrazed or disturbed places.

INDEX TO BOTANICAL NAMES